AF581738

Advances in Soil Pollution and Geotechnical Environment

Advances in Soil Pollution and Geotechnical Environment

Editor

Bing Bai

MDPI • Basel • Beijing • Wuhan • Barcelona • Belgrade • Manchester • Tokyo • Cluj • Tianjin

Editor
Bing Bai
School of Civil Engineering
Beijing Jiaotong University
Beijing
China

Editorial Office
MDPI
St. Alban-Anlage 66
4052 Basel, Switzerland

This is a reprint of articles from the Special Issue published online in the open access journal *Applied Sciences* (ISSN 2076-3417) (available at: https://www.mdpi.com/journal/applsci/special_issues/soil_pollution_geotechnical_environment).

For citation purposes, cite each article independently as indicated on the article page online and as indicated below:

LastName, A.A.; LastName, B.B.; LastName, C.C. Article Title. *Journal Name* **Year**, *Volume Number*, Page Range.

ISBN 978-3-0365-6177-6 (Hbk)
ISBN 978-3-0365-6178-3 (PDF)

Contents

About the Editor

Bing Bai

Bing Bai is currently a professor working at the Beijing Jiaotong University, Beijing, China. He holds a Ph.D. in Geotechnial Engineering from Wuhan University, China. His research interests include geo-environmental engineering, thermal consolidation theory, and contaminant transport theory and control methods. Recently, he devoted himself to research on advances in soil pollution and geotechnical environments. He developed a theory describing the cotransport of heavy metals and suspended particles at different temperatures in porous media and proposed a nonlinear attachment–detachment model with hysteresis suitable for substances with sizes ranging from ions to large particles, which is of great significance in understanding groundwater pollution mechanisms and in the development of purification technology. He has published more than 150 academic papers and five scientific books. He serves as an editorial board member of the Journal of Geotechnical Engineering and Rock and Soil Mechanics of China. He is the member of several professional committees, such as the Special Committee on Soil Constitutive Relationship, the Strength of China Society of Civil Engineering, and the Special Committee on Energy Underground Structure and Engineering of the Chinese Society of Rock Mechanics

Preface to "Advances in Soil Pollution and Geotechnical Environment"

Soil pollution and disposal technology is currently a hot topic in geo-environmental engineering that involves the interaction of mechanisms, migration processes, and treatment measures of various types of pollutants (e.g., heavy metals, organic pollutants, acid or alkali substances, and radioactive materials) with geotechnical media, as well as heat flow processes and thermal damage. Application areas include industrial pollutant treatment in shallow seepage, groundwater exploitation, geothermal resource development, thermal energy storage, waste disposal in metal mining, and landfill leachate barriers.

This Special Issue on: "Advances in Soil Pollution and the Geotechnical Environment" addresses the most recent developments in soil pollution and restoration, contaminant hydrology, and ground disturbance to stimulate fruitful technical and scientific interactions between professionals.

A total of thirteen papers in various fields of soil pollution and the geotechnical environment, including soil pollution, soil restoration, and heat transfer in soils, are presented in this Special Issue. Although the submissions for this Special Issue have been closed, more in-depth research in the field of soil pollution and the geotechnical environment continues to address the challenges we face today, such as soil pollution mechanisms, pollutant migration, and the remediation efficiency of pollutants.

Thanks to all the authors and peer reviewers for their valuable contributions to this the Special Issue, "Advances in Soil Pollution and the Geotechnical Environment". I would also like to express my gratitude to all the staff and people involved in the creation of this Special Issue.

Bing Bai
Editor

Editorial

Special Issue on Advances in Soil Pollution and the Geotechnical Environment

Bing Bai

School of Civil Engineering, Beijing Jiaotong University, Beijing 100044, China; bbai@bjtu.edu.cn

Soil pollution and disposal technology is currently a hot topic in geo-environmental engineering that involves the interaction mechanisms, migration processes, and treatment measures of various types of pollutants (e.g., heavy metals, organic pollutants, acid or alkali substances, and radioactive materials) with geotechnical media, as well as heat flow processes and thermal damage. Application areas include industrial pollutant treatment in shallow seepage, groundwater exploitation, geothermal resource development, thermal energy storage, waste disposal in metal mining, and landfill leachate barriers.

This Special Issue on *Advances in Soil Pollution and the Geotechnical Environment* addresses the most recent developments in soil pollution and restoration, contaminant hydrology, and ground disturbance to stimulate fruitful technical and scientific interactions between professionals.

A total of thirteen papers in various fields of soil pollution and the geotechnical environment, including soil pollution, soil restoration, and heat transfer in soils, are presented in this Special Issue. To comprehend the effect of tidal actions on the nitrogen cycle in silty-clay riparian hyporheic zones, Cai et al. [1] carried out the synchronous monitoring of water level and water quality along a test transect during a spring tidal period and measured the permeability and chemical composition of soil samples from drilled holes. Wang et al. [2] presented a systemic laboratory experimental investigation on the liquid–plastic limit, moisture retention, hydraulic conductivity, and gas diffusion barrier properties of compacted clay amended with attapulgite and diatomite to control desiccation cracks and the migration of water and volatile organic compounds. Tian et al. [3] proposed a calculation method for thermal pore water pressure while considering the overconsolidation effect of saturated clay, which was verified by the relevant experimental data. Zhang et al. [4] explored the effect and mechanism of xanthan gum treatment on the water retention and shear strength characteristics of silt during the wetting process. Wu et al. [5] proposed a novel subgrade using a capillary barrier to reduce the different settlements and stabilities based on seepage theory for unsaturated soils. Qu et al. [6] investigated the migration, adsorption, and desorption characteristics of Pb^{2+} on fine, medium, and coarse sand in a water table fluctuation zone using several laboratory methods, including the kinetic aspects of Pb^{2+} adsorption/desorption and water table fluctuation experiments. Li et al. [7] carried out physical experiments on the nitrogen migration and transformation processes in a groundwater level fluctuation zone and constructed a numerical model for nitrogen migration in a vadose zone and a saturated zone using the software HydrUS-1D. Yang et al. [8] conducted a long-term F–T study of S/S Pb–Zn–Cd composite HM-contaminated soil under six conditions (0, 3, 7, 14, 30, and 90 cycles), with each F–T cycle process lasting 24 h. Kuang et al. [9] simulated the nanoscale elastic properties of hydrated Na-, Cs-, and Ca-MMT with unconstrained system atoms using a molecular dynamics (MD) method. Yu et al. [10] discussed the soil–water characteristic curve and microstructure evolution of unsaturated expansive soil improved by microorganisms in Nanning, Guangxi by means of a filter paper method and scanning electron microscope imaging (SEM). Liu et al. [11] investigated the removal effect of nitrobenzene in an aquifer through a series of two-dimensional sandbox experiments with different stratigraphic structures. Wu et al. [12] proposed a

Citation: Bai, B. Special Issue on Advances in Soil Pollution and the Geotechnical Environment. *Appl. Sci.* **2022**, *12*, 12000. https://doi.org/10.3390/app122312000

Received: 19 November 2022
Accepted: 23 November 2022
Published: 24 November 2022

sedimentation ratio calculation method through the analysis of the settlement load ratio to calculate the roadbed replacement thickness. Chen et al. [13] investigated a contaminated site in Xinxiang City, Henan Province and collected 92 groundwater samples from the site to identify the groundwater chemical characteristics of the sites and their control mechanisms for the remediation of pollutants.

Although the submissions for this Special Issue have been closed, more in-depth research in the field of soil pollution and the geotechnical environment continues to address the challenges we face today, such as soil pollution mechanisms, pollutant migration, and the remediation efficiency of pollutants [14].

Funding: This research received no external funding.

Acknowledgments: Thanks to all the authors and peer reviewers for their valuable contributions to this the Special Issue Advances in Soil Pollution and the Geotechnical Environment. I would also like to express my gratitude to all the staff and people involved with this Special Issue.

Conflicts of Interest: The author declares no conflict of interest.

References

1. Cai, Y.; Xing, J.; Huang, R.; Ruan, X.; Zhou, N.; Yi, D. Occurrence Characteristics of Inorganic Nitrogen in Groundwater in Silty-Clay Riparian Hyporheic Zones under Tidal Action: A Case Study of the Jingzi River in Shanghai, China. *Appl. Sci.* **2022**, *12*, 7704. [CrossRef]
2. Wang, M.; Wen, J.; Zhuang, H.; Xia, W.; Jiang, N.; Du, Y. Screening Additives for Amending Compacted Clay Covers to Enhance Diffusion Barrier Properties and Moisture Retention Performance. *Appl. Sci.* **2022**, *12*, 7341. [CrossRef]
3. Tian, G.; Zhang, Z. A Calculation Method of Thermal Pore Water Pressure Considering Overconsolidation Effect for Saturated Clay. *Appl. Sci.* **2022**, *12*, 6325. [CrossRef]
4. Zhang, J.; Meng, Z.; Jiang, T.; Wang, S.; Zhao, J.; Zhao, X. Experimental Study on the Shear Strength of Silt Treated by Xanthan Gum during the Wetting Process. *Appl. Sci.* **2022**, *12*, 6053. [CrossRef]
5. Wu, Y.; Zhou, X.; Liu, J. Long-Term Performance of the Water Infiltration and Stability of Fill Side Slope against Wetting in Expressways. *Appl. Sci.* **2022**, *12*, 5809. [CrossRef]
6. Qu, J.; Yan, T.; Zhang, Y.; Li, Y.; Tian, R.; Guo, W.; Jiang, J. An Experimental Study on the Migration of Pb in the Groundwater Table Fluctuation Zone. *Appl. Sci.* **2022**, *12*, 3870. [CrossRef]
7. Li, Y.; Wang, L.; Zou, X.; Qu, J.; Bai, G. Experimental and Simulation Research on the Process of Nitrogen Migration and Transformation in the Fluctuation Zone of Groundwater Level. *Appl. Sci.* **2022**, *12*, 3742. [CrossRef]
8. Yang, Z.; Chang, J.; Li, X.; Zhang, K.; Wang, Y. The Effects of the Long-Term Freeze–Thaw Cycles on the Forms of Heavy Metals in Solidified/Stabilized Lead–Zinc–Cadmium Composite Heavy Metals Contaminated Soil. *Appl. Sci.* **2022**, *12*, 2934. [CrossRef]
9. Kuang, L.; Zhu, Q.; Shang, X.; Zhao, X. Molecular Dynamics Simulation of Nanoscale Elastic Properties of Hydrated Na-, Cs-, and Ca-Montmorillonite. *Appl. Sci.* **2022**, *12*, 678. [CrossRef]
10. Yu, X.; Xiao, H.; Li, Z.; Qian, J.; Luo, S.; Su, H. Experimental Study on Microstructure of Unsaturated Expansive Soil Improved by MICP Method. *Appl. Sci.* **2022**, *12*, 342. [CrossRef]
11. Liu, R.; Yang, X.; Xie, J.; Li, X.; Zhao, Y. Experimental Investigation on the Effects of Ethanol-Enhanced Steam Injection Remediation in Nitrobenzene-Contaminated Heterogeneous Aquifers. *Appl. Sci.* **2021**, *11*, 12029. [CrossRef]
12. Wu, Y.; Liu, H.; Liu, J. Use of Foamed Cement Banking for Reducing Expressways Embankment Settlement. *Appl. Sci.* **2021**, *11*, 11959. [CrossRef]
13. Chen, W.; Zhang, Y.; Shi, W.; Cui, Y.; Zhang, Q.; Shi, Y.; Liang, Z. Analysis of Hydrogeochemical Characteristics and Origins of Chromium Contamination in Groundwater at a Site in Xinxiang City, Henan Province. *Appl. Sci.* **2021**, *11*, 11683. [CrossRef]
14. Bai, B.; Bai, F.; Li, X.; Nie, Q.; Jia, X.; Wu, H. The remediation efficiency of heavy metal pollutants in water by industrial red mud particle waste. *Environ. Technol. Innov.* **2022**, *28*, 102944. [CrossRef]

Article

Occurrence Characteristics of Inorganic Nitrogen in Groundwater in Silty-Clay Riparian Hyporheic Zones under Tidal Action: A Case Study of the Jingzi River in Shanghai, China

Yi Cai [1,2,*], Jingwen Xing [1], Ruoyao Huang [1], Xike Ruan [1], Nianqing Zhou [1,*] and Dongze Yi [1]

1 Department of Hydraulic Engineering, College of Civil Engineering, Tongji University, Shanghai 200092, China; 2032262@tongji.edu.cn (J.X.); 2032261@tongji.edu.cn (R.H.); xkruan007@163.com (X.R.); 2132630@tongji.edu.cn (D.Y.)

2 The Yangtze River Water Environment Key Laboratory of the Ministry of Education, Tongji University, Shanghai 200092, China

* Correspondence: caiyi@tongji.edu.cn (Y.C.); nq.zhou@tongji.edu.cn (N.Z.)

Abstract: For comprehending the effect of tidal action on nitrogen cycle in silty-clay riparian hyporheic zones, the synchronous monitoring of water level and water quality was carried out along a test transect during a spring tidal period from 21 to 23 October 2021. Moreover, the permeability and chemical composition of soil samples from drilled holes were measured. Subsequently, the spatiotemporal variation of inorganic nitrogen concentrations in the groundwater in the riparian hyporheic zone was investigated during the study period, and the potential reason was discussed. It is shown that the delayed response time of groundwater level in the silty-clay riparian zone to the tide-driven fluctuation of the river stage increased with distance from the shore and reached 3.0 h at the position 3.83 m away from the shore. The continuous infiltration of the river water under tide action contributed to the aerobic and neutral riparian hyporheic zone conductive to nitrification. Within 4 m away from the bank, the dominant inorganic nitrogen form changed from NO_3^--N to NH_4^+-N, upon increasing the distance from the bank. Additionally, the removal of nitrogen could occur in the riparian hyporheic zone with aerobic and neutral environment under the conjoint control of nitrification, microbial assimilation, and aerobic denitrification.

Keywords: tidal action; silty-clay soil; riparian hyporheic zone; inorganic nitrogen; occurrence characteristics; influencing factors

Citation: Cai, Y.; Xing, J.; Huang, R.; Ruan, X.; Zhou, N.; Yi, D. Occurrence Characteristics of Inorganic Nitrogen in Groundwater in Silty-Clay Riparian Hyporheic Zones under Tidal Action: A Case Study of the Jingzi River in Shanghai, China. *Appl. Sci.* **2022**, *12*, 7704. https://doi.org/10.3390/app12157704

Academic Editor: Bing Bai

Received: 20 May 2022
Accepted: 28 July 2022
Published: 30 July 2022

1. Introduction

In order to ensure the security of the global food supply, a large amount of nitrogen fertilizer has been utilized in the past few decades. For example, the average nitrogen utilization has achieved the rate of 305 kg/hm^2 in China [1]. Excessive bioavailable nitrogen such as nitrate nitrogen (NO_3^-) and ammonia nitrogen (NH_4^+) in rivers will lead to water eutrophication and, consequently, damage the ecological balance of river systems. Particularly in coastal areas, nitrogen pollution in rivers is generally serious due to dense population and active industry and agriculture production. If it cannot be maintained under effective control, the excessive nitrogen will continue to enter the ocean and further affect the health of the coastal ecosystem and the sustainable utilization of marine resources.

Hyporheic zones of river ecosystem are water-saturated sediment (known as aquifer) below the riverbed and extending to the riparian areas on both sides. It is not only a key area for water exchange and solute migration between river and groundwater, but also an important place for microbial growth and metabolism. Mass exchange and energy transfer are frequent and biogeochemical reactions are complex within a hyporheic zone, which is of great significance to the structure, function, and health of river ecosystem [2–4]. Nitrogen is transported between rivers and their riparian zones through hyporheic exchange and transformed between different chemical forms (e.g., NH_4^+, NO_3^-, nitrite nitrogen (NO_2^-), and N_2) through biogeochemical reactions [5]. Therefore, to realize the effective control of nitrogen pollution in rivers, it is necessary to fully comprehend nitrogen migration and transformation in groundwater in hyporheic zones.

Nitrogen in groundwater in a hyporheic zone varies geographically, and its form changes temporally, which forms a complex dynamic cycle [6–9]. Some hydrological events (e.g., rainstorm, flooding, reservoir discharge, and tidal action) can change the recharge relationship between river water and groundwater, consequently influencing nitrogen migration in hyporheic zones [10–13]. Shibata et al. carried out onsite monitoring to study the water exchange and the change in nitrogen concentration in the hyporheic zone during the rainstorm. It was found that the increase in nitrogen concentration in the hyporheic zone was mainly caused by the rainwater infiltration through the overlying unsaturated zone rather than the lateral seepage from the river [14]. The research conducted by Singh et al. indicated that the flood process extended the mixing range of surface water and groundwater and had a significant impact on the mass cycle in the riverbed hyporheic zone [15]. Sawyer et al. studied the impact of the reservoir operation on the lateral hyporheic exchange, and the results showed that the impact range could extend to 1–5 m away from the shore [16]. Unlike the hydrological events mentioned above, tides exhibit periodicity, and the tidal period is usually about 12 or 24 h. This makes the hyporheic exchange more frequent and the biogeochemical process in the hyporheic zone more complex [17]. In recent years, there has been growing concern over the hyporheic zones of tidal rivers. Musial et al. quantified the tide-driven hyporheic exchange fluxes across the bank and the bed and deduced that the tidal bank storage in the sandy hyporheic zone might remove nutrients from rivers [18]. The geochemical measurements and numerical simulation performed by Barnes et al. showed that it was possible that the hyporheic zone with permeable sediment and low organic matter content could be a source of nitrate to the tidal river [19].

Hydrogeological conditions and ecological factors could affect the nitrogen cycle in hyporheic zones [20–22]. McGarr et al. investigated the hyporheic exchange process in a sand–gravel mixed riverbed and pointed it out that the pore water flowed preferentially through the sediment of high-permeability, which affected nitrogen transport pathways [23]. The sediment heterogeneity could result in the existence of tiny anoxic blocks within the aerobic zone near the shore, thus affecting nitrogen processing [24]. The retention time of river water is usually extended in the low-permeability riverbed sediment, which is beneficial to the full reaction between nitrogen and dissolved oxygen (DO) [25]. In addition, the migration ability of NH_4^+ in the hyporheic zone may be limited to a certain extent due to the electrostatic attraction by soil particles [26]. Eco-environmental factors mainly include the reactant concentration, redox potential (Eh), temperature (T), and pH [27–29]. The nitrogen-related biogeochemical reactions such as nitrification and denitrification are greatly governed by DO concentration and Eh in hyporheic zones [30]. Dissolved organic carbon (DOC) is another key factor controlling the nitrogen cycle, and increasing DOC concentration could promote denitrification [31]. Zarnetske et al. found that DOC and DO could be preferentially consumed in the near-shore hyporheic zone [32]. Additionally, temperature could affect the microbial activity, reaction rate, and DO concentration, consequently influencing the process of nitrogen cycle in the hyporheic zone [33,34]. Some studies demonstrated that pH showed a positive correlation with NO_3^- concentration and a negative one with NH_4^+ concentration [35].

Although the nitrogen cycle in hyporheic zones has been extensively studied, there is little work aiming at the case of low-permeability riparian hyporheic zones of tidal rivers. At present, the mechanism underlying how tides drive nitrogen migration and transformation in riparian hyporheic zones with low permeability sediment is not very clear. The main objective of the study was to characterize the spatiotemporal variations of various inorganic nitrogen concentrations in groundwater in poorly permeable hyporheic zones of tidal rivers and discuss the potential reason why such variations existed. A representative transect was chosen to carry out the drilling, monitoring, and sampling. By means of onsite monitoring and lab analysis, the response of the riparian groundwater level to the tide-driven fluctuation of the river stage was studied, the spatiotemporal variations of inorganic nitrogen concentrations in the groundwater in the riparian hyporheic zone were analyzed, and the potential influencing factors were discussed. The results are expected to enrich the theory of river hyporheic zones and provide a scientific basis for nitrogen pollution treatment of tidal rivers.

2. Methodology

2.1. Field Monitoring

The test site was located on the riverbank of the Jingzi River, a tributary of the Huangpu River in Shanghai, China, as illustrated in Figure 1. The Huangpu River, with a total length of approximately 114 km, is the last tributary of the Yangtze River before it empties into the East China Sea. The Huangpu River estuary forms a medium-tidal shallow-water environment where the tides are of irregular semidiurnal features. The tides can penetrate up to the Dianshan Lake and the provincial boundary between Zhejiang and Shanghai. On an average, the tidal period is about 12.5 h, where 4.5 h involves flooding and the remainder involves ebbing [36]. At the Wusong tide gauge station, the average tidal prism is about 5.8×10^7 m^3, the flood current velocity is up to 1.2×10^4 m^3/s, and the mean tidal range is about 2.27 m [37]. The test site was located about 29 km upstream of the Wusong tide gauge station. According to the preliminary research, there was a good response relationship between the river stages at the test site and the sea levels at Wusong station. The flooding at the test site was about 2.5 h behind that at Wusong station. Furthermore, the fluctuation range of the river stage at the site was up to approximately 0.5 m.

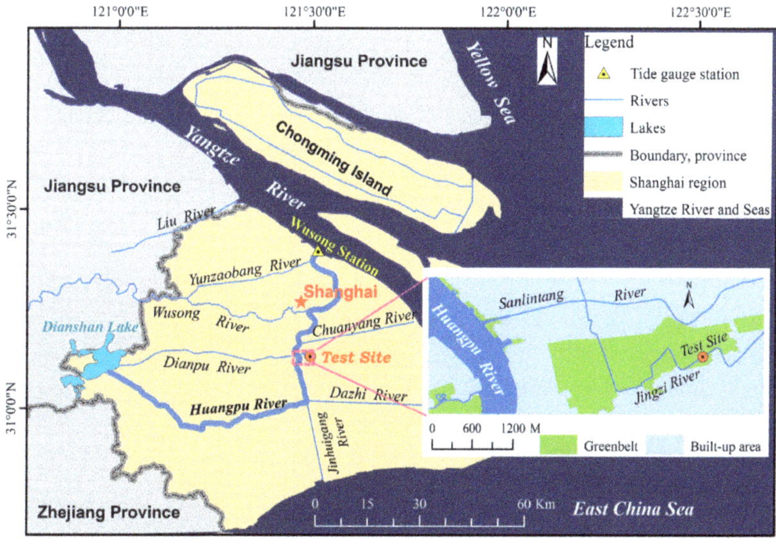

Figure 1. Location of the test site.

The land use is mainly green field and residential in the surrounding area of the test site. The green field is unused and covered with bushes and stunted trees. The Jingzi River flows through the greenbelt, serving as a sewage channel of the vicinity. The terrain around the test site is relatively flat, and the slope degree of the bank is less than 9°. The riverbed sediment and the shallow aquifer materials at the site consist of silty-clay soils extensively distributed on the floodplain of the Huangpu River. The site with gentle slope, poorly permeable sediment, and less interference was considered suitable for the investigation.

The site was instrumented with a perpendicular transect of observation wells and sampling points, as shown in Figure 2. One observation well marked as WR0 was set in the river channel, and four other wells (i.e., WB1, WB2, WB3, and WB4) were laid on the riverbank, 0.71 m, 1.11 m, 2.43 m, and 3.85 m away from the shore, respectively (Figure 2a). The well pipes, made of polyvinyl chloride (PVC) material, were equipped with pipe covers at both ends. A hole punch was employed to make scattered tiny holes with a diameter of about 5 mm in the wall of the pipes. A nylon cloth with a mesh size of 0.15–0.20 mm was wrapped around each pipe. The wells on the riverbank were 200 cm in length and 11 cm in diameter, while WR0 was 80 cm in length and 6 cm in diameter. Each well tube was inserted vertically into the sediment. A high-precision GPS (Z-Survey i70, Manufacturer: CHCNAV Co. Ltd., Shanghai, China) and a leveling instrument (B30, Manufacturer: SOKKIA Co. Ltd., Tokyo, Japan) were used to survey the top elevation of each well. One water-level sensor (HOBO Kit-D-U20-04, Manufacturer: ONSET Co. Ltd., Bourne, Cape Cod, Centerville, MA, USA) was installed at the bottom of each well, with a data collection interval of 5 min.

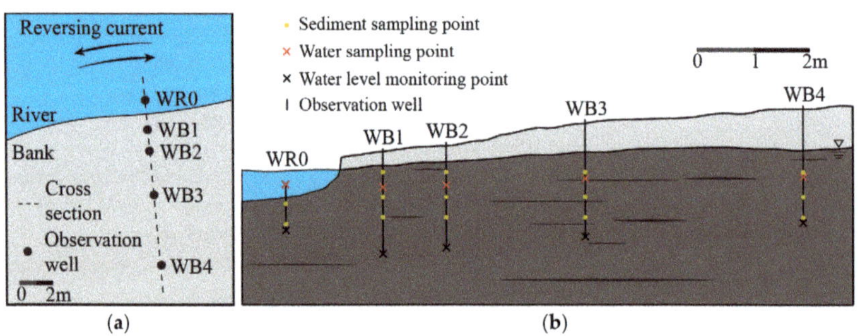

Figure 2. The transect for monitoring and sampling: (**a**) the top view; (**b**) the cross-section.

According to the preliminary work, the fluctuation range of the Jingzi River stage was mostly less than 0.5 m, and the change in groundwater table could not be observed as the river stage fluctuation was smaller than 10 cm (usually occurred during neap tides). Thus, a spring tide period without rain should be chosen to focus on the impact of tides. According to the tide prediction table and the weather forecast information released in Shanghai, the 3 day spring tide period from 21 to 23 October 2021 was selected to carry out the monitoring of water level and water quality. The data recorded by the water-level sensors were exported and processed through atmospheric pressure correction. A portable dissolved oxygen meter (LDO™, Manufacturer: HACH Co. Ltd., Loveland, CO, USA) and a portable multiparameter water quality analyzer (sensION+ MM156, Manufacturer: HACH Co. Ltd., Loveland, CO, USA) were employed to synchronously determine DO concentration, temperature, and pH in the river channel and the four riparian wells.

2.2. Sample Collection

During the drilling of five wells, 14 columnar sediment samples with a diameter of 70 mm and height of 200 mm were collected using a stainless-steel soil sampler. The riverbed sediment samples were collected at the depths of 10–30 cm and 40–60 cm below the riverbed at WR0. Considering the depth distribution of the groundwater table in the riparian zone, the riparian sediment samples were individually collected at the depth of 40–120 cm below the surface at WB1, at the depth of 50–130 cm below the surface at WB2, at the depth of 100–180 cm below the surface at WB3, and at the depth of 120–200 cm below the surface at WB4. The three sediment samples were collected from each riparian borehole at equal intervals. All sediment samples were stored in the separate bags in a cooling box keeping the temperature below 5 °C and then taken back to the lab for testing within 24 h. Sediment samples collected were divided into two aliquots: one for determining the contents of total nitrogen (TN) and total carbon (TC), and the other for analyzing permeability coefficient (K), particle size, and sediment type.

Water samples were collected from the river and the four wells using a liquid sampling tube with a total capacity of 500 mL. Water sampling was carried out at 2 h intervals from 6:00 a.m. to 6:00 p.m. on any particular day of 21–23 October 2021. The water sampling points were 50 cm below the free water table. Figure 2b presents the positions of the sediment and water sampling points. The tube was cleaned with deionized water before sampling to avoid cross-contamination between water samples [38]. Water discharged from the tube was stored in both polyethylene vials of 200 mL and amber glass vials of 20 mL. The former vials were used for the concentration analysis of inorganic nitrogen (NH_4^+, NO_3^-, and NO_2^-), while the latter vials were used for the DOC determination. All water samples were stored in the cooling box and subsequently taken back to the lab for testing within 24 h. Each water sample was filtered through a 0.45 μm filter membrane before testing.

2.3. Sample Analysis

TN content in each sediment sample was determined using a continuous flow analyzer (QuAAtro 39, Manufacturer: SEAL Co. Ltd., Norderstedt, Germany), while TC content was determined using a total organic carbon analyzer (TOC-VCPN, Manufacturer: Shimadzu Co. Ltd., Kyoto, Japan). Particle size was analyzed using a laser particle size meter (Mastersizer 3000, Manufacturer: Malvern Instrument Co. Ltd., Malvern, UK) with the measurement range of 0.01–3500 μm, after larger grains were removed through sieving. On the basis of the data from particle size analysis, the sediment type was determined. The sediment density (ρ) was quantified by the cutting ring method, and the permeability coefficient (K) was determined through the consolidation test.

For any water sample, NH_4^+ concentration was determined using a continuous flow analyzer (QuAAtro 39, Manufacturer: SEAL Co. Ltd., Norderstedt, Germany) with a quantification limit of 0.02 mg/L, NO_3^- concentration was determined using an ion chromatograph (ICS-90, Manufacturer: DIONEX Co. Ltd., Sunnyvale, CA, USA) with a quantification limit of 0.02 mg/L, NO_2^- concentration was determined using a spectrophotometer (U-3010, Manufacturer: HITACHI Co. Ltd., Tokyo, Japan) with a quantification limit of 0.002 mg/L, and DOC concentration was determined using a total organic carbon analyzer (TOC-VCPN, Manufacturer: Shimadzu Co. Ltd., Kyoto, Japan).

2.4. Determination of Periodicityand Response Time of Water-Level Fluctuation

2.4.1. Periodicity

The frequency analysis of the river stage and the riparian groundwater level was carried out using the fast Fourier transform algorithm (FFT) in MATLAB. FFT is applied to transform signals from time domain to frequency domain for quickly analyzing the frequency characteristics of stationary or nonstationary signals. FFT is an optimization algorithm of discrete Fourier transform (DFT). The DFT value $X(k)$ can be defined as follows [39]:

$$X(k) = \sum_{n=0}^{N-1} x(n) W_N^{nk}, \ (k = 0,\ 1,\ 2 \ldots N-1), \tag{1}$$

where $x(n)$ is the value in the n-th place of a sampled timeseries, N is the length of the series, W_N is the rotation factor equal to $exp(-j2\pi/N)$, and j is the imaginary unit, $j^2 = -1$.

Through the DFT mentioned above, spectrum values can be calculated, and the spectrum sequence can be obtained. In order to optimize the calculation efficiency, a sampled timeseries with the length of N is usually divided into two subsequences with the length of $N/2$ according to the symmetry and periodicity of W_N. As a result, a pair of DFT units can reduce the computation amount from N^2 to $Nlog_2N$, which is the reason why FFT is performed quickly. After the FFT calculation, the spectrum diagram can be plotted with signal frequency as the horizontal axis and amplitude $X(k)$ as the vertical axis. The dominant oscillation frequency of the river stage or the groundwater level corresponded to the maximum spectrum value in the diagram; accordingly, the dominant period could be determined.

2.4.2. Delayed Response Time and Correlation

The response of the groundwater level to the rise and fall of the river stage was analyzed using the cross-correlation analysis method. Through cross-correlation analysis, the response relationship between the two timeseries could be estimated, and the time shift could be quantified. For two discrete timeseries signals $Y_1(m)$ and $Y_2(m)$, their cross-correlation function $R(\tau)$ is expressed in Equation (2) [40].

$$R(\tau) = \sum_{m=0}^{M} Y_1(m) \cdot Y_2(m + \tau), \tag{2}$$

where m is the moment, M is the length of the timeseries, τ is the delayed time, and $Y_1(m)$ and $Y_2(m)$ are the timeseries of the river water level and the groundwater level, respectively. A larger $R(\tau)$ value indicates a greater correlation between the two sequences. If $Y_1(m)$ is equal to $Y_2(m)$, then $R(\tau)$ represents the self-correlation function value of the two sequences at τ. Normalization is usually performed on $R(\tau)$, so that the self-correlation is equal to one at the delayed time of zero. The delayed response time of groundwater level to the rise and fall of the river stage corresponded to the maximum $R(\tau)$.

2.5. Statistical Analysis of Water Quality Parameters Data

The seven water quality parameters (NH_4^+, NO_3^-, NO_2^-, DO, DOC, pH, and temperature) were analyzed using analysis of variance (ANOVA) test and Welch's test. Levene's test was employed initially to determine whether the data met the hypothesis of homogeneity of variance. If the result rejected the hypothesis, Welch's test was chosen to analyze the significance of differences in various water quality parameters of the groundwater among different riparian observation wells. Otherwise, the ANOVA test was used. The level of significance was set to $p < 0.05$ for both Welch's test and the ANOVA test.

3. Results

3.1. Variation and Hysteresis of Groundwater Table in the Riparian Hyporheic Zone under Tidal Effect

The river stage fluctuated twice a day and was of irregular semidiurnal characteristics (Figure 3). During the monitoring period, the mean tidal duration was about 12.2 h. The maximum level was 1.196 m a.s.l., and the minimum level was 0.850 m a.s.l., which contributed to a fluctuation range of 34.6 cm.

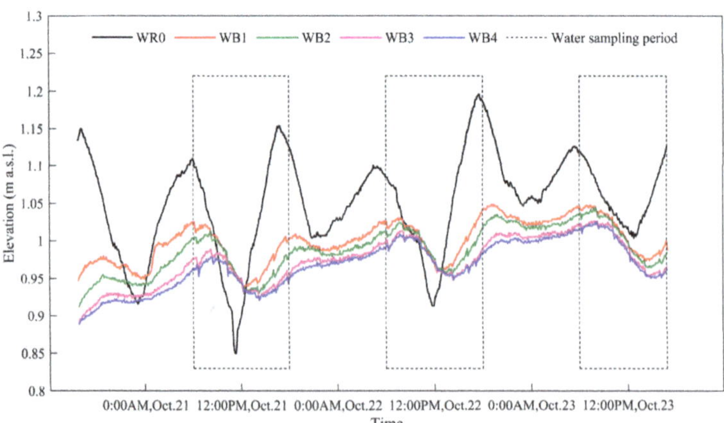

Figure 3. Fluctuation curve of water level in the river channel (WR0) and the four riparian observation wells (WB1–WB4) during the monitoring period from 21 to 23 October 2021.

The groundwater level in any riparian well fluctuated twice a day, just like the river water stage (Figure 3). Regardless of the river channel or the riparian wells, the dominant oscillation period of the water level time series was 12.2 h. The fluctuation range of the water level gradually decreased upon increasing the distance from the shore on any monitoring day, as shown in Table 1. Over the 3 day study period, the fluctuation range of the daily mean river stage declined from 22.7 to 12.9 cm, while that of the daily mean groundwater level in each riparian well changed a little. Furthermore, the groundwater level in the riparian hyporheic zone exhibited a slow response to the rise and fall movement of the river stage. The delayed response time extended from 90 to 180 min, and the correlation coefficient decreased from 0.75 to 0.52, when the distance increased from 0.71 to 3.85 m away from the shore.

Table 1. Fluctuations and the response relationship of the river stage and the groundwater level during the monitoring period from 21 to 23 October 2021.

Observation Well	Mean Water Level Fluctuation Range (cm)			Time Period (h)	Response to the River Stage	
	21 October	22 October	23 October		Delayed Time (min)	Correlation Coefficient
WR0	22.7	17.9	12.9	12.2	—	—
WB1	6.4	5.8	5.2	12.2	90	0.75
WB2	5.6	4.9	5.2	12.2	135	0.66
WB3	4.3	4.0	4.4	12.2	175	0.56
WB4	3.7	3.5	4.3	12.2	180	0.52

During most of the monitoring period, the river stage was higher than the groundwater level and, hence, the river water continued to penetrate into the riparian hyporheic zone under such a hydraulic gradient. As a result, the groundwater level showed an overall upward trend (Figure 3). The daily mean groundwater levels in different riparian wells showed upward trends with the increase of 3.1–3.9 cm over the monitoring period (Figure 4a). Since the temperature of the river water was lower than that of the groundwater during the study period, the infiltration of the river water to the groundwater contributed to the drop in groundwater temperature. Although the river water temperature increased, the groundwater temperature in the riparian hyporheic zone declined (Figure 4b). Moreover, the decline range had a negative correlation with the distance from the shore. In view of the variations in water level and temperature of the groundwater at WB4, the migration range of water moisture and heat in the riparian hyporheic zone could reach 4 m away from the shore during the study period.

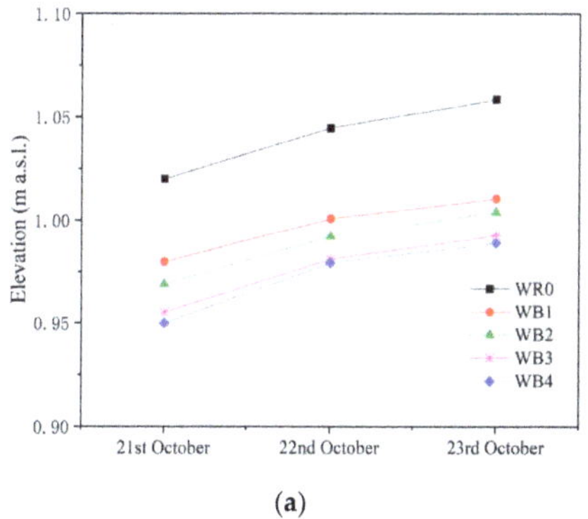

(a)

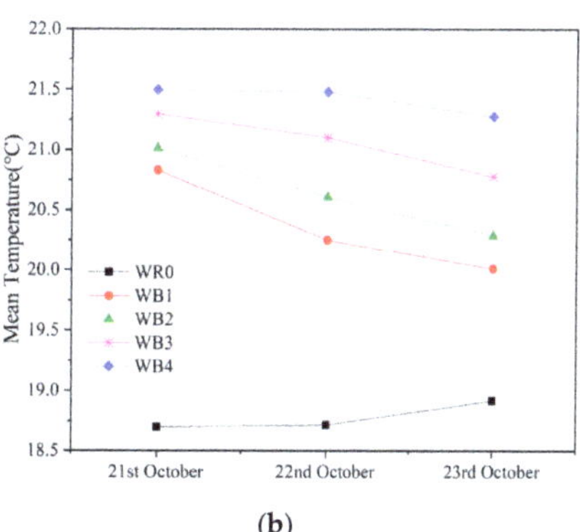

(b)

Figure 4. Variations in (**a**) daily mean water level and (**b**) daily mean temperature of the river water and the groundwater over the monitoring period from 21 to 23 October 2021.

3.2. Spatial Distribution of Inorganic Nitrogen in the Riparian Hyporheic Zone

According to the statistical analysis, the concentration of three species of inorganic nitrogen (NH_4^+, NO_3^-, and NO_2^-) in the groundwater showed significant differences among different riparian wells ($p < 0.001$). The mean NH_4^+ concentration was 0.540 mg/L in the river channel, while those in the riparian observation wells ranged from 0.085 to 0.247 mg/L. This indicated that the NH_4^+ concentration in the river water was higher than that in the riparian groundwater (Figure 5a). The mean NH_4^+ concentrations in the groundwater in WB1, WB2, WB3, and WB4 were 0.085, 0.099, 0.247, and 0.105 mg/L, respectively. It is, thus, clear that the NH_4^+ concentration in the groundwater in the riparian hyporheic zone was lower in the near-shore zone where WB1 and WB2 existed than in the offshore zone where WB3 and WB4 existed.

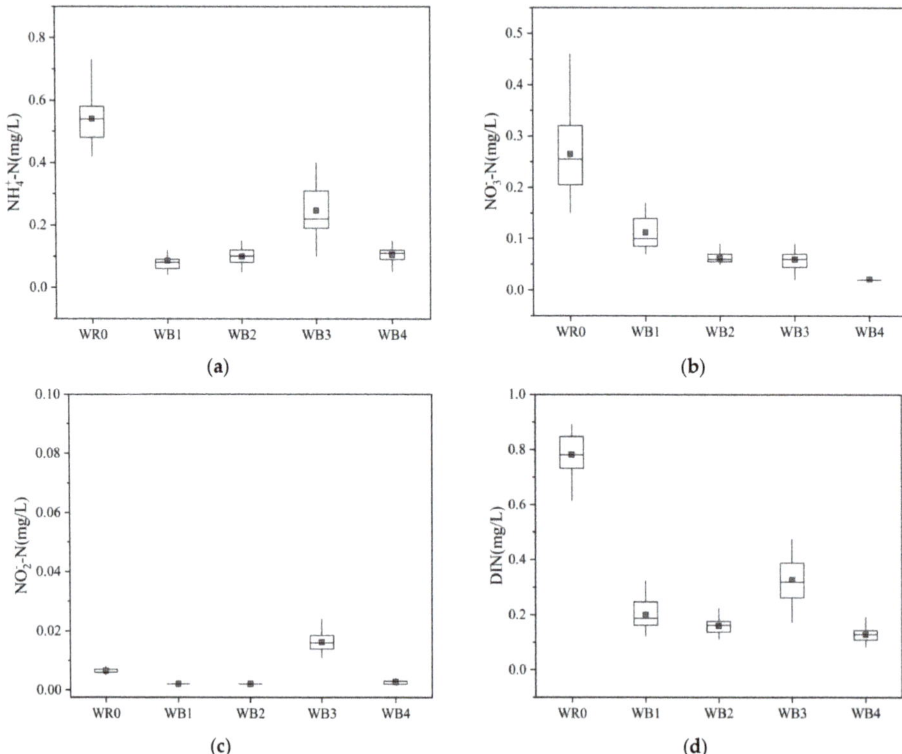

Figure 5. Boxplots of the observed concentrations of inorganic nitrogen in the river water and the riparian groundwater during the study period from 21 to 23 October 2021: (**a**) ammonia nitrogen, NH_4^+; (**b**) nitrate nitrogen, NO_3^-; (**c**) nitrite nitrogen, NO_2^-; (**d**) dissolved inorganic nitrogen, DIN.

The mean NO_3^- concentration was 0.265 mg/L in the river water and varied from 0.022 to 0.112 mg/L in the groundwater in the riparian wells (Figure 5b). This implies that the NO_3^- concentration in the river water was higher than that in the groundwater in the riparian hyporheic zone. The mean NO_3^- concentrations in the groundwater were 0.112, 0.062, 0.060, and 0.022 mg/L in WB1, WB2, WB3, and WB4, respectively. The mean NO_3^- concentrations in the riparian groundwater decreased gradually with increasing distance from the shore. The mean NO_2^- concentrations in both the river water and the groundwater in the four wells were below 0.02 mg/L (Figure 5c). This was mainly due to its chemical instability.

Dissolved inorganic nitrogen (DIN) was approximate to the sum of the concentrations of NH_4^+, NO_3^-, and NO_2^-. The mean DIN concentration was 0.782 mg/L in the river water and varied from 0.129 to 0.327 mg/L in the groundwater in the riparian wells (Figure 5d). The DIN concentration in the river water was higher than that in the riparian groundwater, and the offshore groundwater had a higher DIN concentration than the near-shore groundwater.

The content of NO_3^- in the groundwater accounted for about 56% of the total DIN content at WB1 on average and decreased with increasing distance from the shore (Figure 6). In contrast, the mean NH_4^+ proportion gradually increased from 43% at WB1 to 81% at WB4. For any riparian well, the mean NO_2^- proportion kept the low value less than 5%.

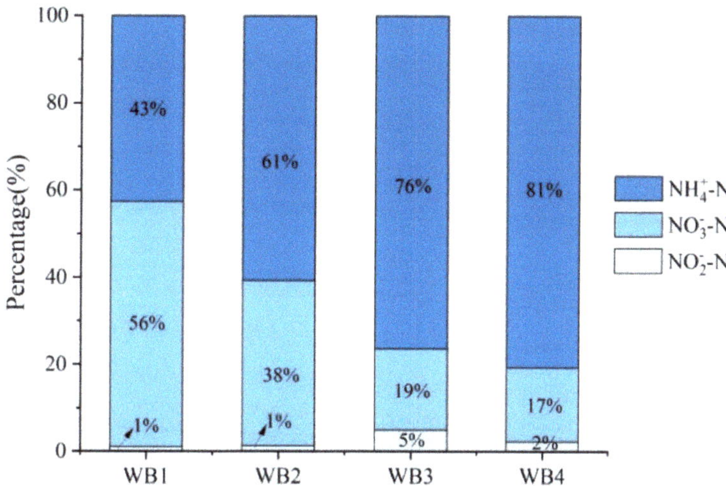

Figure 6. Mean proportions of different forms of inorganic nitrogen in the groundwater in the riparian observation wells during the study period from 21 to 23 October 2021.

3.3. Temporal Variation of Inorganic Nitrogen Concentrations in the Riparian Hyporheic Zone

Over the study period, the daily mean NH_4^+ concentration in the groundwater increased from 0.070 to 0.106 mg/L at WB1 and rose from 0.087 to 0.114 mg/L at WB2 (Figure 7a). In contrast, it decreased from 0.337 to 0.193 mg/L at WB3 and fell from 0.110 to 0.097 mg/L at WB4. It is distinctly clear that the daily mean NH_4^+ concentration showed an upward trend in the near-shore zone and a downward one in the offshore zone.

Meanwhile, the daily mean NO_3^- concentration in the groundwater increased from 0.106 to 0.120 mg/L at WB1 and rose from 0.057 to 0.068 mg/L at WB2 (Figure 7b). However, it changed little in the offshore groundwater. The daily mean NO_2^- concentration in the riparian groundwater showed little change (Figure 7c). This is mainly due to the NO_2^- concentration having a very low concentration in both the river water and the riparian groundwater.

The daily mean DIN concentration increased from 0.178 to 0.228 mg/L at WB1 and ascended from 0.146 to 0.185 mg/L at WB2 (Figure 7d). It decreased from 0.408 to 0.271 mg/L at WB3 and fell from 0.132 to 0.123 mg/L at WB4. It is clear that the variation of the DIN concentration was roughly the same as that of the NH_4^+ concentration in the groundwater; hence, NH_4^+ was generally the dominant form of inorganic nitrogen in the riparian hyporheic zone within 4 m from the shore.

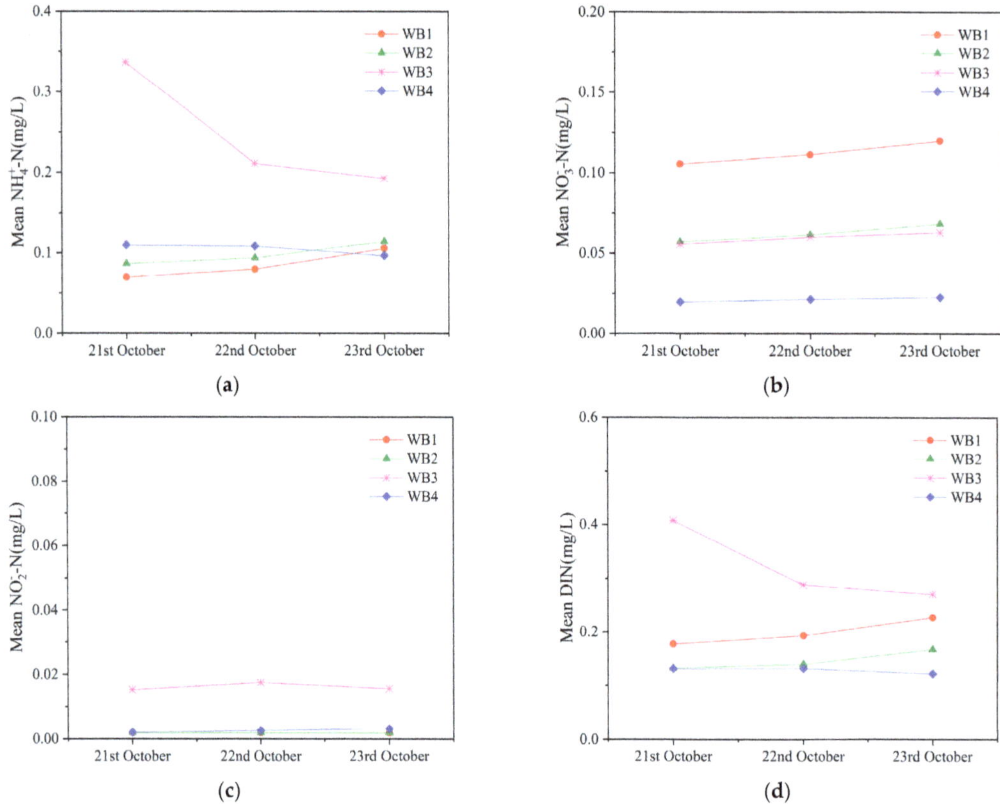

Figure 7. Variations in daily mean inorganic nitrogen concentrations in the river water and the riparian groundwater over the study period from 21 to 23 October 2021: (**a**) ammonia nitrogen, NH_4^+; (**b**) nitrate nitrogen, NO_3^-; (**c**) nitrite nitrogen, NO_2^-; (**d**) dissolved inorganic nitrogen, DIN.

3.4. Physicochemical Characteristics of Sediment and Water in the River Channel and the Riparian Hyporheic Zone

There was a great difference between the horizontal and vertical permeability coefficients of the sediment in the riparian hyporheic zone (Table 2). For the sediment at any specific position, the horizontal permeability coefficient was 10 times higher than the vertical one, which implied that the lateral migrations of both the hyporheic flow and the solutes were dominant along the transect. According to the grain size distribution, the riparian sediment was determined as silty-clay soil.

Table 2. Physical properties of the sediment in the riparian hyporheic zone.

Observation Well	Density (g/cm^3)	Coefficient of Permeability (cm/s)				Granularity Content (%)		Soil Type
		Horizontal		Vertical				
		Average	Standard Deviation	Average	Standard Deviation	Clay ($d < 0.005$ mm)	Silt (0.005 mm $\leq d \leq 0.075$ mm)	
WB1	1.89–1.97	4.40×10^{-6}	6.26×10^{-7}	8.71×10^{-8}	0.98×10^{-9}	27–31	53–57	Silty-clay
WB2	1.92–1.97	4.19×10^{-6}	5.78×10^{-7}	9.14×10^{-8}	0.67×10^{-9}	32–34	54–58	Silty-clay
WB3	1.97–1.99	3.11×10^{-6}	8.37×10^{-7}	8.13×10^{-8}	1.05×10^{-8}	34–41	53–59	Silty-clay
WB4	1.94–1.96	3.97×10^{-6}	2.42×10^{-7}	8.44×10^{-8}	1.13×10^{-8}	32–35	55–57	Silty-clay

Both the TN and the TC contents of the sediments gradually increased with depth, and the low contents of TN and TC appeared within the extent (0.9–1.05 m a.s.l.) where the riparian groundwater fluctuated (Figure 8). This could provide a clue that the extent where the groundwater level varied might be the hotspot of biogeochemical reactions in the riparian hyporheic zone, since large amounts of organic mass such as organic nitrogen and organic carbon were consumed in the course of the reactions involving micro-organisms. Furthermore, the mean TN content of the sediment at WB3 was the highest at 674 mg/kg, compared with that at the three other riparian wells.

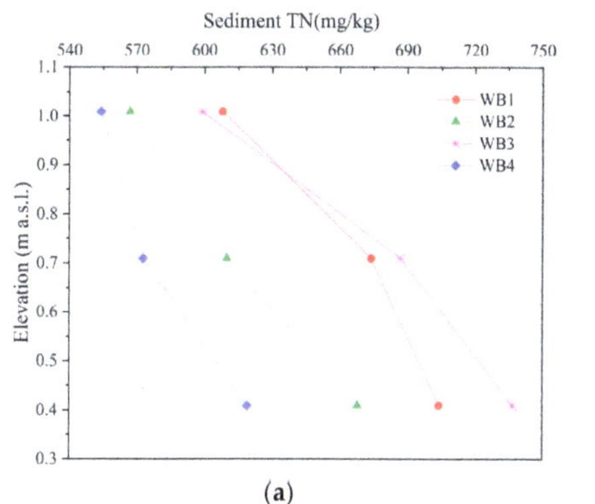

(a)

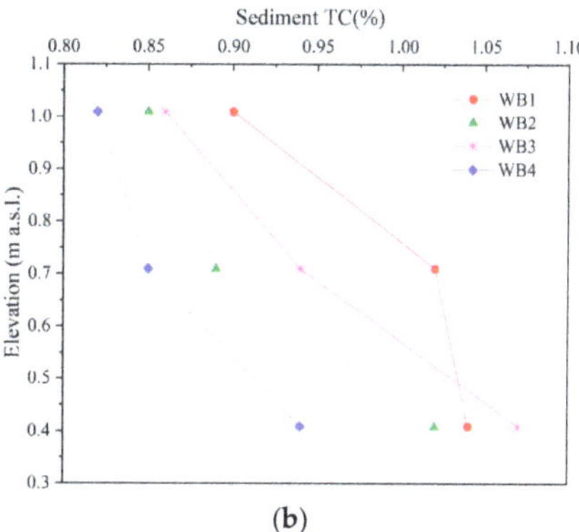

(b)

Figure 8. Vertical distribution of the nitrogen and carbon contents of the riparian sediments: (**a**) total nitrogen, TN; (**b**) total carbon, TC.

The DO concentration in the groundwater showed a significant difference among different riparian wells ($p < 0.05$). The mean DO concentration in the river water was 6.24 mg/L, higher than that in the riparian groundwater (Figure 9a). Moreover, the mean DO concentration in the riparian groundwater decreased gradually with an increase in distance from the shore. The DO concentration of the riparian groundwater ranged from 5.2 to 6.8 mg/L. The difference of the DOC concentration in the groundwater was not significant among the riparian wells ($p > 0.05$). The data distribution range of the DOC concentration in the river water was less than that in the riparian groundwater (Figure 9b). This implies that the river water was not the only source of the DOC in the riparian groundwater.

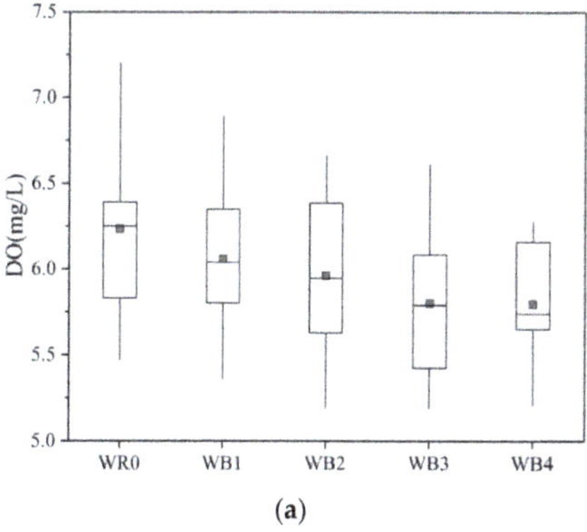

(a)

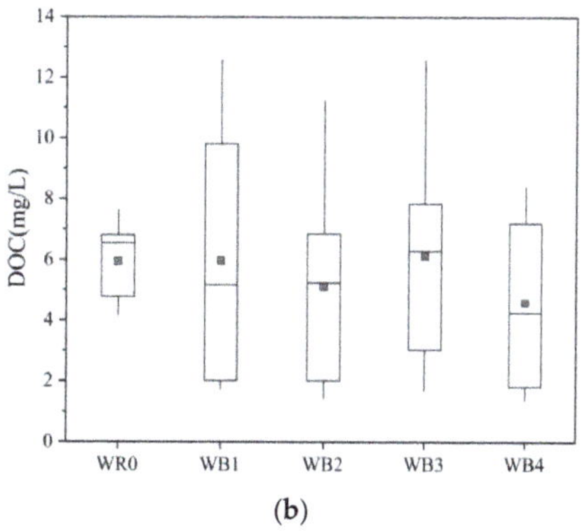

(b)

Figure 9. *Cont.*

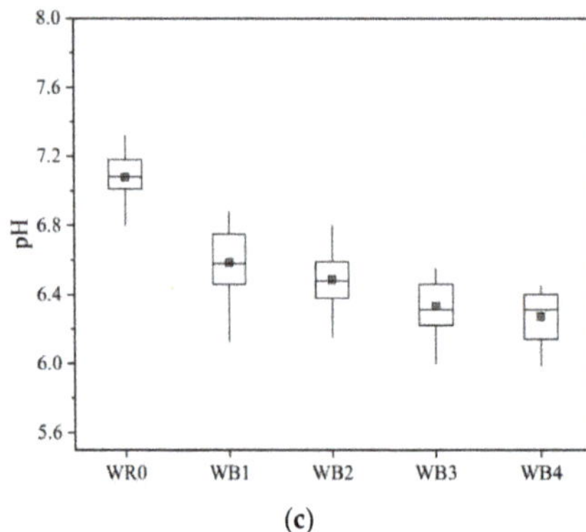

(c)

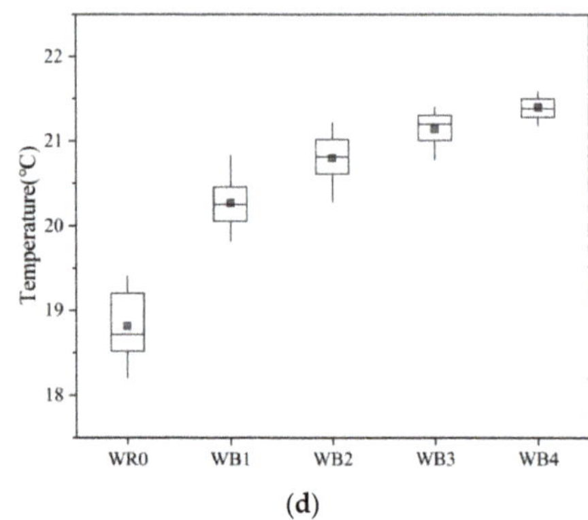

(d)

Figure 9. Spatial variations of eco-environmental factors along the transect: (**a**) dissolved oxygen, DO; (**b**) dissolved organic carbon, DOC; (**c**) pH; (**d**) temperature.

There were significant differences in both the pH and the temperature in the groundwater among different riparian wells ($p < 0.001$). The mean pH of the river water was about 7.08 (Figure 9c), higher than that of the groundwater in any riparian well (ranging from 6.27 to 6.59). Along the transect, the pH decreased with increasing the distance from the shore. The mean temperature of the river water was 18.8 °C, while that of the groundwater in the riparian wells, ranging from 20.3 to 21.4 °C, increased gradually with an increase in distance from the shore (Figure 9d).

4. Discussion

During the study period, the pH and the DO concentrations in the groundwater in the four riparian wells varied from 6.27 to 6.59 (neutral environment) and from 5.2 to 6.8 mg/L, respectively. It is generally considered that an environment is aerobic if DO concentration is greater than 3 mg/L [41]. Thus, the riparian hyporheic zone within 4 m from the bank was an aerobic and neutral environment in favor of nitrification. In past studies, denitrification was regarded as an anaerobic process, and it was considered that denitrification would no longer occur if DO concentration was greater than 4.5 mg/L [42,43]. However, recent research results showed that denitrification could also occur in aerobic zones in soils due to the existence of aerobic denitrifying bacteria which could live in the environment with a DO concentration of 3–9 mg/L and a neutral pH [44–46]. It was, therefore, deduced that aerobic denitrification and nitrification might coexist in the riparian hyporheic zone studied, and the combination of these two effects was conducive to the removal of nitrogen from the riparian zone.

In the investigation, the NH_4^+ concentration in the near-shore groundwater was lower than that in the offshore groundwater, and the lowest value found at WB1 closest to the bank. It was reported by Pan et al. that the NH_4^+ concentration in the riparian groundwater decreased with increasing the distance from the bank of the nitrogen-rich river [47]. This is inconsistent with the results of our research. By comparison, it was found that there were significant differences in NH_4^+ concentration in the river water and type of the sediment. The NH_4^+ concentration in the river water was 0.9–1.5 mg/L, and the sediment mainly consisted of coarse sand in their research, while the former was 0.4–0.7 mg/L and the latter was silty-clay soil in our research. Some studies indicated that a low permeability of the sediment could reduce flow velocity of the pore water and, consequently, prolong the solute residence time, which provided the sufficient reaction time for nitrogen in the hyporheic zone [48,49]. Thus, the lower-permeability sediment and the river water with the lower nitrogen concentration contributed to more consumption and less recharge of the NH_4^+ in the near-shore groundwater, respectively. This could explain why groundwater with less NH_4^+ was observed in the near-shore hyporheic zone. However, upon increasing the distance from the bank, the DO concentration in the groundwater deceased gradually. The decline in DO concentration could lower the intensity of nitrification [50]. Furthermore, the mean TN content in the sediment at WB3 was highest among the four observation wells. The higher TN content could promote ammoniation, which is an important pathway for recharging the NH_4^+ in the groundwater in hyporheic zones [51,52]. The greater recharge and lower consumption resulted in more NH_4^+ existing in the groundwater in the offshore hyporheic zone.

The NO_3^- concentration in the riparian groundwater showed a downward trend as the distance from the bank increased along the test transect. This matches with the work of Zhang et al. [53]. Due to the strong nitrification reactions and the recharge from the river water, the NO_3^- was enriched in the near-shore groundwater. However, in the offshore hyporheic zone with the poorly permeable sediment, the recharge of NO_3^- from the river water was very slow and could take several hours or even more according to the delayed response time of the groundwater level to the river stage in the study. Moreover, due to the decline in DO concentration, a lower amount of the NO_3^- in the groundwater was produced by nitrification. Additionally, the riparian zone beyond 2 m away from the bank was covered with undisturbed shrubs. According to the research conducted by Stark and Hart, the microbial communities in the soil under undisturbed forest ecosystems had the capacity to assimilate most of the NO_3^- produced by nitrification [54]. Hence, the microbial assimilation of NO_3^- could be considered as a reason for the NO_3^- in the offshore groundwater being lower than that in the near-shore groundwater.

During most of the study period, the river stage remained higher than the riparian groundwater level under tide action, which led to a continuous infiltration of river water into the riverbank. As the river water steadily flowed into the hyporheic zone, the NH_4^+ and NO_3^- in the near-shore groundwater were continuously recharged; thus, the NH_4^+ and NO_3^- concentrations showed overall upward trends in the near-shore zone over the 3 day study period. Accompanied by the seepage process, the riparian groundwater level had an uplift which made part of the oxygen in the upper vadose zone enter the groundwater [55]. Thus, the DO concentration in the riparian groundwater increased due to the lateral recharge from the river, as well as the vertical recharge from the upper soil, which contributed to a whole aerobic environment within 4 m from the bank. The increase in DO concentration in the offshore groundwater promoted nitrification [56]. Meanwhile, the fine particles of the silty clay sediment had strong adsorption capacity, which restricted the further migration of the NH_4^+ in the near-shore groundwater to some extent [57]. For these reasons, the NH_4^+ concentration in the offshore groundwater decreased over the study period. Although the further migration of the NO_3^- in the near-shore groundwater and nitrification could have caused an increase in NO_3^- concentration in the offshore groundwater, little change was actually observed. This might be due to the fact that the microbial assimilation of NO_3^- was considerable in the offshore hyporheic zone under the shrubs.

The variation range of the DOC concentration in the groundwater in each riparian well was significantly larger than that in the river. The DOC in the riparian groundwater mainly depended on the dissolution of particulate organic carbon (POC) in the sediment rather than the infiltration of the river water, which is consistent with the research of Peyrard et al. [58]. Some previous studies reported the effect of DOC, pH, and temperature on the nitrogen cycle in the hyporheic zone. The DOC could play an important role in denitrification processes [59]. The reaction rate of nitrogen could be inhibited in an acidic environment (pH < 5.5), but there is an insignificant impact on nitrogen cycle in a neutral environment [60,61]. The nitrification rate in soils gradually increases with temperature within the range of 15–35 °C [62]. However, in our research, there was an insignificant difference in DOC concentration in the riparian groundwater. Moreover, the pH remained neutral, and the temperature had a small change of 1–2 °C in the riparian hyporheic zone. Hence, DOC concentration, pH, and temperature may not be the dominant factors causing the spatiotemporal difference in inorganic nitrogen concentration in the groundwater in the riparian zone studied.

Additionally, it should be noted that the time period involved in the research was a spring tidal period in October, and recharging the riparian aquifers from the river water was dominant during the period. However, the tide-driven hyporheic exchange process varies with tide strength and season. Since the hyporheic exchange greatly influences the nitrogen occurrence, the findings of the study have to be seen in light of some limitations. Future research will be performed to analyze nitrogen occurrence in groundwater in silty-clay hyporheic zones under the action of tides of different strength and its seasonal variation. Moreover, the variations of nitrogen forms and the influencing factors with depth will need to be investigated to reveal the vertical nitrogen cycle in hyporheic zones of tidal rivers.

5. Conclusions

It was confirmed from the study that the groundwater level in the silty-clay riparian zone presented a slow response to the tide-driven fluctuation of the river stage. The delayed response time reached a few hours within 1–4 m away from the bank, which could help to extend the hydraulic and solute retention time. The continuous infiltration of the river water under tide action contributed to the aerobic and neutral riparian hyporheic zone conductive to nitrification. Within 4 m away from the bank, the dominant inorganic nitrogen form changed from NO_3^--N to NH_4^+-N, upon increasing the distance from the bank. In addition, the removal of nitrogen could occur in the riparian hyporheic zone with an aerobic and neutral environment under the joint control of nitrification, microbial assimilation, and aerobic denitrification. It is important to consider that nitrogen occurrence in riparian hyporheic zones of tidal rivers is governed by many factors including tidal strength, permeability and chemical compositions of sediments, DO concentration, vegetation coverage, and even nitrogen concentration in river water.

Author Contributions: Conceptualization, Y.C. and J.X.; methodology, Y.C. and J.X.; software, J.X.; validation, J.X.; formal analysis, Y.C. and J.X.; investigation, Y.C., J.X., N.Z., R.H., X.R. and D.Y.; resources, Y.C. and J.X.; data curation, J.X., R.H. and X.R.; writing—original draft preparation, Y.C. and J.X.; writing—review and editing, Y.C. and N.Z.; funding acquisition, Y.C. and N.Z. All authors have read and agreed to the published version of the manuscript.

Funding: This research was financially supported by the National Natural Science Foundation of China (Grant No. 42077176) and the Natural Science Foundation of Shanghai (Grant No. 20ZR1459700).

Institutional Review Board Statement: Not applicable.

Informed Consent Statement: Not applicable.

Data Availability Statement: Data are contained within the article.

Acknowledgments: The authors specially thank the Shanghai Institute of Geological Survey for providing assistance with sample testing.

Conflicts of Interest: The authors declare no conflict of interest.

References

1. Wang, Y.C.; Lu, Y.L. Evaluating the potential health and economic effects of nitrogen fertilizer application in grain production systems of China. *J. Clean. Prod.* **2020**, *264*, 121635. [CrossRef]
2. Liang, D.; Song, J.X.; Xia, J.; Chang, J.B.; Kong, F.H.; Sun, H.T.; Wu, Q.; Cheng, D.D.; Zhang, Y.X. Effects of heavy metals and hyporheic exchange on microbial community structure and functions in hyporheic zone. *J. Environ. Manag.* **2022**, *303*, 114201. [CrossRef] [PubMed]
3. Ward, A.S. The evolution and state of interdisciplinary hyporheic research. Wiley Interdiscip. *Wiley Interdiscip. Rev. Water* **2016**, *3*, 83–103.
4. Cai, Y.; Shi, T.; Yao, J.L.; Liu, S.G.; Ruan, X.K.; Xu, J. Advance in field monitoring research of hydrodynamic process of hyporheic exchange in rivers. *Adv. Water Sci.* **2021**, *32*, 638–648. (In Chinese)
5. Bernal, S.; Lupon, A.; Ribot, M.; Sabater, F.; Marti, E. Riparian and in-stream controls on nutrient concentrations and fluxes in a headwater forested stream. *Biogeosciences* **2015**, *12*, 1941–1954. [CrossRef]

6. Lamontagne, S.; Cosme, F.; Minard, A.; Holloway, A. Nitrogen attenuation, dilution and recycling in the intertidal hyporheic zone of a subtropical estuary. *Hydrol. Earth Syst. Sci.* **2018**, *22*, 4083–4096. [CrossRef]
7. Hampton, T.B.; Zarnetske, J.P.; Briggs, M.A.; Singha, K.; Harvey, J.W.; Day-Lewis, F.D.; Dehkordy, F.M.; Lane, J.W. Residence time controls on the fate of nitrogen in flow-through lakebed sediments. *J. Geophys. Res.-Biogeosci.* **2019**, *124*, 689–707. [CrossRef]
8. Naranjo, R.C.; Niswonger, R.G.; Davis, C.J. Mixing effects on nitrogen and oxygen concentrations and the relationship to mean residence time in a hyporheic zone of a riffle-pool sequence. *Water Resour. Res.* **2015**, *51*, 7202–7217. [CrossRef]
9. Zhou, N.Q.; Zhao, S.; Shen, X.P. Nitrogen cycle in the hyporheic zone of natural wetlands. *Chin. Sci. Bull.* **2014**, *59*, 2945–2956. [CrossRef]
10. Wen, H.; Perdrial, J.; Abbott, B.W.; Bernal, S.; Dupas, R.; Godsey, S.E.; Harpold, A.; Rizzo, D.; Underwood, K.; Adler, T.; et al. Temperature controls production but hydrology regulates export of dissolved organic carbon at the catchment scale. *Hydrol. Earth Syst. Sci.* **2020**, *24*, 945–966. [CrossRef]
11. Ledesma, J.L.J.; Ruiz-Perez, G.; Lupon, A.; Poblador, S.; Futter, M.N.; Sabater, F.; Bernal, S. Future changes in the Dominant Source Layer of riparian lateral water fluxes in a subhumid Mediterranean catchment. *J. Hydrol.* **2021**, *595*, 126014. [CrossRef]
12. Newcomer, M.E.; Hubbard, S.S.; Fleckenstein, J.H.; Maier, U.; Schmidt, C.; Thullner, M.; Ulrich, C.; Flipo, N.; Rubin, Y. Influence of hydrological perturbations and riverbed sediment characteristics on hyporheic zone respiration of CO_2 and N_2. *J. Geophys. Res.-Biogeosci.* **2018**, *123*, 902–922. [CrossRef]
13. Cai, Y.; Xing, J.W.; Ruan, X.K.; Zhong, N.Q.; Huang, R.Y.; Yi, D.Z. Advances in the numerical simulation of the migration and transformation of nitrogen in hyporheic zones of rivers. *Water Resour. Prot.* **2022**. Available online: http://kns.cnki.net/kcms/detail/32.1356.TV.20220419.1738.004.html (accessed on 20 April 2022). (In Chinese).
14. Shibata, H.; Sugawara, O.; Toyoshima, H.; Wondzell, S.M.; Nakamura, F.; Kasahara, T.; Swanson, F.J.; Sasa, K. Nitrogen dynamics in the hyporheic zone of a forested stream during a small storm, Hokkaido, Japan. *Biogeochemistry* **2004**, *69*, 83–103. [CrossRef]
15. Singh, T.; Wu, L.W.; Gomez-Velez, J.D.; Lewandowski, J.; Hannah, D.M.; Krause, S. Dynamic hyporheic zones: Exploring the role of peak flow events on bedform-induced hyporheic exchange. *Water Resour. Res.* **2019**, *55*, 218–235. [CrossRef]
16. Sawyer, A.H.; Cardenas, M.B.; Bomar, A.; Mackey, M. Impact of dam operations on hyporheic exchange in the riparian zone of a regulated river. *Hydrol. Process* **2009**, *23*, 2129–2137. [CrossRef]
17. Bernhardt, E.S.; Blaszczak, J.R.; Ficken, C.D.; Fork, M.L.; Kaiser, K.E.; Seybold, E.C. Control points in ecosystems: Moving beyond the hot Spot hot moment concept. *Ecosystems* **2017**, *20*, 665–682. [CrossRef]
18. Musial, C.T.; Sawyer, A.H.; Barnes, R.T.; Bray, S.; Knights, D. Surface water-groundwater exchange dynamics in a tidal freshwater zone. *Hydrol. Process* **2016**, *30*, 739–750. [CrossRef]
19. Barnes, R.T.; Sawyer, A.H.; Tight, D.M.; Wallace, C.D.; Hastings, M.G. Hydrogeologic controls of surface water-groundwater nitrogen dynamics within a tidal freshwater zone. *J. Geophys. Res.-Biogeosci.* **2019**, *124*, 3343–3355. [CrossRef]
20. Wallace, C.D.; Sawyer, A.H.; Soltanian, M.R.; Barnes, R.T. Nitrate removal within heterogeneous riparian aquifers under tidal influence. *Geophys. Res. Lett.* **2020**, *47*, e2019GL085699. [CrossRef]
21. Wallace, C.D.; Sawyer, A.H.; Barnes, R.T.; Soltanian, M.R.; Gabor, R.S.; Wilkins, M.J.; Moore, M.T. A model analysis of the tidal engine that drives nitrogen cycling in coastal riparian aquifers. *Water Resour. Res.* **2020**, *56*, e2019WR025662. [CrossRef]
22. Pryshlak, T.T.; Sawyer, A.H.; Stonedahl, S.H.; Soltanian, M.R. Multiscale hyporheic exchange through strongly heterogeneous sediments. *Water Resour. Res.* **2015**, *51*, 9127–9140. [CrossRef]
23. McGarr, J.T.; Wallace, C.D.; Ntarlagiannis, D.; Sturmer, D.M.; Soltanian, M.R. Geophysical mapping of hyporheic processes controlled by sedimentary architecture within compound bar deposits. *Hydrol. Process* **2021**, *35*, e14358. [CrossRef]
24. Briggs, M.A.; Day-Lewis, F.D.; Zarnetske, J.P.; Harvey, J.W. A physical explanation for the development of redox microzones in hyporheic flow. *Geophys. Res. Lett.* **2015**, *42*, 4402–4410. [CrossRef]
25. Krause, S.; Tecklenburg, C.; Munz, M.; Naden, E. Streambed nitrogen cycling beyond the hyporheic zone: Flow controls on horizontal patterns and depth distribution of nitrate and dissolved oxygen in the upwelling groundwater of a lowland river. *J. Geophys. Res.-Biogeosci.* **2013**, *118*, 54–67. [CrossRef]
26. Zhao, L.; Li, Y.L.; Wang, S.D.; Wang, X.Y.; Meng, H.Q.; Luo, S.H. Adsorption and transformation of ammonium ion in a loose-pore geothermal reservoir: Batch and column experiments. *J. Contam. Hydrol.* **2016**, *192*, 50–59. [CrossRef] [PubMed]
27. Shan, J.; Yang, P.P.; Shang, X.X.; Rahman, M.M.; Yan, X.Y. Anaerobic ammonium oxidation and denitrification in a paddy soil as affected by temperature, pH, organic carbon, and substrates. *Biol. Fertil. Soils* **2018**, *54*, 341–348. [CrossRef]
28. Bai, B.; Jiang, S.C.; Liu, L.L.; Li, X.; Wu, H.Y. The transport of silica powders and lead ions under unsteady flow and variable injection concentrations. *Powder Technol.* **2021**, *387*, 22–30. [CrossRef]
29. Helton, A.M.; Ardon, M.; Bernhardt, E.S. Thermodynamic constraints on the utility of ecological stoichiometry for explaining global biogeochemical patterns. *Ecol. Lett.* **2015**, *18*, 1049–1056. [CrossRef]
30. Pescimoro, E.; Boano, F.; Sawyer, A.H.; Soltanian, M.R. Modeling influence of sediment heterogeneity on nutrient cycling in streambeds. *Water Resour. Res.* **2019**, *55*, 4082–4095. [CrossRef]
31. Zarnetske, J.P.; Haggerty, R.; Wondzell, S.M.; Baker, M.A. Labile dissolved organic carbon supply limits hyporheic denitrification. *J. Geophys. Res.-Biogeosci.* **2011**, *116*, G04036. [CrossRef]
32. Zarnetske, J.P.; Haggerty, R.; Wondzell, S.M.; Baker, M.A. Dynamics of nitrate production and removal as a function of residence time in the hyporheic zone. *J. Geophys. Res.-Biogeosci.* **2011**, *116*, G01025. [CrossRef]

33. Herrman, K.S.; Bouchard, V.; Moore, R.H. Factors affecting denitrification in agricultural headwater streams in Northeast Ohio, USA. *Hydrobiologia* **2008**, *598*, 305–314. [CrossRef]
34. Bai, B.; Nie, Q.K.; Zhang, Y.K.; Wang, X.L.; Hu, W. Cotransport of heavy metals and SiO_2 particles at different temperatures by seepage. *J. Hydrol.* **2021**, *597*, 125771. [CrossRef]
35. Zhao, S.; Zhou, N.Q.; Shen, X.P. Driving mechanisms of nitrogen transport and transformation in lacustrine wetlands. *Sci. China-Earth Sci.* **2016**, *59*, 464–476. [CrossRef]
36. Zong, H.C.; Zhang, W.S.; Zhang, J.S. Study on influence of sea level rise on storm tidal levels of Huangpu River. *Yangtze River* **2014**, *45*, 1–3. (In Chinese)
37. Huang, X.Y. Hydrological characteristics of Shanghai area. *J. China Hydrol.* **1987**, *32*, 43–46. (In Chinese)
38. Harvey, J.W.; Bohlke, J.K.; Voytek, M.A.; Scott, D.; Tobias, C.R. Hyporheic zone denitrification: Controls on effective reaction depth and contribution to whole-stream mass balance. *Water Resour. Res.* **2013**, *49*, 6298–6316. [CrossRef]
39. Marcotte, D. Fast variogram computation with FFT. *Comput. Geosci.* **1996**, *22*, 1175–1186. [CrossRef]
40. Qian, W.M.; Liang, H.Y.; Yang, G.Q. *Applied Stochastic Processes*, 1st ed.; Higher Education Press: Beijing, China, 2014; pp. 101–109. (In Chinese)
41. Anderson, T.H.; Taylor, G.T. Nutrient pulses, plankton blooms, and seasonal hypoxia in western Long Island Sound. *Estuaries* **2001**, *24*, 228–243. [CrossRef]
42. Xue, D.M.; Botte, J.; De Baets, B.; Accoe, F.; Nestler, A.; Taylor, P.; Van Cleemput, O.; Berglund, M.; Boeckx, P. Present limitations and future prospects of stable isotope methods for nitrate source identification in surface- and groundwater. *Water Res.* **2009**, *43*, 1159–1170. [CrossRef]
43. Patureau, D.; Bernet, N.; Delgenes, J.P.; Moletta, R. Effect of dissolved oxygen and carbon-nitrogen loads on denitrification by an aerobic consortium. *Appl. Microbiol. Biotechnol.* **2000**, *54*, 535–542. [CrossRef] [PubMed]
44. Huang, T.L.; Zhou, S.L.; Zhang, H.H.; Bai, S.Y.; He, X.X.; Yang, X. Nitrogen removal characteristics of a newly isolated indigenous aerobic denitrifier from oligotrophic drinking water reservoir, Zoogloeasp N299. *Int. J. Mol. Sci.* **2015**, *16*, 10038–10060. [CrossRef] [PubMed]
45. Kim, M.; Jeong, S.Y.; Yoon, S.J.; Cho, S.J.; Kim, Y.H.; Kim, M.J.; Ryu, E.Y.; Lee, S.J. Aerobic denitrification of pseudomonas putida AD-21 at different C/N ratios. *J. Biosci. Bioeng.* **2008**, *106*, 498–502. [CrossRef] [PubMed]
46. Hu, J.A.; Yang, X.Y.; Deng, X.Y.; Liu, X.M.; Yu, J.X.; Chi, R.; Xiao, C.Q. Isolation and nitrogen removal efficiency of the heterotrophic nitrifying-aerobic denitrifying strain K17 from a rare earth element leaching site. *Front. Microbiol.* **2022**, *13*, 905409. [CrossRef]
47. Pan, J.; Li, R.F.; Meng, X.T.; Ye, M.X. Biogeochemical characteristics of nitrogen migrogen and transformation in subsurface flow belt driven by river collection. *Environ. Eng.* **2021**, *39*, 62–68. (In Chinese)
48. Yabusaki, S.B.; Wilkins, M.J.; Fang, Y.L.; Williams, K.H.; Arora, B.; Bargar, J.; Beller, H.R.; Bouskill, N.J.; Brodie, E.L.; Christensen, J.N.; et al. Water table dynamics and biogeochemical cycling in a shallow, variably-saturated floodplain. *Environ. Sci. Technol.* **2017**, *51*, 3307–3317. [CrossRef]
49. Krause, S.; Blume, T.; Cassidy, N.J. Investigating patterns and controls of groundwater up-welling in a lowland river by combining Fibre-optic Distributed Temperature Sensing with observations of vertical hydraulic gradients. *Hydrol. Earth Syst. Sci.* **2012**, *16*, 1775–1792. [CrossRef]
50. Zarnetske, J.P.; Haggerty, R.; Wondzell, S.M.; Bokil, V.A.; Gonzalez-Pinzon, R. Coupled transport and reaction kinetics control the nitrate source-sink function of hyporheic zones. *Water Resour. Res.* **2012**, *48*, W11508.
51. Marumoto, T.; Anderson, J.P.E.; Domsch, K.H. Decomposition of ^{14}C- and ^{15}N-labelled microbial cells in soil. *Soil Biol. Biochem.* **1982**, *14*, 461–467. [CrossRef]
52. Weatherill, J.J.; Krause, S.; Ullah, S.; Cassidy, N.J.; Levy, A.; Drijfhout, F.P.; Rivett, M.O. Revealing chlorinated ethene transformation hotspots in a nitrate-impacted hyporheic zone. *Water Res.* **2019**, *161*, 222–231. [CrossRef]
53. Zhang, Y.; Yang, P.H.; Wang, J.L.; Xie, S.Y.; Chen, F.; Zhan, Z.J.; Ren, J.; Zhang, H.Y.; Liu, D.W.; Meng, Y.K. Geochemical characteristics of lateral hyporheic zone between the river water and groundwater, a case study of Maanxi in Chongqing. *Environ. Sci.* **2016**, *37*, 2478–2486. (In Chinese)
54. Stark, J.M.; Hart, S.C. High rates of nitrification and nitrate turn-over in undisturbed coniferous forests. *Nature* **1997**, *385*, 61–64. [CrossRef]
55. Williams, M.D.; Oostrom, M. Oxygenation of anoxic water in a fluctuating water table system: An experimental and numerical study. *J. Hydrol.* **2000**, *230*, 70–85. [CrossRef]
56. Covatti, G.; Grischek, T. Sources and behavior of ammonium during riverbank filtration. *Water Res.* **2021**, *191*, 116788. [CrossRef]
57. Wu, F.; Wang, Y.; Yan, X.; Xia, C.; Xiao, Y.; Huo, J.; Zhou, X. Adsorption characteristics of suspended sediment to ammonia nitrogen in Lanzhou section of the Yellow River. *China J. Environ. Eng.* **2014**, *8*, 3201–3207.
58. Peyrard, D.; Delmotte, S.; Sauvage, S.; Namour, P.; Gerino, M.; Vervier, P.; Sanchez-Perez, J.M. Longitudinal transformation of nitrogen and carbon in the hyporheic zone of an N-rich stream: A combined modelling and field study. *Phys. Chem. Earth* **2011**, *36*, 599–611. [CrossRef]

59. Bohlke, J.K.; Antweiler, R.C.; Harvey, J.W.; Laursen, A.E.; Smith, L.K.; Smith, R.L.; Voytek, M.A. Multi-scale measurements and modeling of denitrification in streams with varying flow and nitrate concentration in the upper Mississippi River basin, USA. *Biogeochemistry* **2009**, *93*, 117–141. [CrossRef]
60. Lehtovirta-Morley, L.E.; Stoecker, K.; Vilcinskas, A.; Prosser, J.I.; Nicol, G.W. Cultivation of an obligate acidophilic ammonia oxidizer from a nitrifying acid soil. *Proc. Natl. Acad. Sci. USA* **2011**, *108*, 15892–15897. [CrossRef]
61. Kemmitt, S.J.; Wright, D.; Goulding, K.W.T.; Jones, D.L. pH regulation of carbon and nitrogen dynamics in two agricultural soils. *Soil Biol. Biochem.* **2006**, *38*, 898–911. [CrossRef]
62. Hu, R.; Wang, X.P.; Pan, Y.X.; Zhang, Y.F.; Zhang, H. The response mechanisms of soil N mineralization under biological soil crusts to temperature and moisture in temperate desert regions. *Eur. J. Soil Biol.* **2014**, *62*, 66–73. [CrossRef]

Article

Screening Additives for Amending Compacted Clay Covers to Enhance Diffusion Barrier Properties and Moisture Retention Performance

Min Wang [1], Jiaming Wen [2], Heng Zhuang [1], Weiyi Xia [3], Ningjun Jiang [1] and Yanjun Du [1,*]

1 Jiangsu Key Laboratory of Urban Underground Engineering & Environmental Safety, Institute of Geotechnical Engineering, Southeast University, Nanjing 210096, China; wang-min@seu.edu.cn (M.W.); zhuangheng@seu.edu.cn (H.Z.); jiangn@seu.edu.cn (N.J.)
2 Zijin (Changsha) Engineering Technology Co., Ltd., Changsha 410006, China; wenjiaming@seu.edu.cn
3 Jiangsu Environmental Engineering Technology Co., Ltd., Nanjing 210019, China; xiaweiyi@seu.edu.cn
* Correspondence: duyanjun@seu.edu.cn

Featured Application: The dual-additives-amended compacted clay covers in this article could apply to the cover system of industrial contaminated sites, municipal solid waste landfill, and industrial solid waste landfill. The dual-additives could enhance the anti-cracking, moisture retention, gas barrier, and hydraulic performance of compacted clay covers.

Abstract: The cover systems in contaminated sites have some problems, including desiccation cracks, which would lead to degradation of the barrier performance. This study presented a systemic laboratory experimental investigation on the liquid–plastic limit, moisture retention, hydraulic conductivity (k), and gas diffusion barrier properties of amended compacted clay by attapulgite and diatomite for controlling desiccation cracks and migration of water and volatile organic compounds (VOCs). The results showed that the attapulgite could enhance the moisture retention and liquid limit of amended compacted clay. Diatomite could reduce the gas diffusion coefficient (D_θ) significantly. The compacted clay amended by the dual-additives component of attapulgite and diatomite could enhance the liquid limit, moisture retention percent, gas barrier property, and hydraulic performance compared with the unamended clay. Based on the experimental data obtained, the dosage of additives was targeted to be 5%. The moisture retention percent of dual-additives (attapulgite 4% and diatomite 1%) amended clay increased by 82%, the k decreased by 25%, and the D_θ decreased by 42% compared with unamended clay. Scanning electron microscopy (SEM), BET-specific surface area test method (BET), Mercury Intrusion Porosimetry (MIP), and thermogravimetric analysis (TGA) indicated the enhancement mechanism of additives-amended compacted clay.

Keywords: additives; compacted clay cover; moisture retention; gas diffusion barrier; hydraulic conductivity

Citation: Wang, M.; Wen, J.; Zhuang, H.; Xia, W.; Jiang, N.; Du, Y. Screening Additives for Amending Compacted Clay Covers to Enhance Diffusion Barrier Properties and Moisture Retention Performance. *Appl. Sci.* **2022**, *12*, 7341. https://doi.org/10.3390/app12147341

Academic Editor: Bing Bai

Received: 9 June 2022
Accepted: 18 July 2022
Published: 21 July 2022

1. Introduction

The cover system, including compacted clay cover (CCC) and geomembrane (GM), applied in contaminated sites could prevent the migration of volatile or semi-volatile organic compounds (VOCs/SVOCs) and limit the movement of precipitation into the underlying waste [1–5]. However, the published literature [6–8] have shown that the cover system has some problems, including geomembrane defects, and desiccation cracks of the CCC, which would lead to the degradation of the barrier performance (Figure 1). Rowe et al. [6] presented that the geomembrane defects are inevitable. The monitoring results of 205 sites show that the percent of complete geomembrane is less than 30%. In addition, more than 50 percent of geomembranes have more than five loopholes per ha. The barrier performance of the cover system would be degraded by geomembrane defects. Then, the CCC contributes to the role of VOCs/SVOCs and the precipitation barrier. However, desiccation cracks in CCC act as preferential flow paths and affect the barrier performance of CCC, eventually breaking through the cover system [9]. There are various issues resulting in cracks of CCC [7,10], including differential settlements, extreme drought [11], and dry–wet cycle [9]. The resistance to cracking, i.e., the tensile strength or fracture toughness of the soil, also changes upon drying. Finally, the adhesion at interfaces, which is essential in providing the restraint for desiccation cracks to form, changes with moisture content [12,13]. Omidi et al. [8] showed that the hydraulic conductivity increases nearly two orders of magnitude due to the desiccation cracks of CCC, which would have a great impact on the cover system. Albrecht and Benson [14] presented that the hydraulic conductivity of cracked soils is typically several orders of magnitude greater than that of intact soils.

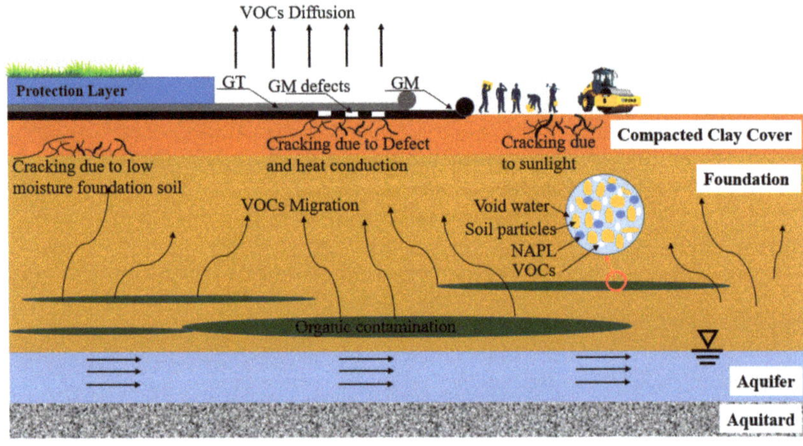

Figure 1. Problems of compacted clay cover desiccation and possible migration of VOCs [6–14].

To restrain the desiccation cracks of CCC, moisture retention additives were added to the soil to improve its moisture retention and desiccation cracks inhibition performance. Moisture retention additives could inhibit moisture evaporation and regulate soil temperature by adsorbing water hundreds or even thousands of times its weight [15]. According to different synthetic materials, the moisture retention additives can be divided into four kinds: modified starch, synthetic polymer, modified cellulose, and other natural compounds and their derivatives, blends, and composites [16]. It is cumbersome to prepare the synthetic polymer moisture retention additives. In addition, it is not suitable for practical engineering because the synthetic polymer would be completely degraded in soil within some years [17]. Therefore, natural materials and their derivatives could be selected as amended materials, namely attapulgite, diatomite, and zeolite. However, the natural zeolite cannot meet the requirements due to its small pore size and being easy to block [18]. Finally, attapulgite and diatomite are proposed as additives for amending CCC.

Usually, the migration modes of gases in the soil included advection and diffusion [5]. Conant et al. [19] found that diffusion is the main migration mode of trichloroethylene (TCE) vapor through a detailed, field-scale analysis of the transport behavior of solvent vapors within the unsaturated zone. The unsaturated zone at the site is approximately 3.5 m thick and comprises the upper portion of the sequence of glaciolacustrine sands and silts of the Borden aquifer. You et al. [20] found that although the transient advective flux can be greater than the diffusive flux; under most of the field conditions the net contribution of the advective flux is one to three orders of magnitude less than the diffusive flux. The advective flux contributes comparably with the diffusive flux only when the gas-filled porosity is less than 0.05. The advective transport of VOCs can be induced by the discrepancy in density between gas phase VOCs and clean air [21–23], water table fluctuation at coastal sites [21,23–25], and atmospheric pressure fluctuation [26,27]. Therefore, in most non-coastal sites and unsaturated zones with low permeability, the VOCs' migration mode in the organic contaminated sites is mainly diffusion [28–30]. Rowe [31] found that the gas barrier should not only prevent water migration, but also the organic contaminants transmission. Therefore, the development of amended CCC with excellent moisture retention and VOCs/SVOCs diffusion barrier performance is significant.

A systematic study of the moisture retention, hydraulic conductivity, and gas barrier properties of amended CCC was investigated through laboratory experiments. In this paper, the optimal dosage and ratio of dual-additives of the amended CCC were studied through the laboratory test. The hydraulic conductivity and gas barrier properties of amended CCC with optimal dosage and ratio were evaluated. Then, the mechanism behind the phenomenon was demonstrated through SEM, BET, MIP, and TGA test analyses.

2. Materials and Methods

2.1. Preparation of Amended Compacted Clay

Three types of amended compacted clay were prepared in this study: (1) the compacted clay amended by attapulgite with the dosage of 0%, 1%, 3%, 5%, and 10% (dry weight of attapulgite to dry weight of mixture powder); (2) the compacted clay amended by diatomite with the dosage of 0%, 1%, 3%, 5%, and 10% (dry weight of attapulgite to dry weight of mixture powder); and (3) the compacted clay amended by attapulgite and diatomite, here referred to as dual-additives, with the ratio of 4, 2, 1, 0.5, and 0.25 (dry weight of attapulgite to dry weight of diatomite); the dosage of dual-additives was attapulgite 4% and diatomite 1%, attapulgite 3.3% and diatomite 1.7%, attapulgite 2.5% and diatomite 2.5%, attapulgite 1.7% and diatomite 3.3%, and attapulgite 1% and diatomite 4% (dry weight of dual-additives to dry weight of mixture powder). The powdered clay used was obtained from Jining (Shandong, China), the powdered attapulgite and diatomite were manufactured by Zhengzhou (Henan, China). The optimum moisture content (w_{op}) and maximum dry density (p_{dmax}) were obtained by compaction test as per JTG E40 T0131-2007. This laboratory compaction method was used to determine the compaction curve compacted in a 152 mm diameter mold with a 45 N rammer dropped from a height of 450 mm, producing a compactive effort of 2677 kN-m/m^3 with 3 layers and 98 blows per layer. While, the ASTM D1557 compaction method was used to determine the compaction curve compacted in a 152.4 mm diameter mold with a 44.48 N rammer dropped from a height of 457.2 mm, producing a compactive effort of 2700 kN-m/m^3 with 5 layers and 56 blows per layer. The compaction curve of clay and dual-additives (attapulgite 4% and diatomite 1%)-amended clay are shown in Figure 2. The physical properties and main oxide content of these constituent materials used to prepare amended compacted clay are shown in Tables 1 and 2, respectively. The powdered materials used in this study were air dried and passed through a No. 200 (0.075 mm) sieve as specified in ASTM D 698 [32]. The initial moisture content of the specimens was determined to be 30% (weight of water to dry weight of solid). The liquid–plastic limit test and specific gravity test were conducted as per ASTM D 4318 [33] and ASTM D 854 [34], respectively, using distilled water.

Table 1. The physical properties of constituent materials used for preparing amended compacted clay.

Property	Standard	Constituent Material		
		Clay	Attapulgite	Diatomite
Specific gravity, (G_s)	ASTM D854 (2014)	2.73	-	-
Liquid limit, *LL* (%)	ASTM D4318 (2018)	54	197	121
Plasticity index, *PI* (%)		26	105	35
Classification	ASTM D2487 (2018)	CH	MH	MH
Optimum moisture content, w_{op} (%)	JTG E40 T0131-2007	27.6	-	-
Maximum dry density, p_{dmax} (g/cm^3)		1.7	-	-

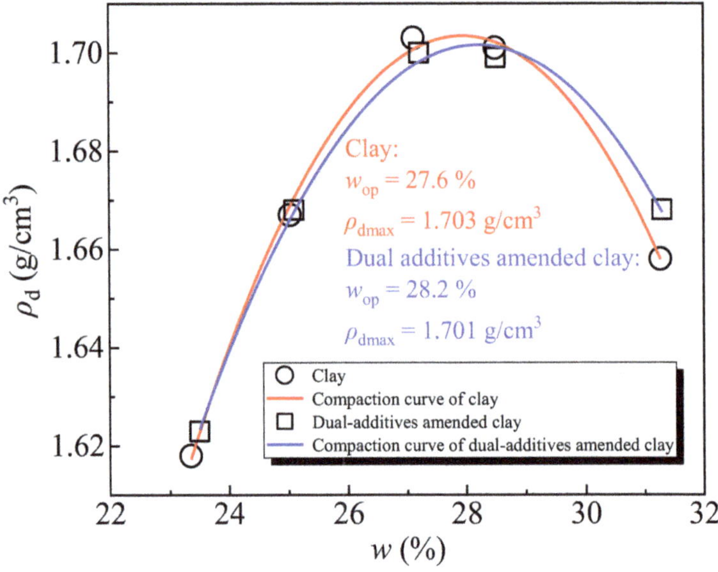

Figure 2. Compaction curve of the clay and dual-additives-amended clay.

Table 2. The main oxide content of attapulgite and diatomite.

Component		SiO_2	Al_2O_3	Fe_2O_3	Na_2O	K_2O	CaO	MgO	MnO	TiO_2	Loss
Content (%)	Attapulgite	58.05	9.55	6.2	0.14	1.13	1.18	11.02	0. 61	0.47	11.65
	Diatomite	90.2	3.4	1.2	0.5	0.5	0.4	0.4	0. 61	0.2	2.59

The amended compacted clay in this study was prepared by mixing predetermined amounts of powdered clay, attapulgite, and diatomite directly. The dosage and ratio (dry weight basis) of additives are shown in Table 3. Then, the specimens were prepared by (1) mixing the air-dried clay, attapulgite, and diatomite thoroughly to prepare the dry mixture (Table 3); (2) adding predetermined amounts of tap water incrementally to the dry mixture, then mixing thoroughly and sealing for 24 h to assure homogeneity; (3) compacting by hydraulic pressure and sealing for 14 d until the specimens reached moisture balance. The number of parallel specimens was three. The coefficient of compaction was targeted to be 85%. The diameter of the specimens in the moisture retention and flexible-wall hydraulic conductivity test was set as 50 mm and the height was 50 mm. While, the diameter of the specimens in the gas diffusion test was set as 61.88 mm and the height was 20 mm.

Table 3. The experimental scheme of the moisture retention, liquid limit, and gas diffusion.

Specimens	The Initial Moisture Content	The Additives Dosage	Ratio (Attapulgite: Diatomite)	Test Parameter
Attapulgite-amended	30%	0%, 1%, 3%, 5%, 10%	-	Moisture retention percent (W_t); Liquid limit (LL); Plastic limit (PL); Gas diffusion coefficient (D_θ)
Diatomite-amended		0%, 1%, 3%, 5%, 10%	-	
Dual-additives-amended		5%	4, 2, 1, 0.5, 0.25	

2.2. Moisture Retention Test

In the moisture retention test, the effects of attapulgite dosage, diatomite dosage, and dual-additives ratio on the moisture retention capacity of amended compacted clay were studied respectively. It was characterized by its moisture retention percent, as shown in Equation (1).

$$W_t = \frac{m_{wt}}{m_{w0}} \cdot 100\% = \frac{m_s \cdot w_0 - m_0 - m_t}{m_s \cdot w_0} \cdot 100\% \quad (1)$$

where, W_t is the moisture retention percent of the specimens at time t; m_{wt} is the mass of moisture in the samples at time t; m_{w0} is the mass of moisture in the specimens at the initial time; m_s is the total mass of soil during specimens preparation; w_0 is the initial moisture content during specimens preparation; m_0 is the total mass of the specimens at the initial time; m_t is the total mass of the specimens at time t.

The detailed test scheme of the moisture retention test is shown in Table 3. The dosage and ratio of dual-additives were determined through the orthogonal test. The test was conducted according to the following procedure: (1) placing the specimens in a 70 × 55 mm perforated aluminum box after sealing for 14 d; (2) placing the specimens in the oven, the temperature was targeted to be 60 °C, recording the mass change per three hours, and calculating the moisture retention percent of the specimens according to Equation (1) [35–38].

2.3. Gas Diffusion Testing Procedures and Apparatus

In the gas diffusion test, the effects of the attapulgite dosage, the diatomite dosage, and the ratio of dual-additives on the gas diffusion barrier performance of the amended compacted clay were conducted. Oxygen was used in this study, because when the gas dissolves easily in water or reacts with the components in the soil, the measured gas diffusion coefficient of the soil was low, resulting in the test error. The gas barrier performance of amended compacted clay was characterized by gas diffusion coefficient (D_θ) per the research results of Taylor et al. [39–41]. D_θ was calculated based on the first Fick's law and the change rate of oxygen concentration in the diffusion chamber with time:

The first Fick's law is Equation (2):

$$\frac{dq}{dt} = -D_\theta \cdot A \cdot \frac{\Delta C_t}{h_s} \quad (2)$$

where: q is the volume of gas diffusing into the chamber, cm^3; t is the diffusion time, s; A is the diffusion area of specimens, 30 cm^2 in this study; h_s is the height of specimens, 2 cm in this study; D_θ is the gas diffusion coefficient of specimens, cm^2/s; ΔC_t is the gas concentration gradient of both ends of specimens, g/cm^3.

The rate of gas diffusion volume into the chamber with time can be also presented as Equation (3):

$$\frac{dq}{dt} = \frac{d(\Delta C_t)}{dt} \cdot h_c \cdot A' \quad (3)$$

where: h_c is the height of the chamber, 15 cm in this study; A' is the diffusion area of the chamber, 133 cm^2 in this study.

Then, Equation (4) is obtained by combining Equations (2) and (3).

$$-D_\theta \cdot A \cdot \frac{\Delta C_t}{h_s} = \frac{d(\Delta C_t)}{dt} \cdot h_c \cdot A' \quad (4)$$

At the initial time (t = 0), the oxygen concentration in the diffusion chamber is 0, in the atmosphere is C_0, ΔC_0 was the difference between the oxygen concentration in the atmosphere and the diffusion chamber before the test, $\Delta C_0 = C_0$. With initial conditions $t = 0$, $\Delta C_t = \Delta C_0 = C_0$, Equation (4) is integrated from 0 to t, obtaining results as Equation (5):

$$\ln\left(\frac{\Delta C_t}{\Delta C_0}\right) = -\frac{D_\theta}{h_s \cdot h_c} \cdot \frac{A}{A'} \cdot t \tag{5}$$

ΔC_0, h_s, h_c, A, and A' are the constants. Therefore, D_θ can be calculated through the relationship of ΔC_t and t.

The optimal dosage and ratio of additives were determined through the orthogonal test. The test scheme of the gas diffusion test is shown in Table 3. The gas diffusion test was conducted as per the single chamber method recommended by Taylor et al. [39–41]. The testing schematic apparatus used for this study is shown in Figure 3. The apparatus consisted of a gas diffusion chamber, an oxygen transducer (KE-25, HELM AG., Hamburg, Germany), and a data acquisition (NL-115, HELM AG., Hamburg, Germany).

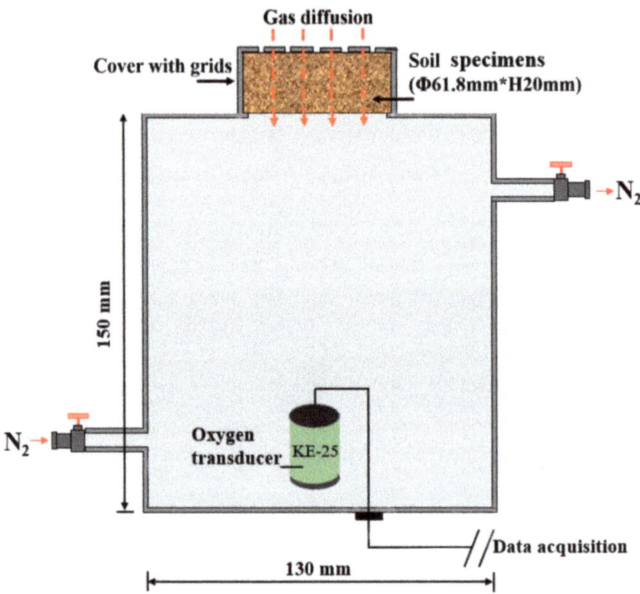

Figure 3. The chamber of soil gas diffusion test.

All experiments were conducted at a room temperature of 25 ± 1 °C and relative humidity of 64% ± 2% as per the following procedures: (1) The vacuum grease was applied to the inner wall of the cover (with 85% grids on the top) and outer wall of chamber top for lubrication and sealing. Then the prepared specimens were placed on the top of the chamber, and capped with the cover. (2) The inlet valve connected with the nitrogen cylinder and outlet valve was opened. It was deemed that the air in the chamber was discharged completely by nitrogen until oxygen concentration in the chamber decreased to 0.3–0.6%. Then, the inlet and outlet valves were closed after continuing to supply nitrogen for 10–15 s. (3) The data acquisition and oxygen transducer were used to automatically collect and record the oxygen concentration (C_t) in the chamber per 5 min until the oxygen concentration in the chamber was the same as the atmosphere (C_0) and reached the stable state ($C_t = C_0$). Combining the C_t and C_0 obtained from the above, the gas diffusion coefficient D_θ was calculated based on Equation (5).

2.4. Flexible-Wall Hydraulic Conductivity Test

The unamended, attapulgite-amended, diatomite-amended, and dual-additives (attapulgite 4% and diatomite 1%)-amended compacted clay were prepared with moisture content of 30% and additives dosage of 5%. The degree of compaction was set as 85%. The specimens of 50 mm in diameter and 50 mm in height were prepared, assembled in a flexible-wall permeameter, and saturated using deaired tap water. Hydraulic conductivity test was conducted in the permeameter using the constant head method as per ASTM D 5084 [42]. All hydraulic conductivity tests in the study were conducted under hydraulic gradients (i) of 40 and effective confining stress of 36.7 kPa calculated by Equation (6) as follows [43–45].

$$\sigma_c{}' = \sigma_3 - \frac{1}{3}(p_2 + 2p_1) \tag{6}$$

where $\sigma_c{}'$= average effective confining stress; σ_3 = cell pressure (50 kPa in this study); p_1 = bottom seepage pressure (20 kPa in this study); p_2 = top seepage pressure (0 kPa in this study).

2.5. Scanning Electron Microscopy (SEM) and Mercury Intrusion Porosimetry (MIP) Tests

Image analyses using SEM (SIGMA 500, ZEISS-Tech, Jena, Germany) and pore size distribution using MIP (AutoPore V 9620, Micromeritics Instrument Corporation, Atlanta, America) were performed on the amended compacted clay specimens to evaluate the microstructural-amended mechanism. First, the attapulgite-amended, diatomite-amended, dual-additives-amended, and unamended compacted clay specimens were cut into 1 cm^3 pieces, the dosages of additives were all 5% (dry weight basis), and the ratio of dual-additives was 4 (dry weight of attapulgite 4% to dry weight of diatomite 1%). Then, the SEM and MIP specimens were frozen using liquid nitrogen for 5 min at a temperature of −120 °C. Sublimation of water was conducted at a temperature of −80 °C for 24 h in a vacuum freeze-drying unit at −18 N (Nanjing Xianou Instruments Manufacture Co., Ltd., Nanjing, China). After freeze-drying, the frozen SEM specimens were cut into small blocks with a natural fracture surface area of approximately 0.25 cm^2, and these blocks were coated with a thin gold layer and then subjected to SEM analyses. The freeze-dried MIP specimens were subjected to MIP analyses in compliance with ASTM D 4404-18.

2.6. BET Specific Surface Area Test Method (BET) and Thermogravimetric Analysis (TGA)

To evaluate the mechanisms of additives treatment, BET and TGA (DSC200F3, NETZSCH-Gerätebau GmbH, Germany), analyses were conducted on attapulgite-amended, diatomite-amended, dual-additives-amended, and unamended compacted clay specimens. The specimens were prepared according to the procedures as follows. First, the attapulgite-amended, diatomite-amended, dual-additives-amended, and unamended clay powders were mixed thoroughly to prepare the dry mixture. The dosages of additives were all designed as 5% (dry weight basis) and the ratio of dual-additives was 4 (dry weight of attapulgite 4% to dry weight of diatomite 1%). Then, the powder mixtures were frozen using liquid nitrogen for 5 min at a temperature of −120 °C. Sublimation of water was conducted at a temperature of −80 °C for 24 h in a vacuum freeze-drying unit at −18 N (Nanjing Xianou Instruments Manufacture Co., Ltd., Nanjing, China). Then, the specimens were subjected to BET and TG analyses, respectively.

3. Results and Discussion

3.1. Liquid–Plastic Limit and Moisture Retention Tests Results

Figure 4 presents the plasticity chart of the clay amended by additives with different dosages and ratios. The dosage of dual-additives-amended compacted clay was 5%. As the dosage of attapulgite increased from 0% to 10%, the plastic limit of attapulgite-amended specimens increased from 28% to 41%, and the liquid limit increased from 54% to 74%. Its liquid limit increased by 36% compared with unamended clay. With the dosage of diatomite increasing from 0% to 10%, the plastic limit of clay increased from 28% to 29%, and the liquid limit increased from 54% to 57%. Its liquid limit increased by about 5% compared with the unamended clay. The effect of diatomite on the liquid limit of amended compacted clay was not significant. When the mass ratio of attapulgite to diatomite was 0.25, the plastic limit of the dual-additives-amended compacted clay decreased from 32% to 30%, and the liquid limit decreased from 64% to 60% compared with the mass ratio of 4. The decreasing ranges of plastic limit and liquid limit were 9% and 6% respectively. When the dosage is 5%, the liquid limit of the compacted clay amended by attapulgite and dual-additives (ratio of 4) were 62% and 64% respectively.

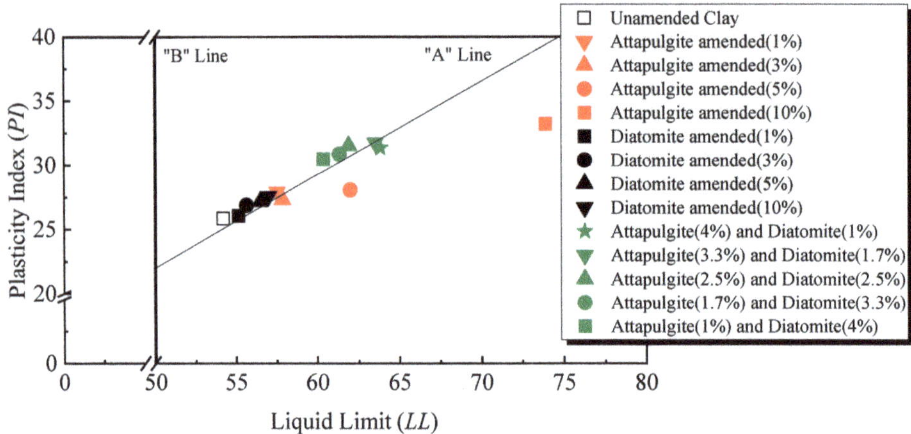

Figure 4. The plasticity chart of the clay amended by additives with different dosages and ratios.

Figure 5 shows how moisture retention percent varies with the dosage of additives. In general, the moisture retention percent increases with the dosage of the attapulgite increasing, while the dosage of diatomite has little effect on its moisture retention percent. The moisture retention percent with the 10% dosage of attapulgite was similar to 5%. When the attapulgite dosage was 5%, the moisture retention percent was about 66% higher than that of unamended clay. Attapulgite could effectively increase the liquid limit of clay and enhance its moisture retention capacity. When the diatomite dosage was 5%, the moisture retention percent was the largest, which was 7% higher than that of unamended clay. Therefore, the dosage of attapulgite and diatomite was found to be 5% considering the cost.

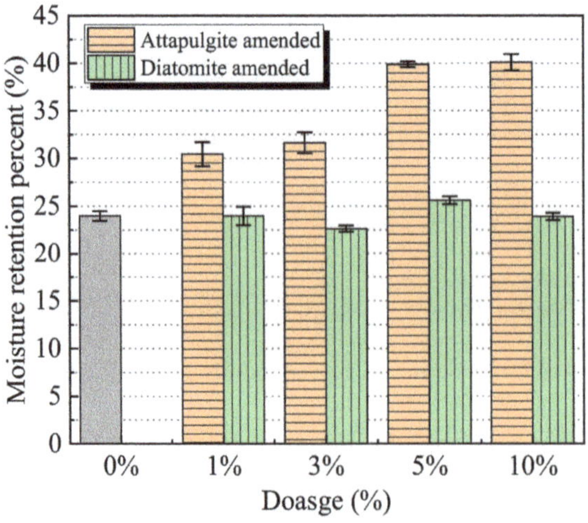

Figure 5. The moisture retention percent with different dosages of attapulgite and diatomite.

Figure 6 presents the variation of moisture retention percent of the dual-additives-amended clay with the ratio of attapulgite to diatomite. The moisture retention capacity of dual-additives-amended compacted clay was the best when the dry weight of attapulgite to the dry weight of diatomite was 4, which was 82% higher than that of unamended clay. It shows that the dual-additives have a great effect on enhancing the moisture retention capacity of amended compacted clay. The liquid limit of compacted clay amended by dual-additives was 3% higher than attapulgite with the dosage of 5%. The moisture retention percent of compacted clay amended by dual-additives (ratio of 4) was 9% higher than attapulgite with the dosage of 5%. Therefore, the increase in the liquid limit of compacted clay amended by additives is one of the reasons for the improvement of its moisture retention capacity.

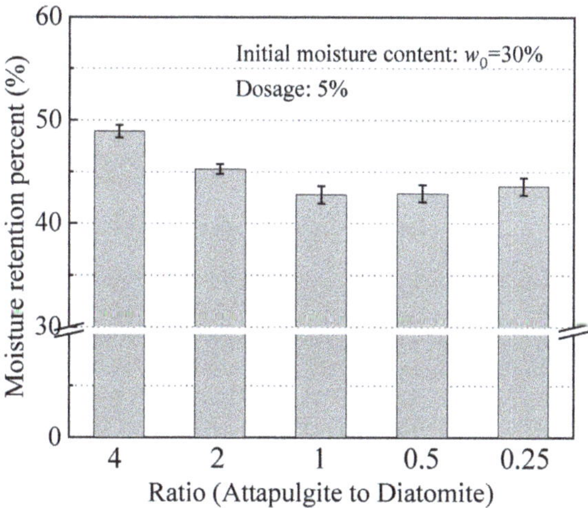

Figure 6. The moisture retention percent with different ratios of attapulgite to diatomite.

The results showed that the attapulgite could greatly increase the moisture retention percent, whereas the diatomite has a limited effect. It is attributed to the molecule's structure [46–48]. There are a lot of pores in the crystal of attapulgite, which could adsorb most cations, water molecules, and organic molecules of a certain size with the Van der Waals Forces, similar to a "zeolite molecular sieve". At the same time, the crystalline water molecules combined with Mg^{2+} at the edge could form a hydrogen bond, which belonged to the synergistic effect of physical adsorption and chemical adsorption. Whereas, the adsorption mechanism of diatomite is mainly chemical adsorption [49–51]. It could adsorb water molecules due to the hydrogen bonds which are combined with the water molecules and the hydroxyl belonged to the Silanol group and silanediol group. However, it has a limited effect on enhancing the moisture retention percent of clay due to the weak chemical adsorption. Therefore, attapulgite could be screened as an additive to enhance the moisture retention capacity.

3.2. Gas Diffusion Test Results

The results of the gas diffusion test of the compacted clay amended by attapulgite or diatomite are shown in Figure 7. The oxygen concentration in the diffusion chamber increased with the time of all dosages. At which time the oxygen concentrations inside and outside the chamber were the same, it reached equilibrium. Figure 7 presents that the D_θ decreased with the increase of attapulgite dosage and its gas barrier performance gradually improved. With the increase in dosage of diatomite, the diffusion rate of oxygen decreased. It shows that diatomite has a good effect on gas barrier performance. It also indicates that diatomite could reduce the D_θ of clay better.

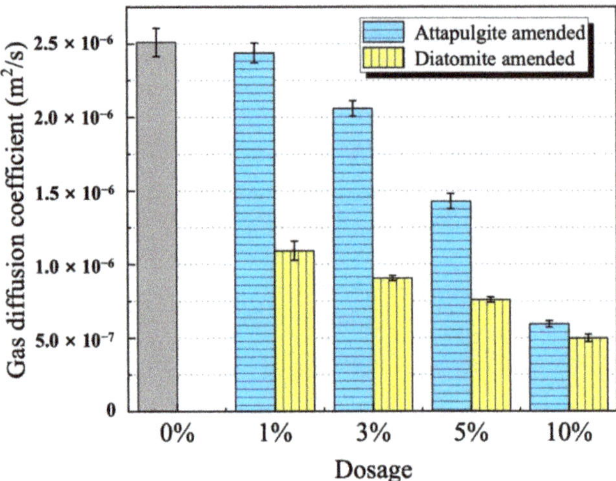

Figure 7. The gas diffusion coefficient with different dosages of attapulgite and diatomite.

The results of the gas diffusion test with different ratios of dual-additives are shown in Figure 8. The initial moisture content was 30% and the dosage of dual-additives was 5% of all the specimens. Among that, the D_θ of the amended clay was the lowest at 4.4×10^{-7} m^2/s when the ratio was 1. After considering the economic cost, the dosage of diatomite should be reduced as much as possible. Therefore, it is suggested that the mass ratio of attapulgite to diatomite be 4, the dosage 5%, and the D_θ of the amended compacted clay 8.5×10^{-7} m^2/s, reduced by about 42% compared with the unamended clay.

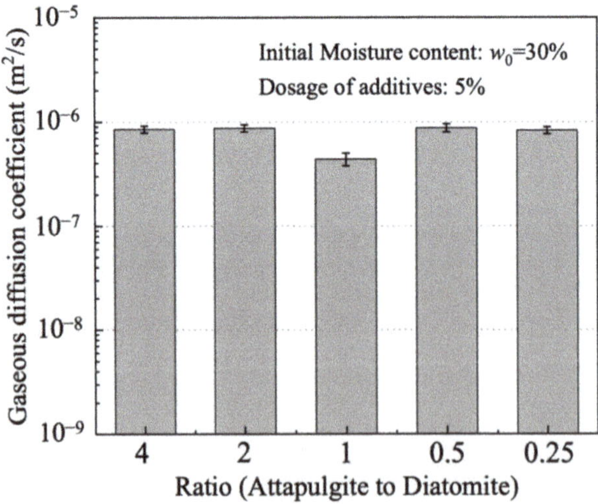

Figure 8. The effect of ratios of additives on the gas diffusion coefficient.

Comparing the gas diffusion test results of the clay amended by attapulgite and diatomite, it can be obtained that the effect of diatomite on the gas barrier performance of the clay is more obvious compared with that of attapulgite, which can be attributed to those crystal structures. Nevertheless, the attapulgite crystals are chain structures or fibrous structures [48], meaning the gas could still migrate through the gap of fibrous or chain structure between the crystals after mixing and compaction. While diatomite, as a porous layered siliceous rock of biological genesis, is composed of diatom wall shells, which are distributed with microporous structure [49–51]. There are few and complex channels for gas migration owing to its wall shells. Therefore, diatomite could be screened as one of the additives to improve the gas barrier performance of amended compacted clay.

In conclusion, it is apparent that the optimal dosage of dual-additives was determined to be 5% and the mass ratio of attapulgite to diatomite 4, based on the experimental results of moisture retention percent and gas diffusion coefficient. In addition, the economic cost is also a factor to be considered.

3.3. Flexible-Wall Hydraulic Conductivity Test Results

Figure 9 shows the results of hydraulic conductivity of amended compacted clay. During permeation with tap water, the k of the unamended, attapulgite amended, diatomite amended, and dual-additives-amended compacted clay were 8.3×10^{-9} m/s, 6.3×10^{-9} m/s, 7.4×10^{-9} m/s, and 5.9×10^{-9} m/s, respectively. Importantly, the k of compacted clay amended by attapulgite, diatomite, and dual-additives were lower than unamended compacted clay, which decreased by 25%, 11%, and 29%, respectively. These results indicate that the dual-additives-amended clay possessed better hydraulic performance than unamended clay.

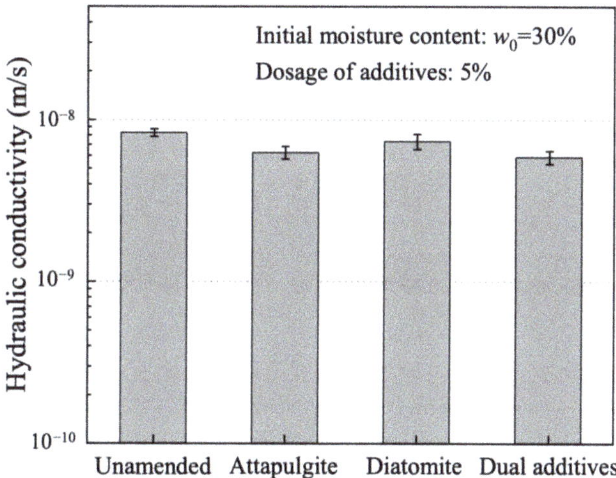

Figure 9. The flexible wall penetration test results of amended compacted clay.

3.4. SEM, BET, MIP, and TGA Tests Results

SEM images for the compacted clay specimens (a) Unamended; (b) Attapulgite-amended; (c) Diatomite-amended; and (d) Dual-additives-amended are shown in Figure 10. It is seen from Figure 10a that the flaky, unamended compacted clay particles polymerize to form agglomerates. There are gaps between adjacent agglomerates for gas and water molecules to pass through. Figure 10b shows that the acicular attapulgite particles attached to the surface of the clay and filled the gaps, which reduced the gaps between clay particles and increased the water retention percent. Figure 10c presents that the size of diatomite particles with micropores was larger than clay particles. The pores between the clay particles decreased, and the micropores on the surface of the diatomite made the gas migration path more complex. However, the microporous structure was larger than those of "zeolite molecular sieve", which had a limited effect on moisture molecules adsorption. Therefore, it had a limited effect on the enhancement of moisture retention performance. Figure 10d shows that the dual-additives filled and decreased the pores between clay particles and the micropores structure made the gas migration more complex and difficult.

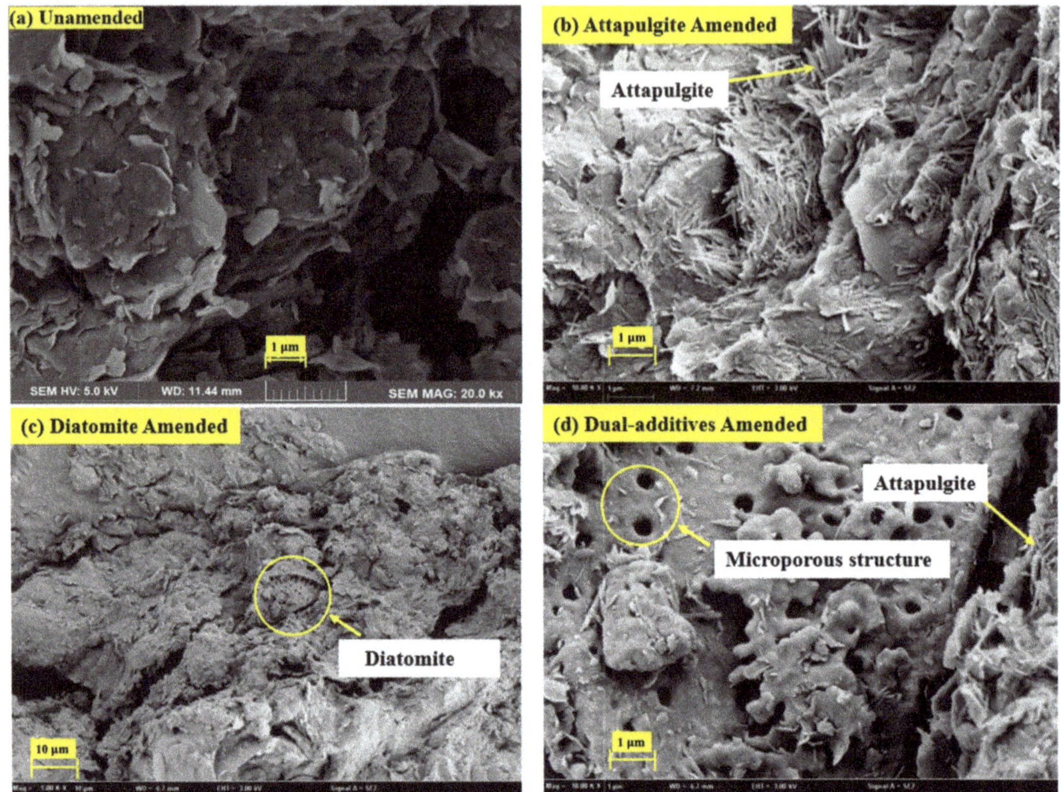

Figure 10. SEM images of compacted clay specimens.

Figure 11 shows the cumulative intruded pore volume and incremental intruded pore volume vs. pore size diameter (PSD) of amended compacted clay specimens with different additives (dosage 5%). The distribution is shown for a diameter range of 3 nm to 425 μm and a pressure range of 0.1 pasi to 61,000 pasi. The average PSD of specimens amended by dual-additives, attapulgite, diatomite, and unamended were 30.6, 31.19, 35.02, and 47.28 nm respectively. Porosities were 19.4%, 21.5%, 27.2%, and 47.3% respectively. The average PSD and porosity of dual-additives-amended compacted clay decreased by 55% and 144% respectively compared with unamended ones. Figure 11a presents that the cumulative intruded pore volume of all amended specimens was lower than unamended ones, regardless of PSD. The cumulative intruded pore volume of dual-additives-amended was the lowest, which decreased by 60% at PSD of 5 nm compared with unamended. It illustrates that the dual-additives filled the gaps and pores of compacted clay significantly. Figure 11b shows that the incremental intruded pore volume of unamended compacted clay increased significantly at a PSD of 100 nm. The peak of curves of amended compacted clay were in the range of 5–100 nm, especially at 15 and 40 nm. The incremental intruded pore volume of all specimens was almost zero in the range of 500–100,000 nm. It indicates that almost all of the pores in amended compacted clay had a diameter range of 5–500 nm. Therefore, the dual-additives could enhance moisture retention and gas barrier performance by decreasing the gaps and pores of compacted clay.

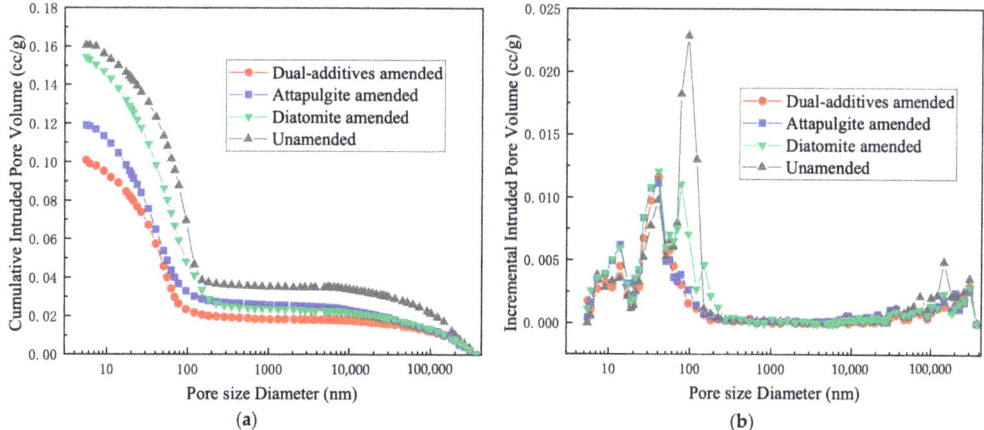

Figure 11. MIP results of amended compacted clay specimens: (**a**) cumulative intruded pore volume vs. pore size diameter; and (**b**) incremental intruded pore volume.

The BET test results of clay powder amended by different additives are shown in Figure 12. The PSDs of less than 2 nm, 2–50 nm, and larger than 50 nm were defined as microporous, mesoporous, and macroporous respectively. It is seen that the microporous amount of amended clay increased significantly compared with unamended clay, and the mesoporous and microporous volume increases were not significant. The microporous volume of attapulgite-amended, diatomite-amended, and dual-additives-amended clay were 0.00721 $cm^3/g{\cdot}nm$, 0.00707 $cm^3/g{\cdot}nm$, and 0.00724 $cm^3/g{\cdot}nm$ respectively when the pore size diameter was 2 nm, increases of 9%, 7%, and 9% respectively compared with unamended clay. So, the dual-additives could enhance the moisture retention performance of compacted clay. The specific surface area and Density Functional Theory (DFT) pore distribution results are presented in Table 4. It was seen that the specific surface area of clay amended by additives increased 13–15% compared with unamended clay. The volume of microporous and mesoporous attapulgite and dual-additives-amended clay increased significantly. Therefore, the dual-additives decreased the microporous size diameter of clay powder and increased its specific surface area.

Table 4. The BET test results.

Specimens	Dosage	Ratio (Attapulgite to Diatomite)	Specific Surface Area (m^2/g)	DFT Pore Distribution (cm^3/g)		Average Microporous Size Diameter (nm)
				Less than 1.863 nm	Less than 19.577 nm	
Unamended	-	-	43.5843	0.01119	0.05508	0.7847
Attapulgite-amended	5%	-	49.7271	0.01284	0.07191	0.7829
Diatomite-amended	5%	-	45.4255	0.01201	0.05729	0.7803
Dual-additives-amended	5%	4	49.3243	0.01284	0.06779	0.7811

The TGA test results of clay amended by different additives are shown in Figure 13. The water in the clay was classified into tightly bound water (TBW) connected with clay minerals by hydrogen bond, loosely bound water (LBW) connected with clay minerals by molecular force [52,53], and free water; the limit of water decomposition temperature was 120~230 °C, 75~120 °C, and 25~75 °C respectively [54]. The DTG curve was obtained from the differentiation of the TG curve. Each minimum point at the valley on the DTG curve represents the water decomposition point. Figure 13a shows that the temperature limits of TBW, LBW, and free water of unamended clay were 29.4 °C, 101.55 °C, and 184.41 °C respectively. The LBW temperature limit of attapulgite-amended clay was 106.68 °C, an increased of 5.05% compared with unamended clay, and its TBW temperature limit was 210.46 °C, an increase of 14%. The LBW temperature limit of diatomite-amended clay was

105.10 °C, an increase of 3.5%, and that of dual-additives-amended clay increased by 8.9%. The TBW and free water have no relationship with the moisture retention performance of clay. The LBW is double-diffuse layers (DDLs) of clay minerals [55,56], which is the critical factor in the moisture retention performance of clay. Figures 4,5 and 13, present that the trend of the temperature limit of clay amended by different additives is consistent with the results of moisture retention percent and liquid limit. It illustrates that the moisture retention percent of amended clay increases with the increasing the temperature limit of LBW. That is attributed to the fact that the thermal energy required to lose the same amount of LBW is greater with the increasing temperature limit; therefore, its moisture retention performance is enhanced at the same temperature (60 °C).

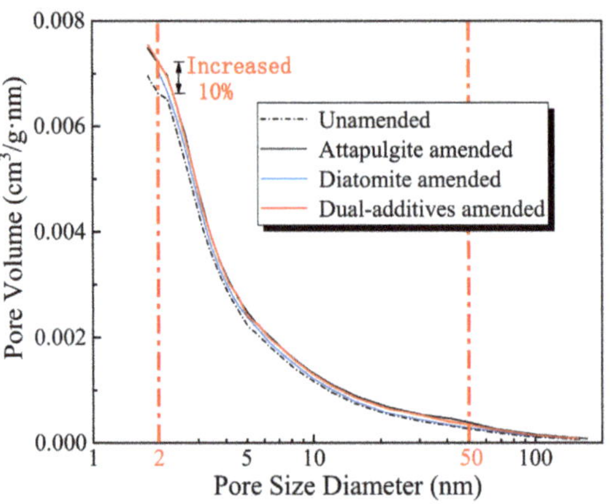

Figure 12. BET tests results of clay specimens amended by attapulgite, diatomite, dual-additives, and unamended.

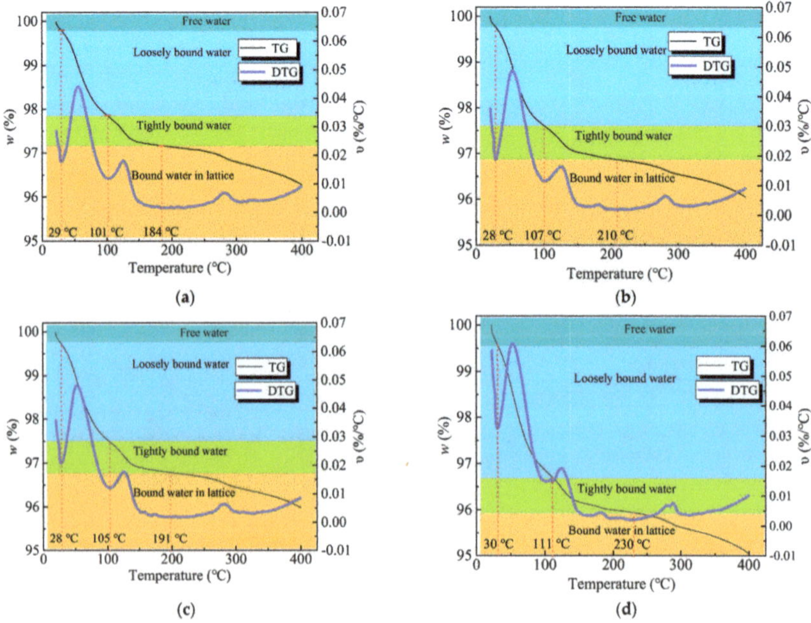

Figure 13. TG tests results of specimens: (**a**) Unamended; (**b**) Attapulgite-amended; (**c**) Diatomite-amended; and (**d**) Dual-additives-amended.

4. Conclusions

The optimum dosage and ratio of additives for amending compacted clay were screened and determined through moisture retention and gas diffusion test. The principal reason for the enhancement of its moisture retention capacity was investigated by the liquid–plastic limit test. Subsequently, the hydraulic performance of the compacted clay amended by additives and unamended was conducted. Based on the results, the following conclusions can be drawn:

1. The attapulgite could enhance the moisture retention performance of the clay, but its effect on gas barrier performance was limited. When its dosage was 5%, its moisture retention percent was 66% higher than that of unamended clay. The enhancement mechanism behind the phenomenon is attributed to increasing the liquid limit of clay;
2. The diatomite could effectively decrease the gas diffusion coefficient of clay, while it had a limited effect on the moisture retention performance. As its dosage was 5%, the moisture retention percent of clay only increased by 7%;
3. Considering the moisture retention capacity, gas barrier performance, and cost, the optimal dosage of dual-additives was targeted to be 5%, and the optimal ratio of attapulgite to diatomite was 4. The moisture retention percent of the dual-additives-amended compacted clay increased by 82%, and the gas diffusion coefficient decreased by 42% compared with unamended clay. The k of enhanced compacted clay amended by attapulgite and dual-additives was 6.3×10^{-9} m/s and 5.9×10^{-9} m/s respectively, decreases of 25% and 29% compared with the unamended clay;
4. The SEM and MIP analyses presented that dual-additives effectively filled the intergranular pores of the amended clay and micropore structure increased the gas migration path. The BET test results showed that dual-additives increased the amount of microporous to enhance the moisture retention performance of amended clay. Meanwhile, the dual-additives increased the specific surface area and decreased the average PSD. The TGA results demonstrated that dual-additives increased the temperature limit of loosely bound water to enhance the moisture retention performance of amended clay.

Author Contributions: Conceptualization, M.W., J.W. and Y.D.; methodology, M.W., Y.D. and J.W.; software, M.W.; validation, M.W. and N.J.; formal analysis, M.W. and J.W.; investigation, M.W., J.W. and H.Z.; resources, M.W.; data curation, M.W. and J.W.; writing—original draft preparation, M.W., J.W., H.Z. and W.X.; writing—review and editing, M.W., Y.D., H.Z. and N.J.; visualization, M.W.; supervision, M.W., Y.D. and W.X.; project administration, Y.D.; funding acquisition, Y.D. All authors have read and agreed to the published version of the manuscript.

Funding: Financial support for this project was provided by the National Key R&D Program of China (Grant No. 2018YFC1803100), the National Natural Science Foundation of China (Grant Nos. 41877248, 42177133, and 41902276).

Data Availability Statement: Some or all data, models, or code that support the findings of this study are available from the corresponding author upon reasonable request.

Conflicts of Interest: The authors declare that they have no known competing financial interests or personal relationships that could have appeared to influence the work reported in this paper.

References

1. Chen, Z.; Kamchoom, V.; Chen, R. Landfill gas emission through compacted clay considering effects of crack pathway and intensity. *Waste Manag.* **2022**, *143*, 215–222. [PubMed]
2. Albright, W.H.; Benson, C.H.; Gee, G.W.; Abichou, T.; McDonald, E.V.; Tyler, S.W.; Rock, S.A. Field performance of a compacted clay landfill final cover at a humid site. *J. Geotech. Geoenviron. Eng.* **2006**, *132*, 1393–1403.
3. Ng, C.W.W.; Chen, Z.K.; Coo, J.L.; Chen, R.; Zhou, C. Gas breakthrough and emission through unsaturated compacted clay in landfill final cover. *Waste Manag.* **2015**, *44*, 155–163. [PubMed]
4. Mahmoodlu, M.G.; Hassanizadeh, S.M.; Hartog, N.; Raoof, A.; van Genuchten, M.T. Evaluation of a horizontal permeable reactive barrier for preventing upward diffusion of volatile organic compounds through the unsaturated zone. *J. Environ. Manag.* **2015**, *163*, 204–213.

5. Verginelli, I.; Capobianco, O.; Hartog, N.; Baciocchi, R. Analytical model for the design of in situ horizontal permeable reactive barriers (HPRBs) for the mitigation of chlorinated solvent vapors in the unsaturated zone. *J. Contam. Hydrol.* **2017**, *197*, 50–61.
6. Rowe, R.K.; Brachman, R.W.I. Assessment of equivalence of composite liners. *Geosynth. Int.* **2004**, *11*, 273–286.
7. Hewitt, P.J.; Philip, L.K. Problems of clay desiccation in composite lining systems. *Eng. Geol.* **1999**, *53*, 107–113.
8. Omidi, G.H.; Thomas, J.C.; Brown, K.W. Effect of desiccation cracking on the hydraulic conductivity of a compacted clay liner. *Water Air Soil Pollut.* **1996**, *89*, 91–103.
9. Julina, M.; Thyagaraj, T. Combined effects of wet-dry cycles and interacting fluid on desiccation cracks and hydraulic conductivity of compacted clay. *Eng. Geol.* **2020**, *267*, 105505.
10. Aldaeef, A.A.; Rayhani, M.T. Hydraulic performance of compacted clay liners under simulated daily thermal cycles. *J. Environ. Manag.* **2015**, *162*, 171–178.
11. Tang, C.S.; Shi, B.; Liu, C.; Suo, W.B.; Gao, L. Experimental characterization of shrinkage and desiccation cracking in thin clay layer. *Appl. Clay Sci.* **2011**, *52*, 69–77.
12. Amarasiri, A.L.; Kodikara, J.K.; Costa, S. Numerical modelling of desiccation cracking. *Int. J. Numer. Anal. Methods Geomech.* **2011**, *35*, 82–96.
13. Kodikara, J.K.; Choi, X. A simplified analytical model for desiccation cracking of clay layers in laboratory tests. In *Unsaturated Soils, Proceedings of the UNSAT 2006 Conference, Carefree, AZ, USA, 2–6 April 2006*; Miller, G.A., Zapata, C.E., Houston, S.L., Fredlund, D.G., Eds.; ASCE Geotechnical Special Publication: Carefree, AZ, USA, 2006; Volume 2, pp. 2558–2567.
14. Albrecht, B.A.; Benson, C.H. Effect of desiccation on compacted natural clays. *J. Geotech. Geoenviron. Eng.* **2001**, *127*, 67–75.
15. Kabiri, K.; Omidian, H.; Zohuriaan-Mehr, M.J.; Doroudiani, S. Superabsorbent hydrogel composites and nanocomposites: A review. *Polym. Compos.* **2011**, *32*, 277–289.
16. Farrell, C.; Ang, X.Q.; Rayner, J.P. Water-retention additives increase plant available water in green roof substrates. *Ecol. Eng.* **2013**, *52*, 112–118.
17. Polman, E.M.; Gruter, G.J.M.; Parsons, J.R.; Tietema, A. Comparison of the aerobic biodegradation of biopolymers and the corresponding bioplastics: A review. *Sci. Total Environ.* **2021**, *753*, 141953.
18. Florez, C.; Restrepo-Baena, O.; Tobon, J.I. Effects of calcination and milling pre-treatments on natural zeolites as a supplementary cementitious material. *Constr. Build. Mater.* **2021**, *310*, 125220.
19. Conant, B.H.; Gillham, R.W.; Mendoza, C.A. Vapor transport of trichloroethylene in the unsaturated zone: Field and numerical modeling investigations. *Water Resour. Res.* **1996**, *32*, 9–22.
20. You, K.; Zhan, H. Comparisons of diffusive and advective fluxes of gas phase volatile organic compounds (VOCs) in unsaturated zones under natural conditions. *Adv. Water Resour.* **2013**, *52*, 221–231.
21. Li, H.; Jiao, J.J. One-dimensional airflow in unsaturated zone induced by periodic water table fluctuation. *Water Resour. Res.* **2005**, *41*, 1–10. [CrossRef]
22. Li, J.; Zhan, H.; Huang, G.; You, K. Tide-induced airflow in a two-layered coastal land with atmospheric pressure fluctuations. *Adv. Water Resour.* **2011**, *34*, 649–658.
23. You, K.; Zhan, H. Can atmospheric pressure and water table fluctuations be neglected in soil vapor extraction? *Adv. Water Resour.* **2012**, *35*, 41–54.
24. Jiao, J.J.; Li, H. Breathing of coastal vadose zone induced by sea level fluctuations. *Geophys. Res. Lett.* **2004**, *31*, L11502. [CrossRef]
25. Guo, H.P.; Jiao, J.J. Numerical study of airflow in the unsaturated zone induced by sea tides. *Water Resour. Res.* **2008**, *44*, W06402. [CrossRef]
26. You, K.; Zhan, H.; Li, J. A new solution and data analysis for gas flow to a barometric pumping well. *Adv. Water Resour.* **2010**, *33*, 1444–1455.
27. You, K.; Zhan, H.; Li, J. Gas flow to a barometric pumping well in a multilayer unsaturated zone. *Water Resour. Res.* **2011**, *47*, W05522. [CrossRef]
28. Choi, J.W.; Smith, J.A. Geoenvironmental factors affecting organic vapor advection and diffusion fluxes from the unsaturated zone to the atmosphere under natural conditions. *Environ. Eng. Sci.* **2005**, *22*, 95–108.
29. Smith, J.A.; Chiou, C.T.; Kammer, J.A.; Kile, D.E. Effect of soil moisture on the sorption of trichloroethene vapor to vadose-zone soil at Picatinny Arsenal, New Jersey. *Environ. Sci. Technol.* **1990**, *24*, 676–683.
30. Batterman, S.A.; McQuown, B.C.; Murthy, P.N.; McFarland, A.R. Design and evaluation of a long-term soil gas flux sampler. *Environ. Sci. Technol.* **1992**, *26*, 709–714.
31. Rowe, R.K. Environmental Geotechnics: Past, Present and Future? In Proceedings of the 8th International Congress on Environmental Geotechnics, Hangzhou, China, 28 October–1 November 2018; Springer: Singapore, 2018.
32. *ASTM D 698*; Standard Test Methods for Laboratory Compaction Characteristics of Soil Using Standard Effort (12,400 ft-lbf/ft^3 (600 kN-m/m^3)). ASTM: West Conshohocken, PA, USA, 2012.
33. *ASTM D 4318*; Standard Test Methods for Liquid Limit, Plastic Limit, and Plasticity Index of Soils. ASTM: West Conshohocken, PA, USA, 2018b.
34. *ASTM D 854*; Standard Test Methods for Specific Gravity of Soil Solids by Water Pycnometer. ASTM: West Conshohocken, PA, USA, 2014b.
35. Sun, L.P.; Du, Y.M.; Shi, X.W.; Chen, X.; Yang, J.H.; Xu, Y.M. A new approach to chemically modified carboxymethyl chitosan and study of its moisture-absorption and moisture-retention abilities. *J. Appl. Polym. Sci.* **2006**, *102*, 1303–1309.

36. Barajas-Ledesma, R.M.; Wong, V.N.; Little, K.; Patti, A.F.; Garnier, G. Carboxylated nanocellulose superabsorbent: Biodegradation and soil water retention properties. *J. Appl. Polym. Sci.* **2022**, *139*, 51495.
37. Mi, Y.; Miao, Q.; Cui, J.; Tan, W.; Guo, Z. Novel 2-Hydroxypropyltrimethyl Ammonium Chitosan Derivatives: Synthesis, Characterization, Moisture Absorption and Retention Properties. *Molecules* **2021**, *26*, 4238. [CrossRef]
38. Wang, J.; Jin, W.; Hou, Y.; Niu, X.; Zhang, H.; Zhang, Q. Chemical composition and moisture-absorption/retention ability of polysaccharides extracted from five algae. *Int. J. Biol. Macromol.* **2013**, *57*, 26–29. [PubMed]
39. Taylor, S.A. Oxygen diffusion in porous media as a measure of soil aeration. *Soil Sci. Soc. Am. J.* **1950**, *14*, 55–61.
40. Currie, J.A. Gaseous diffusion in porous media Part 1. A non-steady state method. *Br. J. Appl. Phys.* **1960**, *11*, 314–317.
41. Su, Z.; Wu, B.; Gong, Y. Determination of gas diffusion coefficient in soils with different porosities. *Trans. Chin. Soc. Agric. Eng.* **2015**, *31*, 108–113.
42. *ASTM D5084*; Standard Test Methods for Measurement of Hydraulic Conductivity of Saturated Porous Materials Using a Flexible Wall Permeameter. ASTM: West Conshohocken, PA, USA, 2016.
43. Malusis, M.A.; McKeehan, M.D. Chemical compatibility of model soil-bentonite backfill containing multiswellable bentonite. *J. Geotech. Geoenviron. Eng.* **2013**, *139*, 189–198.
44. Bohnhoff, G.L.; Shackelford, C.D. Hydraulic conductivity of polymerized bentonite-amended backfills. *J. Geotech. Geoenviron. Eng.* **2014**, *140*, 04013028.
45. Yang, Y.L.; Reddy, K.R.; Du, Y.J.; Fan, R.D. Short-term hydraulic conductivity and consolidation properties of soil-bentonite backfills exposed to CCR-impacted groundwater. *J. Geotech. Geoenviron. Eng.* **2018**, *144*, 04018025.
46. Boudriche, L.; Calvet, R.; Hamdi, B.; Balard, H. Effect of acid treatment on surface properties evolution of attapulgite clay: An application of inverse gas chromatography. *Colloid. Surf.* **2011**, *392*, 45–54.
47. Duan, Z.; Zhao, Q.; Wang, S.; Yuan, Z.; Zhang, Y.; Li, X.; Wu, Y.; Jiang, Y.; Tai, H. Novel application of attapulgite on high performance and low-cost humidity sensors. *Sens. Actuators B Chem.* **2020**, *305*, 127534.
48. Liu, D.B.; Li, Y.; Xiao, H.F.; Xu, L.D.; Zhou, Y.Z.; Han, W.Y.; Li, H.R.; Jiang, Y.M.; Qiu, Y.S. Attapulgite Clay and Its Application in Radionuclide Treatment: A Review. *Sci. Adv. Mater.* **2018**, *10*, 1529–1542.
49. de Namor, A.F.D.; El Gamouz, A.; Frangie, S.; Martinez, V.; Valiente, L.; Webb, O.A. Turning the volume down on heavy metals using tuned diatomite. A review of diatomite and modified diatomite for the extraction of heavy metals from water. *J. Hazard. Mater.* **2012**, *241*, 14–31.
50. Sun, M.; Zou, C.; Xin, D. Pore structure evolution mechanism of cement mortar containing diatomite subjected to freeze-thaw cycles by multifractal analysis. *Cem. Concr. Compos.* **2020**, *114*, 103731.
51. Zheng, J.; Shi, J.; Ma, Q.; Dai, X.; Chen, Z. Experimental study on humidity control performance of diatomite-based building materials. *Appl. Therm. Eng.* **2017**, *114*, 450–456.
52. Li, Y.L.; Wang, T.H.; Su, L.J. Determination of bound water content of loess soils by isothermal adsorption and thermogravimetric analysis. *Soil Sci.* **2015**, *180*, 90–96.
53. Li, S.; Wang, C.; Zhang, X.; Zou, L.; Dai, Z. Classification and characterization of bound water in marine mucky silty clay. *J. Soils Sediments* **2019**, *19*, 2509–2519.
54. Wang, Y.; Lu, S.; Ren, T.; Li, B. Bound water content of air-dry soils measured by thermal analysis. *Soil Sci. Soc. Am. J.* **2011**, *75*, 481–487.
55. Fu, X.L.; Shen, S.Q.; Reddy, K.R.; Yang, Y.L.; Du, Y.J. Hydraulic Conductivity of Sand/Biopolymer-Amended Bentonite Backfills in Vertical Cutoff Walls Permeated with Lead Solutions. *J. Geotech. Geoenviron. Eng.* **2022**, *148*, 04021186.
56. Zhuang, H.; Wang, J.; Gao, Z. Anisotropic and Noncoaxial Behavior of Soft Marine Clay under Stress Path Considering the Variation of Principal Stress Direction. *Int. J. Geomech.* **2022**, *22*, 04022062.

Article

A Calculation Method of Thermal Pore Water Pressure Considering Overconsolidation Effect for Saturated Clay

Gailei Tian and Zhihong Zhang *

Key Laboratory of Urban Security and Disaster Engineering of China Ministry of Education, Beijing University of Technology, Beijing 100124, China; tiangailei@emails.bjut.edu.cn
* Correspondence: zhangzh2002@bjut.edu.cn

Abstract: With the increase of soil consolidation degree, the pore water pressure induced by thermal loading drops dramatically. To conveniently and quickly calculate the thermal pore water pressure inside the soil under different overconsolidation states and quantify overconsolidation effect on thermal pore water pressure, a calculation method of thermal pore water pressure considering overconsolidation effect for saturated clay is proposed. The method is verified by the relevant experimental data, and good agreements were achieved. Through analyzing the influence mechanism of *OCR* on the thermal pore water pressure, three important findings were captured. (1) For overconsolidated clay, thermal pore water pressure decreases nonlinearly with the increase of *OCR*. (2) There is a critical threshold of *OCR* 4.3; when $1 < OCR \leq 4.3$ (slightly overconsolidated state), the ratio of compression line slope to recompression line slope (Λ) of overconsolidated clay is consistent with that of the normally consolidated clay, while when $OCR > 4.3$ (highly overconsolidated state), the value of Λ is smaller than that of normally consolidated clay. (3) For highly overconsolidated clay ($OCR > 4.3$), considering the reducing of Λ with *OCR*, the prediction accuracy of the thermal pore pressure calculation method has been greatly improved; especially when *OCR* equals 30, the prediction accuracy improves by 92.7% as temperature change achieves 35 °C.

Keywords: overconsolidation effect; thermal pore water pressure; calculation method; saturated clay

Citation: Tian, G.; Zhang, Z. A Calculation Method of Thermal Pore Water Pressure Considering Overconsolidation Effect for Saturated Clay. *Appl. Sci.* **2022**, *12*, 6325. https://doi.org/10.3390/app12136325

Academic Editor: Bing Bai

Received: 20 May 2022
Accepted: 19 June 2022
Published: 21 June 2022

1. Introduction

During the service period of buffer materials for nuclear waste repository [1–3], thermal prefabricated vertical drains [4,5], geothermal energy piles [6], and buried cables and pipelines [7], the temperature of the surrounding soil may change significantly. However, when the soil temperature changes, thermal pore water pressure will be generated in the soil due to the difference of thermal expansion characteristics between the pore fluid and soil particles. In case of geological formations with low hydraulic conductivity, the thermal pore water can cause the loss of effective stress. Furthermore, with the dissipation of thermal pore water pressure, the soil will produce additional thermal volume change. It is found that the magnitude of thermal pore water pressure mainly depends on the stress history. Knowledge of the volume change of saturated soil under different stress states, which is a key factor that needs to be considered in design of thermo-active structures, must accurately estimate the thermal pore water pressure with different degrees of overconsolidation.

A series of studies have investigated the thermal response of saturated clay soils. To explore the evolution law of thermal pore water pressure under undrained conditions, Campanella and Mitchell [8], Ghaaowd et al. [9], Abuel-Naga et al. [10], and Ghabezloo et al. [11] carried out related experimental investigations. These experimental studies found that the magnitude of thermal pore water pressure depends on the compressibility of the soil, the physicochemical coefficient of structural volume change, initial void ratio, initial effective stress, the change in temperature, and thermal expansion coefficients of pore water and soil particles. Furthermore, Abuel-Naga et al. [10], Ghabezloo et al. [11],

Burghignoli et al. [12], and McCartney [13] explored the thermally induced volume change of soil by conducting drainage heating tests. Among them, Abuel-Naga et al. [10] used the theoretical microstructure mechanism to explain the thermally induced volume change behavior under different stresses. Based on the principle of particle rearrangement in porous granular materials undergoing thermodynamic process, Bai et al. [14,15] established a thermo-hydro-mechanical constitutive model. This model can accurately describe the irreversible consolidation of normally consolidated saturated soils induced under thermal loading and the aging effect caused by cyclic thermal loading. Subsequently, under the framework of granular thermodynamics, Bai et al. [16] derived a generalized effective stress principle, which can automatically consider the influence of the stress path, temperature path, and soil structure.

It should be noted that both the magnitude of thermal pore water pressure and the sign of the thermally induced volume change are affected by the degree of overconsolidation [10,12,14,17]. With the dissipation of the thermal pore water pressure, normally consolidated clays will exhibit irreversible volume shrinkage, highly overconsolidated clays will undergo elastic thermal expansion, and slightly overconsolidated clays will show a volume change trend that expands first and then shrinks [13,18–20]. Meanwhile, for overconsolidated clays, there is a transition temperature. When the soil temperature exceeds the transition temperature, soil deformation will change from expansion to contraction [18,21–23]. The reason for the different deformation laws of soils with different overconsolidation degrees may be due to the difference in thermal pore water pressure [21]. To explore the effect of overconsolidation ratio on the changes of thermal pore water pressure and thermal volume, constitutive models that can consider the influence of stress history are established based on the Cambridge model or modified Cambridge model [3,10,24,25]. However, because the yield caused by temperature and stress is considered, many parameters are involved in these constitutive models, which makes the solution process extremely cumbersome. Furthermore, these constitutive models cannot obtain an explicit expression of thermal pore water pressure [26], which makes engineering applications very inconvenient. To conveniently and quickly predict the thermal pore water pressure, Campanella and Mitchell [8] proposed a thermo-porous-mechanical model by using concepts of thermoelasticity and linear elasticity. However, for highly overconsolidated soils, the thermo-porous-mechanical model may no longer be applicable. Based on the unified hardening model established by Yao and Zhou [25], Wang et al. [26] established a calculation method of thermal pore water pressure including *OCR*, but when the soil is in highly overconsolidated state, there is a large gap between the prediction results and the experimental results. In addition, the acquisition of thermal parameters in this calculation method requires additional thermal tests.

In summary, various methods for calculating thermal pore water pressure have been developed. However, the overconsolidation effect has not been well explored. To conveniently and quickly predict the thermal pore water pressure in overconsolidated clay and quantify overconsolidation effect on the thermal pore water pressure, by introducing the parameter Λ affected by *OCR* and the nonlinear relationship between *OCR* and the thermal pore water pressure, a calculation method of thermal pore water pressure considering overconsolidation effect for saturated clay is proposed. In addition, the calculation method is applied to predict the thermal pore water pressure, and the predicted results are compared with experimental data.

2. Calculation Method of Thermal Pore Water Pressure with Different *OCR*

Campanella and Mitchell [8] developed a calculation model of thermal pore water pressure in saturated clay during an undrained heating test:

$$\Delta u_T = \frac{n(\alpha_w - \alpha_s) + \alpha_{sT}}{m_v}\Delta T \tag{1}$$

in which Δu_T (kPa) is the change of pore pressure caused by thermal loading, n is the porosity, m_v (kPa^{-1}) is the compressibility of soil skeleton, ΔT (°C) is the change in temperature, α_{sT} (°C^{-1}) is the physicochemical coefficient of structural volume change, and α_w (°C^{-1}) and α_s (°C^{-1}) are the volumetric expansion coefficients of pore water and soil particles, respectively.

Under undrained heating conditions, saturated soil will expand along the secondary compression curve, and the compressibility of soil skeleton can be estimated by the isotropic recompression curve and written as

$$m_v \approx (m_v)_r = \frac{1}{1+e_0}\frac{\kappa}{p'} \tag{2}$$

in which e_0 is the initial void ratio, κ is the slope of the isotropic recompression line, p' is the mean effective stress, and $(m_v)_r$ is the compressibility of the soil skeleton determined from the recompression curves.

Substituting Equation (2) into Equation (1) results in

$$\Delta u_T = \frac{n(\alpha_w - \alpha_s) + \alpha_{sT}}{\kappa}(1+e_0)p'\Delta T \tag{3}$$

For most soils, the values of α_w and α_s are usually taken as 1.7×10^{-4} (°C^{-1}) and 3.5×10^{-5} (°C^{-1}), respectively [8,9,12,27].

Since the change of porosity during undrained heating tests is generally negligible and has little effect on the thermal pore water pressure, the porosity (n) is approximately equal to the initial porosity (n_0) [9,28].

The physicochemical coefficient α_{sT} is often used to characterize the volume change caused by soil structural rearrangement under unit thermal loading, which is not straightforward to obtain [8,9]. In fact, α_{sT} can be estimated from experimental data of Δu_T at a given ΔT, and the relationship is rearranged as [9]

$$\alpha_{sT} = [\frac{\Delta u_T}{p_0'}\frac{\kappa}{1+e_0}\frac{1}{\Delta T}] - n(\alpha_w - \alpha_s) \tag{4}$$

where $p_0{}'$ (kPa) is the initial mean effective stress.

Because Equation (4) needs to know the experimental data of Δu_T, the undrained heating test must be carried out. To avoid complicated heating tests, Ghaaowd et al. [9] proposed an empirical expression for the physicochemical coefficient of normally consolidated clay, which can be written as

$$\alpha_{sT} = 1.0 \times 10^{-4} e^{-0.014 I_p} \tag{5}$$

where I_p is soil plasticity index and e is the Napierian base.

In addition, through undrained heating tests, Wang et al. [26] found that the thermal pore water pressure of normally consolidated and overconsolidated soil meets

$$\Delta u_{Tocr} = \ln(1 + \frac{1.717}{OCR})\Delta u_T \tag{6}$$

where Δu_{Tocr} is the thermal pore water pressure of the overconsolidated soil.

Substituting Equation (3) into Equation (6) results in

$$\Delta u_{Tocr} = \ln(1 + \frac{1.717}{OCR})\frac{n(\alpha_w - \alpha_s) + \alpha_{sT}}{\kappa}(1+e_0)p'\Delta T \tag{7}$$

According to Equation (7), it can be seen that κ is a key parameter for estimating thermal pore water pressure. However, compared to the slope of the isotropic recompression line, the slope of the isotropic compression line (Λ) is more commonly adopted in studies to calculate thermal pore water pressure. Therefore, the value of Λ is used to estimate κ. Meanwhile, according to the Cam-Clay model, it can be known that the compression index, C_c, and Λ, and the rebound index, C_s, and κ satisfy, respectively,

$$\kappa = 0.434C_s \quad (8)$$

$$\Lambda = 0.434C_c \quad (9)$$

Assuming that the ratio of Λ and κ is Λ, the below equation can be obtained:

$$\Lambda = \frac{\Lambda}{\kappa} = \frac{C_c}{C_s} \quad (10)$$

For normally consolidated clays, Λ is generally constant, and the range of Λ is about 4.8~10 [9,29–31]. However, the stress history can affect the value of Λ [32–34] and then change the magnitude of the thermal pore water pressure. In order to better reflect the influence of stress history on thermal pore water pressure, the relationship between rebound index C_s and OCR is introduced [32]:

$$C_s = 0.0213\ln(OCR) + 0.0288 \quad (11)$$

Substituting Equation (10) into Equation (7) results in

$$\Delta u_{Tocr} = \ln(1 + \frac{1.717}{OCR})\frac{n(\alpha_w - \alpha_s) + \alpha_{sT}}{\Lambda}(1 + e_0)\Lambda p'\Delta T \quad (12)$$

Equation (12) is the final calculation method of thermal pore water pressure applicable to both normally consolidated and overconsolidated clays. This calculation method can not only consider the direct weakening effect of OCR, but also the influence of the variation Λ with OCR on thermal pore water pressure.

3. Validation of the Calculation Method

The calculation method of thermal pore water pressure, which can consider overconsolidation effect, was applied to predict the variation of thermal pore water pressure, and the predicted results are compared with the experimental data.

3.1. Comparison with the Experimental Data in Undrained Heating Test of Wang et al. (2017)

To investigate the influence of OCR on thermal pore water pressure under undrained conditions, Wang et al. [26] used a temperature-controlled GDS triaxial testing apparatus to conduct undrained heating tests on normally consolidated and overconsolidated kaolin clays. During the test, saturated kaolin clay was consolidated at a pressure of 300 kPa. After consolidation was completed, the consolidation pressure was unloaded to 150 kPa, 100 kPa, 75 kPa, 37.5 kPa, 30 kPa, and 10 kPa to obtain soil samples with OCR of 1, 2, 3, 4, 8, 10, and 30. Subsequently, the obtained soil samples were used to carry out undrained heating tests. Furthermore, the physical parameters of kaolin clay are shown in Table 1.

Table 1. Physical parameters of kaolin clay obtained from the test of Wang et al. (2017).

θ (%)	S_r	G_s	I_P	Λ	κ	n_0
59–61	98	2.72	27	0.17	0.026	0.63

Through calculation, it has been found that when the clay is in slightly overconsolidated state ($1 < OCR \leq 4.3$), the Λ obtained by Equations (10) and (11) is larger than that of normally consolidated clay. In this case, the value of Λ adopts that of normally consolidated clay. However, for highly overconsolidated clays ($OCR > 4.3$), the values of Λ are different from those of normally consolidated clays, and when OCR of soil is 8, 10, and 30, the calculated values of Λ are 5.4, 5.0, and 3.8, respectively. Thus, it can be seen that for saturated clays, there is a critical threshold of *OCR* that determines whether the stress history will change the calculated value of Λ, thereby reducing the thermal pore water pressure. In this paper, the critical threshold of *OCR* is taken as 4.3. Subsequently, Λ is brought into Equation (12) to predict the thermal pore water pressure. The predicted results and their comparison with the experimental results of Wang et al. [26] are presented in Figure 1.

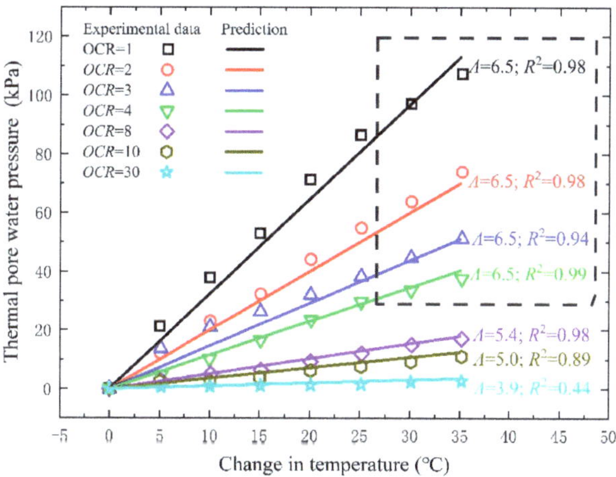

Figure 1. Comparison of predicted thermal pore water pressure against the results of Wang et al. (2017).

It can be seen from Figure 1 that all experimental data points are basically on the predicting lines. Meanwhile, through calculation, it has been found that except for the predicted results with *OCR* equal to 30, the coefficients of determination (R^2) of other predicted results are all greater than or equal to 0.89. Moreover, it should be noted that when *OCR* is equal to 30, although the coefficient of determination of the predicted results is only 0.44, the maximum difference between the predicted results and the experimental data is within 1 kPa, which is acceptable in engineering. Therefore, it can be concluded that the calculation method established in this paper can accurately predict the undrained thermal pore water pressure for both normally consolidated and overconsolidated clays.

3.2. Comparison with the Experimental Data in Undrained Heating Test of Abuel-Naga et al. (2007)

Using modified oedometer and triaxial test apparatus, Abuel-Naga et al. [10] carried out undrained heating tests on soft Bangkok clay to study the thermal pore water pressures under different stress conditions. To obtain soils with different overconsolidation ratios, the consolidation pressure of 200 kPa was unloaded to 100 kPa and 50 kPa. The physical parameters of soft Bangkok clay are shown in Table 2. Meanwhile, although Λ was not reported in this undrained heat test, Ghaaowd et al. [9] found that when Λ = 10, their model-predicted results and the experimental results of Abuel-Naga et al. [10] fit best, so the value of Λ was taken as 10. Furthermore, since the maximum *OCR* is 4 and the critical threshold is not reached, the influence of stress history on Λ is ignored.

Table 2. Physical parameters of soft Bangkok clay obtained from Abuel-Naga et al. (2007).

θ (%)	S_r	G_s	I_P	Λ	n_0
90–95	98	2.68	60	0.46	0.63

The predicted results of Equation (12) are compared with the experimental results of Abuel-Naga et al. [10], and the comparison results are shown in Figure 2.

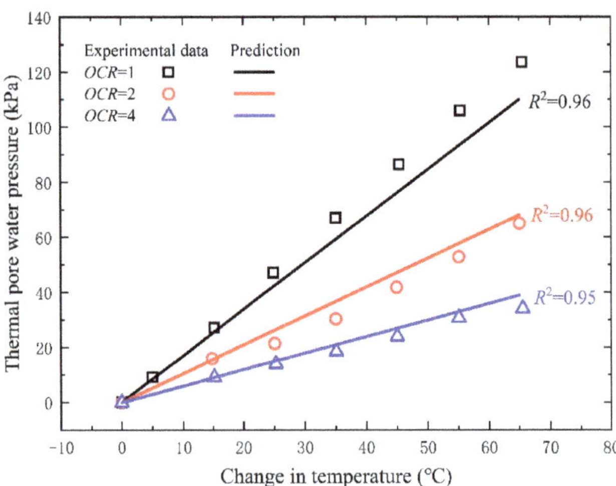

Figure 2. Comparison of predicted thermal pore water pressure against the results of Abuel-Naga et al. (2007).

From Figure 2, it can be observed that all experimental data points are around the predicting lines. In addition, through calculation, it was found that the coefficients of determination (R^2) of all predicted results are 0.95 or above, which indicates that the predicted results are highly consistent with the experimental results. The consistency between the predicted results and the experimental results ensures the accuracy of the calculation method established in this paper.

4. Application of the Calculation Method

In this section, the calculation method is applied to investigate the influence of over-consolidation effect on thermal pore water pressure using MATLAB software. The consolidation pressure is fixed at 300 kPa, and thermal loadings are taken as 35 °C, 50 °C, 65 °C, and 80 °C, respectively. Furthermore, the required parameters for numerical calculation are the same as those in Table 1. The evolution law of thermal pore water pressure with *OCR* under different thermal loadings is shown in Figure 3.

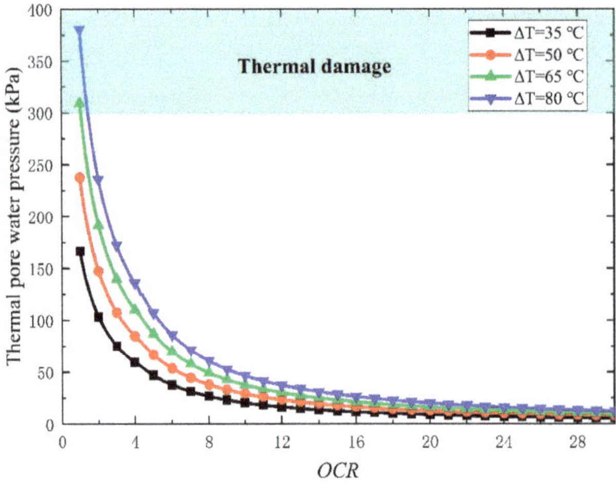

Figure 3. Evolution of the thermal pore water pressure with *OCR* under different thermal loadings.

From Figure 3, it can be observed that the value of thermal pore water pressure drops sharply with increasing *OCR*. Meanwhile, the difference of thermal pore water pressure under different thermal loadings also decreases gradually. This is mainly because the thermal pore water pressure is generated by the difference in thermal expansion characteristics of soil particles and pore water, and the degree of overconsolidation of soil can reduce this difference of thermal expansion. In addition, it can be found from Figure 3 that when thermal loading is large and *OCR* is small, the value of thermal pore water pressure can exceed the effective stress of the soil. In other words, the soil may suffer thermal damage. In this case, if *OCR* is properly increased, the thermal pore water pressure will be significantly reduced, thus avoiding the occurrence of thermal damage.

5. Discussion

As mentioned above, *OCR* can not only directly reduce the thermal pore water pressure (Equation (6)), but it can also change the magnitude of the thermal pore water pressure by affecting the value of Λ (Equations (10) and (11)). However, the effect of *OCR* on Λ is often ignored in existing calculation methods [9,26], which may be the main reason for overestimating the thermal pore water pressure of highly overconsolidated soil. To quantitatively analyze the effect of Λ on the thermal pore water pressure, two cases are given here, namely, Case 1: the predicted results without considering the variation of Λ ($\Lambda = 6.5$), and their comparison with the experimental results; and Case 2: the predicted results considering the variation of Λ ($\Lambda = 5.4$, 5.0, and 3.8), and their comparison with the experimental results. The effect of Λ on the thermal pore water pressure is analyzed by comparing the coincidence between experimental results and predicted results. Figure 4a,b show the comparison of experimental results and predicted results for the two cases.

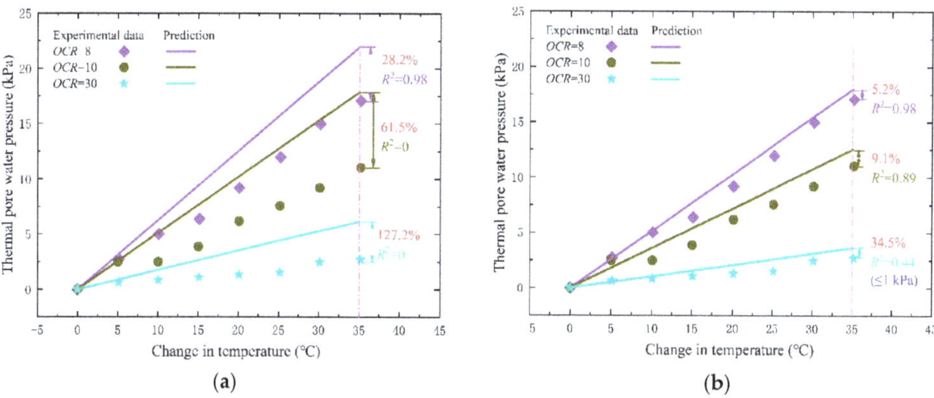

Figure 4. Comparison between experimental results and predicted results of (**a**) Case 1; (**b**) Case 2.

From Figure 4a, it can be observed that the predicted results without considering the variation of Λ will seriously overestimate the thermal pore water pressure. For clays with *OCR* equal to 10 and 30, the coefficients of determination (R^2) of predicted results are both 0. When the change in temperature is 35 °C, the difference between the experimental results and predicted results reaches 61.5% for the clay with *OCR* equal to 10, and the difference reaches 127.2% for the clay with *OCR* equal to 30. However, from Figure 4b, it can be found that the coincidence between the predicted results considering the variation of Λ and experimental results was greatly improved. Especially, when *OCR* is equal to 30, the accuracy of the predicted results was improved by 92.7%, which means that the variation of Λ with *OCR* cannot be ignored when predicting the thermal pore water pressure inside the highly overconsolidated clay.

6. Conclusions

In this study, by considering the effect of *OCR* on the parameter Λ and the nonlinear relationship between *OCR* and the thermal pore water pressure, a calculation method of thermal pore water pressure considering overconsolidation effect for saturated clay is proposed. Compared with previous calculation methods, the proposed calculation method has higher accuracy and the required parameters are easy to obtain. Moreover, through applying the calculation method and comparing the predicted results of the calculation method with the experimental data, the following conclusions can be drawn:

- Based on the proposed calculation method, the critical threshold of *OCR* (4.3) for determining whether the value of Λ will vary with *OCR* is obtained. When the clay is in slightly overconsolidated state, i.e., $1 < OCR \leq 4.3$, the value of Λ is the same as that of normally consolidated clay. However, when the clay is in a highly overconsolidated state, i.e., $OCR > 4.3$, the value of Λ of overconsolidated clay is smaller than that of normally consolidated clay, and gradually decreases with the increase of *OCR*.
- The proposed calculation method can accurately predict the evolution of thermal pore water pressure under undrained conditions, and is supported by some undrained heating test results of overconsolidated saturated clay, which can provide a theoretical basis for the design of thermo-active structures.
- The effect of *OCR* on thermal pore water pressure is related to the heating rate. However, the calculation method proposed in this study does not consider the influence of heating rate. The follow-up work can explore the comprehensive effects of *OCR* and heating rate on the thermal pore water pressure under undrained conditions theoretically and experimentally.

Author Contributions: Conceptualization, Z.Z.; methodology, Z.Z.; validation, Z.Z. and G.T.; formal analysis, Z.Z. and G.T.; investigation, Z.Z. and G.T.; writing—original draft preparation, Z.Z. and G.T. All authors have read and agreed to the published version of the manuscript.

Funding: This research was funded by the Second Tibetan Plateau Scientific Expedition and Research (STEP) Program (No. 2019QZKK0905).

Informed Consent Statement: Not applicable.

Data Availability Statement: Data are contained within the article.

Acknowledgments: Zhihong Zhang is supported by the Second Tibetan Plateau Scientific Expedition and Research (STEP) Program (No. 2019QZKK0905).

Conflicts of Interest: The authors declare no conflict of interest.

References

1. Abuel-Naga, H.M.; Bergado, D.T.; Ramana, G.V.; Grino, L. Experimental evaluation of engineering behavior of soft Bangkok clay under elevated temperature. *J. Geotech. Geoenviron. Eng.* **2006**, *132*, 902–910. [CrossRef]
2. Monfared, M.; Sulem, J.; Delage, P.; Mohajerani, M. On the THM behaviour of a sheared Boom clay sample: Application to the behaviour and sealing properties of the EDZ. *Eng. Geol.* **2012**, *124*, 47–58. [CrossRef]
3. Cheng, W.; Hong, P.Y.; Pereira, J.M.; Cui, Y.J.; Tang, A.M.; Chen, R. Thermo-elasto-plastic modeling of saturated clays under undrained conditions. *Comput. Geotech.* **2020**, *125*, 103688. [CrossRef]
4. Abuel-Naga, H.M.; Lorenzo, G.A.; Bergado, D.T. Current state of knowledge on thermal consolidation using prefabricated vertical drains. *Geotech. Eng. J. SEAGS AGSSEA* **2013**, *44*, 132–141.
5. Pothiraksanon, C.; Bergado, D.T.; Abuel-Naga, H.M. Full-scale embankment consolidation test using prefabricated vertical thermal drains. *Soils Found.* **2010**, *50*, 599–608. [CrossRef]
6. Brochard, L.; Honorio, T. Thermo-poro-mechanics under adsorption applied to the anomalous thermal pressurization of water in undrained clays. *Acta Geotech.* **2021**, *24*, 2713–2727. [CrossRef]
7. Li, H.; Kong, G.; Wen, L.; Yang, Q. Pore pressure and strength behaviors of reconstituted marine sediments involving thermal effects. *Int. J. Geomech.* **2021**, *21*, 06021008. [CrossRef]
8. Campanella, R.G.; Mitchell, J.K. Influence of temperature variations on soil behavior. *J. Soil Mech. Found. Div.* **1968**, *94*, 709–734. [CrossRef]

9. Ghaaowd, I.; Takai, A.; Katsumi, T.; Mccartney, J.S. Pore water pressure prediction for undrained heating of soils. *Environ. Geotech.* **2015**, *4*, 70–78. [CrossRef]
10. Abuel-Naga, H.M.; Bergado, D.T.; Bouazza, A. Thermally induced volume change and excess pore water pressure of soft Bangkok clay. *Eng. Geol.* **2007**, *89*, 144–154. [CrossRef]
11. Ghabezloo, S.; Sulem, J. Stress dependent thermal pressurization of a fluid-saturated rock. *Rock Mech. Rock Eng.* **2009**, *42*, 1. [CrossRef]
12. Burghignoli, A.; Desideri, A.; Miliziano, S. A laboratory study on the thermomechanical behavior of clayey soils. *Can. Geotech. J.* **2000**, *37*, 764–780. [CrossRef]
13. McCartney, J.S. Thermal volume change of unsaturated silt under different stress states and suction magnitudes. *E3S Web Conf. EDP Sci.* **2016**, *9*, 09009. [CrossRef]
14. Bai, B.; Yang, G.C.; Li, T.; Yang, G.S. A thermodynamic constitutive model with temperature effect based on particle rearrangement for geomaterials. *Mech. Mater.* **2019**, *139*, 103180. [CrossRef]
15. Bing, B.; Yan, W.; Dengyu, R.; Fan, B. The effective thermal conductivity of unsaturated porous media deduced by pore-scale SPH simulation. *Front. Earth Sci.* **2022**, *10*, 943853.
16. Bai, B.; Zhou, R.; Cai, G.Q.; Hu, W.; Yang, G.C. Coupled thermo-hydro-mechanical mechanism in view of the soil particle rearrangement of granular thermodynamics. *Comput. Geotech.* **2021**, *137*, 104272. [CrossRef]
17. Liu, Q.; Deng, Y.B.; Mao, W.Y.; Deng, Y.B. The influence of over consolidation ratio on thermal consolidation properties of clay. *J. Disaster. Prev. Mitigation. Eng.* **2019**, *39*, 607–614.
18. Cekerevac, C.; Laloui, L. Experimental study of thermal effects on the mechanical behaviour of a clay. *Int. J. Numer. Anal. Methods Geomech.* **2004**, *28*, 209–228. [CrossRef]
19. Di Donna, A.; Laloui, L. Response of soil subjected to thermal cyclic loading: Experimental and constitutive study. *Eng. Geol.* **2015**, *190*, 65–76. [CrossRef]
20. Houhou, R.; Sutman, M.; Sadek, S.; Laloui, L. Microstructure observations in compacted clays subjected to thermal loading. *Eng. Geol.* **2021**, *287*, 105928. [CrossRef]
21. Baldi, G.; Hueckel, T.; Pellegrini, R. Thermal volume changes of the mineral water system in low porosity clay soils. *Can. Geotech. J.* **1988**, *25*, 807–825. [CrossRef]
22. Towhata, I.; Kuntiwattanakul, P.; Seko, I.; Ohishi, K. Volume change of clays induced by heating as observed in consolidation tests. *Soils Found.* **1993**, *33*, 170–183. [CrossRef]
23. Sultan, N.; Delage, P.; Cui, Y.J. Temperature effects on the volume change behavior of boom clay. *Eng. Geol.* **2002**, *64*, 135–145. [CrossRef]
24. Hueckel, T.; Borsetto, M. Thermoplasticity of saturated soils and shales: Constitutive equations. *J. Geotech. Eng.* **1990**, *116*, 1765–1777. [CrossRef]
25. Yao, Y.P.; Zhou, A.N. Non-isothermal unified hardening model: A thermo-elasto-plastic model for clays. *Geotech.* **2013**, *63*, 1328–1345. [CrossRef]
26. Wang, K.J.; Hong, Y.; Wang, L.Z.; Li, L.L. Effect of heating on the excess pore water pressure of clay under undrained condition. *Chin. J. Rock Mech. Eng.* **2017**, *36*, 2288–2296.
27. Takai, A.; Ghaaowd, I.; McCartney, J.S.; Katsumi, T. Impact of drainage conditions on the thermal volume change of soft clay. *Geo-Chicago* **2016**, *2016*, 32–41.
28. Uchaipichat, A.; Khalili, N. Experimental investigation of thermo-hydro-mechanical behaviour of an unsaturated silt. *Geotech.* **2009**, *59*, 339–353. [CrossRef]
29. Schofield, A.; Wroth, P. *Critical State Soil Mechanics*; McGraw-Hill: London, UK, 1973.
30. Lou, X.M.; Li, D.N.; Yang, M. Statistical analysis for rebound deformation parameters of silty clay at bottom of deep excavation in Shanghai. *J. Tongji Univ.* **2012**, *40*, 535–540.
31. He, P.; Wang, W.D.; Xu, Z.H. Empirical correlations of compression index and swelling index for Shanghai clay. *Rock. Soil. Mech.* **2018**, *39*, 3773–3782.
32. Zhou, K.; Sun, D.A. Experiments on compression characteristics of remoulded soft clay in shanghai. *J. Shanghai Univ.* **2009**, *15*, 99–104.
33. Shi, W.; Wang, J.; Guo, L.; Hu, H.T.; Jin, J.Q.; Jin, F.Y. Undrained cyclic behavior of overconsolidated marine soft clay under a traffic-load-induced stress path. *Mar Georesour. Geotechnol.* **2018**, *36*, 163–172. [CrossRef]
34. Wang, J.; Zhuang, H.; Guo, L.; Wu, T.Y.; Yuan, Z.H.; Sun, H.L. Secondary compression behavior of over-consolidated soft clay after surcharge preloading. *Acta Geotech.* **2022**, *17*, 1009–1016. [CrossRef]

Article

Experimental Study on the Shear Strength of Silt Treated by Xanthan Gum during the Wetting Process

Junran Zhang *, Zhihao Meng, Tong Jiang, Shaokai Wang, Jindi Zhao and Xinxin Zhao

Henan Province Key Laboratory of Geomechanics and Structural Engineering, North China University of Water Resources and Electric Power, Zhengzhou 450045, China; z20201020288@stu.ncwu.edu.cn (Z.M.); jiangtong@ncwu.edu.cn (T.J.); wangshaokai@ncwu.edu.cn (S.W.); x201810207162@stu.ncwu.edu.cn (J.Z.); zhaoxinxinncwu@126.com (X.Z.)

* Correspondence: zhangjunran@ncwu.edu.cn; Tel.: +86-15890092736

Abstract: Traditional materials such as fly ash and lime are generally used to improve soils but can severely pollute the environment. Eco-friendly protocols, such as the application of xanthan gum, are therefore essential for soil treatment. In this study, a series of microscopic tests, water retention characteristics tests, and shear tests were carried out on silt, which are known to have poor engineering properties, to explore the effect and mechanism of xanthan gum treatment on the water retention and shear strength characteristics of silt during the wetting process. The results show that the water retention capacity of the treated silt increases with increasing xanthan gum content, and a hysteresis effect is clearly observed. The cohesion and internal friction angle of the silt strongly decrease with increasing water content, and the strength significantly weakens. However, the strength of the silt treated with xanthan gum is consistently higher than that of the untreated silt. The microscopic tests show that soil pores are gradually filled by xanthan gum with good water-retaining properties, thus significantly enhancing the water retention capacity. Furthermore, the hydrogel that cements the soil particles forms by the bonding effects between xanthan gum and soil particles, which greatly improves the silt strength.

Keywords: xanthan gum; silt; water retention capacity; strength; wetting process; microscopic tests

Citation: Zhang, J.; Meng, Z.; Jiang, T.; Wang, S.; Zhao, J.; Zhao, X. Experimental Study on the Shear Strength of Silt Treated by Xanthan Gum during the Wetting Process. *Appl. Sci.* **2022**, *12*, 6053. https://doi.org/10.3390/app12126053

Academic Editor: Bing Bai

Received: 19 May 2022
Accepted: 13 June 2022
Published: 14 June 2022

1. Introduction

Soil treatment measures (e.g., cement, lime, fly ash) are commonly used to improve the strength and water retention capacity of soils [1]. Traditional materials such as Portland cement have long been used in geotechnical engineering to treat soil, but a large amount of CO_2 is emitted during the production process of cement and lime [2]. One ton of Portland cement production has been shown to release approximately one ton of CO_2, and one ton of lime output releases approximately 0.86 tons of CO_2 [3]. Alternative materials, such as metakaolin and calcium hydroxide mixtures, alkaline aluminosilicate minerals, and fly ash–based inorganic polymer concrete, have therefore emerged to control and reduce carbon emissions [4–6]. Microbial soil treatment not only protects the ecological environment by controlling carbon emissions compared with traditional and inorganic methods but also significantly improves the strength and ductility of treated soils [7]. Microbial technology, namely biopolymers, has been applied to improve soil mass. For example, the unconfined compressive strength of 0.5% biopolymer was shown to be higher than that of 10% cement after treatment. The large-scale commercialization of biopolymers has good economic feasibility due to the high cost of cement in less-developed countries and can help improve the strength and durability of geotechnical engineering [8]. Driven by this huge catalytic potential and differing from traditional geotechnical engineering soil treatment technology, improved microbial soil treatment technology has been explored, including microbial and inanimate microbial improvement technology.

Since the onset of the 21st century, a series of studies have addressed different kinds of microbial technologies for soil treatment: synthetic hydrophilic polyacrylamide additives can improve the strength and stability of different soils [9]; casein and sodium caseinate biopolymers can improve the strength of sand [10]; the Panstrains of Enterobacteriaceae cultivated by cell-free fermentation liquid can stabilize soil [11]; and microbial-induced calcium precipitation technology (MICP) can strengthen and stabilize soil [12]. Recent studies showed that biopolymers could produce hydrogels and induce a pore-blocking effect, which can significantly reduce the permeability, whereas calcite precipitated by MICP does not have such a strong pore-blocking effect [13]. Inanimate microbial technology does not require a strict culture environment and offers greater advantages in enhancing the water stability of soil. Inanimate microbial technology can also produce hydrogel colloid and promote the transport of heavy metals in contaminated soil to improve the soil and protect the environment [14].

Numerous suitable biopolymers for soil improvement have emerged in recent years, and biopolymers have become popularized in engineering. For example, the commercial, large-scale production of composite fiber polymer has been applied [15], and the polymer lignin is widely used in manufacturing industries [16,17]. Polyacrylamide polymers have been widely used in the United States to irrigate land and control sand erosion and runoff protection, as well as to construct helicopter landing pads to reduce dust pollution [10]. Recent studies showed that gellan gum and agar gum could significantly improve soil durability [18], that xanthan gum can maintain water for vegetation growth in the soil to prevent desertification [19], and that both xanthan gum and gellan gum can improve the dynamic characteristics of sand [20]. Adding a small amount of biopolymer can greatly improve the soil strength, and guar gum is more advantageous for treating collapsible soil and clay using the wet mixing method, but xanthan gum is superior when treating silty fine-grained soil [21–23]. Xanthan gum can be used as a stabilizer for slope protection in geotechnical engineering. Xanthan gum also has a low application cost and very competitive price compared with other biopolymers [24].

In the late 20th century, Wallingford and Sanchez pointed out that xanthan gum is more capable of absorbing water than other polysaccharide polymers [25,26]. Xanthan gum can be used to effectively improve the water retention capacity of engineering soil, such as through hydraulic seepage barriers and underground pollution stabilization [14,24,27]. Zhou et al. [28] verified that xanthan gum significantly enhances the water retention of soil. Because xanthan gum can separate soil particles and fill soil pores, the pores of sand become larger [29], and its water retention performance is very strong [30], thus improving the water retention capacity of the soil.

Ayeldeen and Qureshi et al. [21,31] indicated that the addition of xanthan gum could improve the collapsibility resistance of collapsible soil, as well as the water disintegration resistance and strength of sandy soil. Cabalar et al. [32] showed that the compaction degree, viscosity, and strength of clays treated with xanthan gum were enhanced at low water content. Chang et al. [33] found that the unconfined compressive strength (UCS) of drying soil tended to stabilize upon increasing the xanthan gum content to a certain range. According to the UCS test results of Latifi [34], stability can be achieved by adding xanthan gum to bentonite and kaolinite at low water content. Soldo et al. [35] found that the strength of soil with 2% xanthan gum content is close to the maximum at low water content. Soldo and Sujatha et al. [36,37] indicated that water content is an important factor affecting soil strength and that the strength of silty sand and silt treated with xanthan gum is greatly improved at low water content. Engineering soil must usually be cured for a few days to reach a low water content state. Because the water content of engineering soil is low, the soil strength is generally high but easily affected by rainfall infiltration, which can reduce the soil strength and stability. Most soil treatment studies conducted strength tests in the range of high suction, but few tests have focused on soil strength over the entire range of water content of soil treated by xanthan gum. Recent studies showed a significant decrease in the wetting strength of sand after treatment with biopolymer guar gum [38]. However, for sand treated with xanthan gum, the wetting strength was greater than the initial strength [39]. Based on the above research results, further in-depth tests were carried out to systematically study the influence of xanthan gum on the wetting strength characteristics of silt over the full range of water content.

The objective of this study was to investigate the effect and mechanism of the strength weakening characteristics of silt treated with xanthan gum (XG-silt) during the wetting process. A series of microscopic tests, water retention characteristics tests, and direct shear tests were carried out on XG-silt using scanning electron microscopy, mercury porosimeter, a WP4C dew point potential meter, and a Shear Trac-II test system. The experimental results qualitatively and quantitatively reveal the variation law and internal mechanism of the wetting strength characteristics of XG-silt, which provides a useful scientific basis for the design and construction of related geotechnical engineering projects.

2. Materials and Methods

2.1. Materials

Figure 1 shows that Henan Province is located in the North China Plain of the middle and lower reaches of the Yellow River, where silt is widely distributed. The silt sample location is at an engineering project in the Xing-yang area, west of Zhengzhou city in northern Henan Province, which adjoins the south bank of the Yellow River. The experimental alluvial silt has poor early strength and water stability [40,41]. Xanthan gum is therefore used to strengthen the silt in the Yellow River flooding area due to the complex engineering properties of silt.

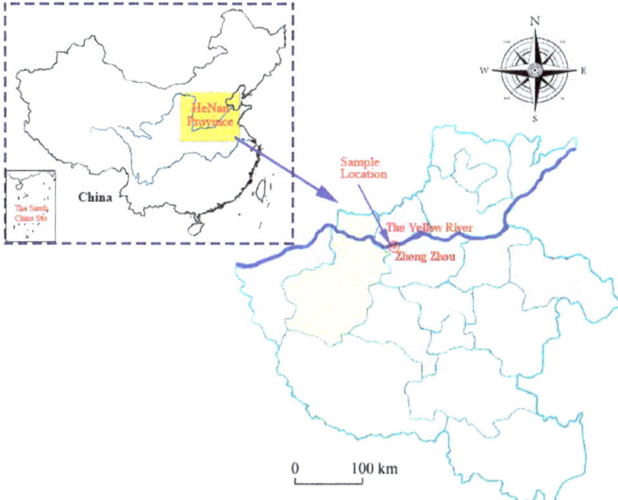

Figure 1. Silt sample location.

Xanthan gum is a high molecular anionic polysaccharide polymer produced by the aerobic fermentation of Xanthomonas and carbohydrates. It is carbon neutral, sustainable and reproducible, stable to acid, alkali, and heat, and has excellent compatibility with a variety of salts. It is made from non-food crops at a low cost and can be prepared in large quantities [26]. The xanthan gum used to treat the silt was purchased from Fu Feng Biotechnology Co., Ltd. in Inner Mongolia. The reagent was of food grade and analytical grade. The storage environment conditions were 25 °C and 35% relative humidity. Pictures of the dry silt and xanthan gum are shown in Figure 2.

Figure 2. Samples of natural soil and xanthan gum: (**a**) silt; (**b**) xanthan gum.

The basic physical properties of natural silt are listed in Table 1. The grading curve determined by hydrometer analyses is shown in Figure 3. The gradation parameters of the soil sample are coefficient of curvature C_c = 0.75, nonuniform coefficient C_u = 4.39, and liquid limit w_l = 21.4%, and the silt shows poor gradation with a low liquid limit.

Table 1. Basic physical properties of silt.

Physical Parameters	Values
Plastic limit, w_P (%)	12.6
Liquid limit, w_L (%)	21.4
Plasticity index, I_P	8.8
Specific gravity of the solid, G_s	2.70
Natural water content, w_0 (%)	5.10
Natural dry weight, ρ_d (g/cm^3)	1.62
Maximum dry density, ρ_{dmax} (g/cm^3)	1.77

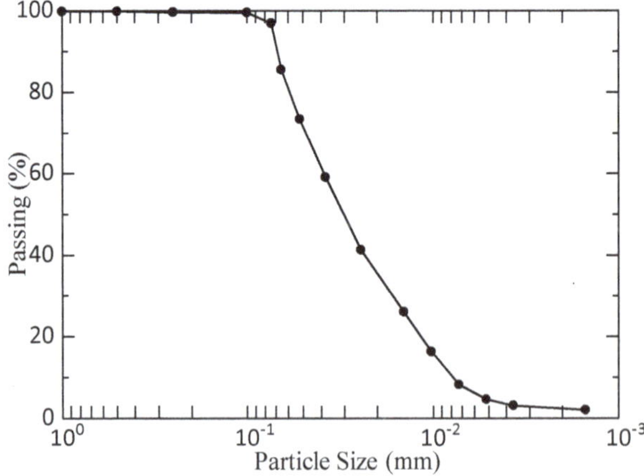

Figure 3. Grading curve of the silt.

The basic parameters of xanthan gum (XG) are shown in Table 2, and its molecular structure is shown in Figure 4.

Table 2. Basic physical properties of xanthan gum.

Product	Grade	Viscosity/CP	Color	State	Shear Performance Value
XG	food grade	1475	light beige	powder	7.7

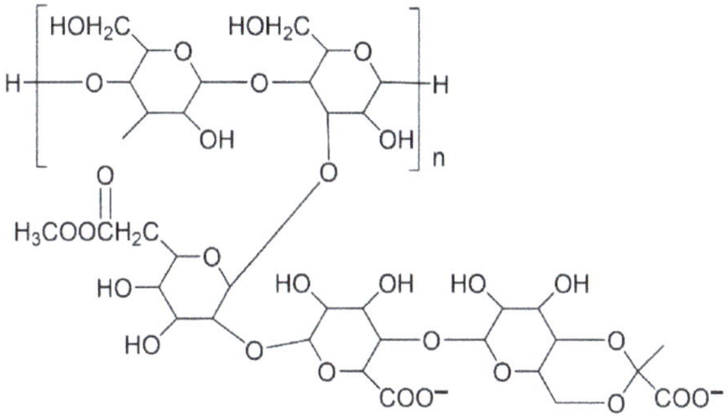

Figure 4. Molecular structure of xanthan gum.

2.2. Methods

2.2.1. Sample Preparation

Soil samples were prepared following the dry method [42], which was commonly used in previous tests. The dried silt powder was fully mixed with dried xanthan gum. Water was then added, and the mixture was stirred evenly. Based on previous experimental studies, the optimal mass ratio (m_{bp}/m_s) of dried XG and dried silt in the water retention characteristics tests was set as 0.0%, 0.5%, 1.0%, 1.5%, and 2.0%, and the optimal mass ratio in the direct shear tests was set as 2.0% [21,33,34,43]. In order to ensure the uniformity of XG-silt, dry soil was stirred and mixed evenly with XG of the corresponding quality. Deionized water was added to form wet soil that avoids the influence of additional chemical components in the water. The soil was then sealed at a constant temperature for 24 h until the water was fully infiltrated. The initial water content of the specimen was 15%, and 92% of the maximum dry density, $\rho_d = 1.63\ g/cm^3$, was selected as the initial dry density. After specimen preparation, a vacuum pump was used to vacuum and saturate the specimen for testing.

2.2.2. Scanning Electron Microscopy Tests

The specimen was rapidly frozen with liquid nitrogen (−190 °C) and vacuum-pumped for 24 h with a vacuum freeze-drying apparatus to ensure the specimen reached the dry state when the water of the specimen was completely sublimated. The dry specimens were then sliced into suitable sheet sizes, and scanning electron microscope (SEM) tests were performed at different magnifications (×50, ×100, ×200, ×500, ×1000, ×2000, ×5000, and ×10,000) using a JBM-7500F field emission SEM (JEOL, Japan). Only four suitable magnification images (×200, ×2000, ×5000, and ×10,000) were selected to fully present the results.

2.2.3. Mercury Intrusion Tests

The pores of the soil sample are assumed to be cylindrical. Mercury is non-infiltrating and therefore does not flow into solid pores. Under low pressure, mercury first intrudes into the gaps and large pores, then gradually intrude into the micropores with increasing mercury pressure:

$$P = -2\sigma \cdot \cos\theta / r \quad (1)$$

where P is the mercury intrusion pressure [Pa], r is the pore radius [m], σ is the surface tension of the intrusion liquid, σ = 0.484 N/m, and θ is the contact angle (130°).

2.2.4. Water Retention Characteristics Test

The saturated specimens were dried and wetted at constant humidity and temperature. The suction was intermittently measured using a WP4C dew point water potential meter (Figure 5). The test specimen's initial size was d_0 = 33 mm and h_0 = 7 mm.

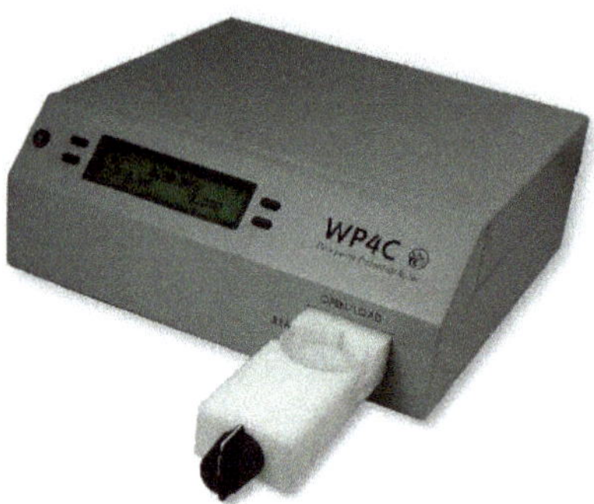

Figure 5. WP4C dewpoint potential meter.

2.2.5. Direct Shear Tests

The size of the test specimen was d_0 = 64 mm and h_0 = 25 mm. Saturated specimens with 0.0% and 2.0% XG content were dried and wetted. The initial water content of the specimens was 22%. The drying and wetting paths of the specimens are shown in Figure 6 The specimens were uniformly dried to a state of 2% water content (drying path AB) and then wetted to a different water content (2%, 4%, 8%, 16%, or 22%). Fast shear tests were then carried out under vertical stresses of 50, 100, and 200 kPa. The automatic direct shear and residual shear test system is shown in Figure 7. The Shear Trac-II test system was operated by computer shear software and the control panel of the shear apparatus to receive immediate feedback on the loads and displacements monitored by the sensors.

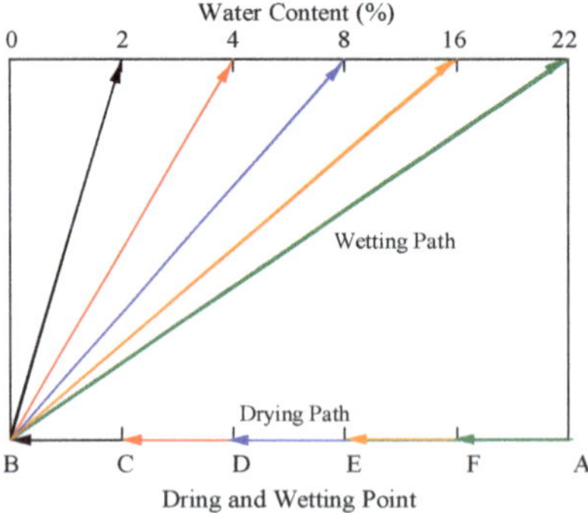

Figure 6. Drying and wetting paths.

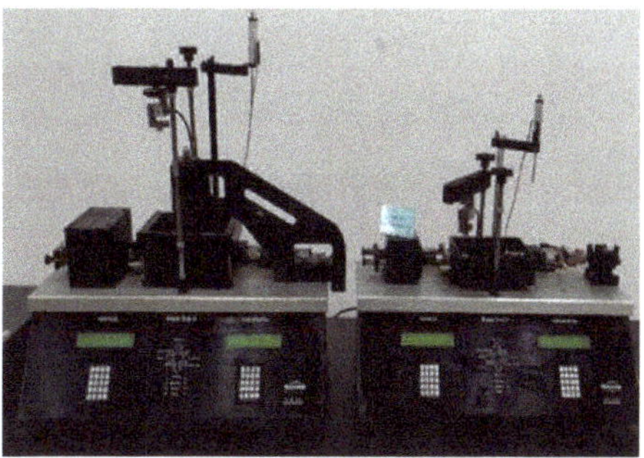

Figure 7. Shear Trac-II test system.

3. Results and Discussion

3.1. SEM Images of Xanthan Gum-Treated Silt

Figures at four suitable magnifications (×200, ×2000, ×5000, and ×10,000) were present for microscopic analysis. At ×200 magnification, the macroscopic structure of soil particle distribution can be seen. At the magnification of ×2000, ×5000, and ×10,000, the internal microstructure of soil can be clearly observed. The following are the conclusions based on SEM pictures.

Figure 8a–c show the SEM test results of XG-silt at ×200 magnification. The mass cementation between the silt particles and XG becomes increasingly large with increasing XG content.

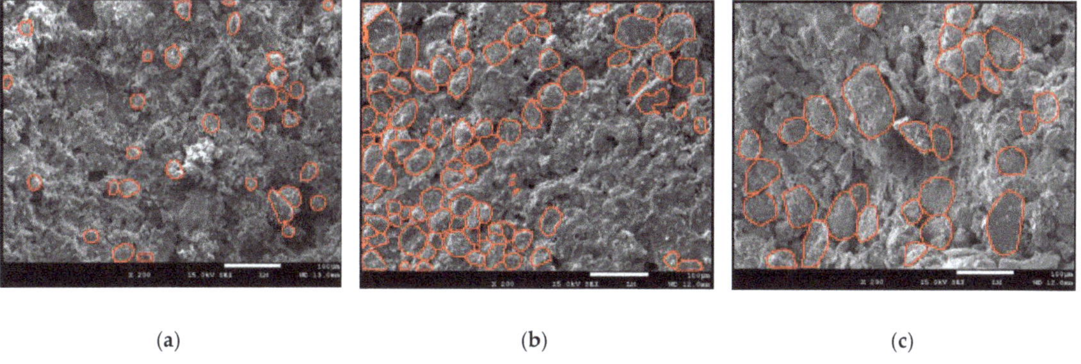

(a) (b) (c)

Figure 8. SEM pictures of XG-silt at ×200 magnification: (**a**) m_{bp}/m_s = 0%; (**b**) m_{bp}/m_s = 1.0%; (**c**) m_{bp}/m_s = 2.0%.

Figure 9a–c show the SEM images of XG-silt at ×2000 magnification. The pore diameter decreases with increasing XG content.

Figure 10a–c show the SEM images of XG-silt at ×5000 magnification. A comparison shows that more soil particles bonded together with increasing XG content, and the small particles became soil mass with cementation [44]. The gaps and pores of the samples were also gradually filled with XG with increasing XG content, showing observable cementation, and the samples became increasingly dense.

(a) (b) (c)

Figure 9. SEM pictures of XG-silt at ×2000 magnification: (**a**) m_{bp}/m_s = 0%; (**b**) m_{bp}/m_s = 1.0%; (**c**) m_{bp}/m_s = 2.0%.

(a) (b) (c)

Figure 10. SEM pictures of XG-silt at ×5000 magnification: (**a**) m_{bp}/m_s = 0%; (**b**) m_{bp}/m_s = 1.0%; (**c**) m_{bp}/m_s = 2.0%.

Figure 11a–c show the SEM images of XG-silt at ×10,000 magnification. Part of the XG formed a biofilm covering the surface of soil particles with increasing XG content, and the other part formed bridge connections (biological polymerization chains) between the aggregate gap [21,37,42]. Additionally, the small particle aggregates became larger with closer connections.

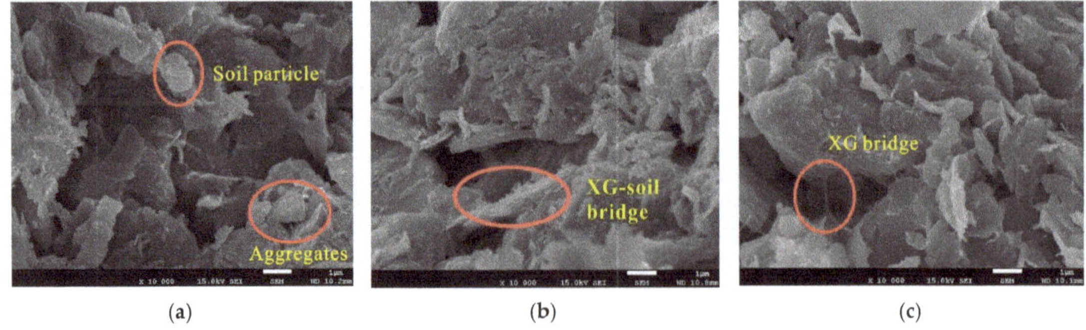

(a) (b) (c)

Figure 11. SEM pictures of XG-silt at ×10,000 magnification: (**a**) m_{bp}/m_s = 0%; (**b**) m_{bp}/m_s = 1.0%; (**c**) m_{bp}/m_s = 2.0%.

3.2. Mercury Intrusion Test

The pore size distribution of the treated specimens with different XG contents was investigated, which is helpful in analyzing the interaction between the silt and XG. Figure 12a,b show the cumulative pore size and differential pore size, respectively.

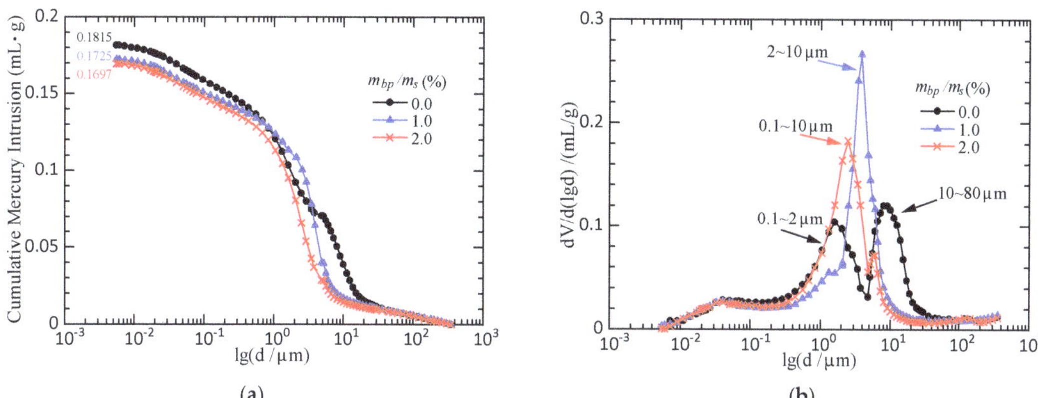

Figure 12. Mercury intrusion pore size distribution: (**a**) cumulative curves; (**b**) differential curves.

Kodikara et al. [45] found that there are aggregates in the soil, which leads to different pore sizes. The pore size distribution of soil is therefore generally bimodal and manifested as micropore and macropore peaks. Figure 12a shows that the cumulative pore intrusion of the mercury intrusion test specimens decreases from 0.1815 mL/g for XG-free silt to 0.1697 mL/g for silt with 2% XG, which implies that the pore size distribution decreases with increasing XG content. Figure 12b shows when 1% XG is added to pure silt, the large pores (10–80 μm) are gradually connected by XG hydrogels and reduced to medium pores (2–10 μm). Small pores (0.1–2 μm) in the silt are then filled with XG, and their number gradually decreases. The number of large and small pores in silt, therefore, strongly decreases, while the number of medium pores significantly increases. The pore size distribution of silt also turns into unimodal pore size distribution. Upon increasing the XG content from 1% to 2%, the medium pores in the silt are gradually connected by the filling of XG and hydrogels, resulting in a significant reduction in the number of medium pores and a large increase in the number of small pores.

3.3. Water Retention Characteristics Test

Figure 13 shows the water retention characteristic curve of the XG-silt specimens. Figure 13a,b show the drying and wetting curves, respectively, and Figure 14 shows the drying–wetting hysteresis curve. Figure 13a,b show that the drying curves all shift to the upper right with increasing XG content, which reflects an increase in the water retention capacity of the treated specimens. Especially in the range of low suction, the water retention capacity of the silt significantly increases with increasing XG content. Because XG carries carboxylic acid (–COOH) and hydroxyl (–OH) with a negative charge, it can interact with cations in the soil particles to generate a capillary force, thus absorbing and retaining more water [46].

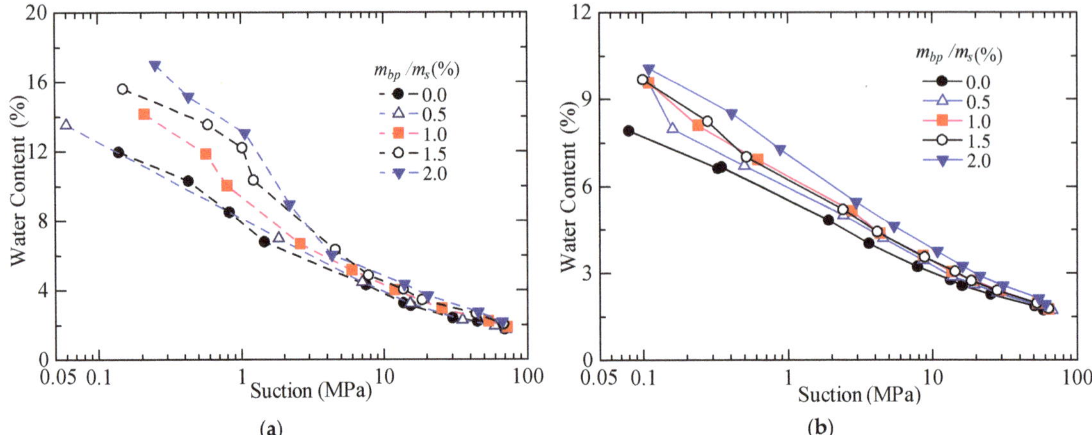

Figure 13. Water retention curves of silt with different XG contents: (**a**) drying process; (**b**) wetting process.

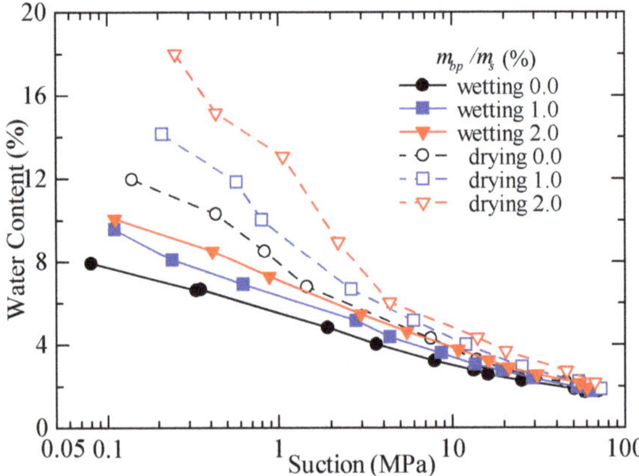

Figure 14. Drying–wetting hysteresis curve of silt with different XG contents.

Figure 14 shows that the water retention capacity of the treated specimens during the drying process is stronger than that during the wetting process, which is related to the water retention hysteretic effect of the soil during drying [47–49]. The hysteretic circle area between the drying and wetting curves of the treated specimens also gradually increases with increasing XG content. The drying and wetting curves of specimens tend to be consistent in the high-suction range [50].

The SEM and mercury intrusion test results show that more biofilms covered the surface of the soil particles with increasing XG content, and the bridging connections are found in the silt pores, which gradually filled with XG. The water retention capacity of the treated silt increased due to the strong water-retaining properties of XG [25,26,29,51], and the hysteretic circle area between the drying and wetting curves also gradually increased. Previous experimental studies show that more water can be retained in soil pores after treatment with XG, which is consistent with the water retention characteristics test results of this study. Additionally, XG has a stronger water retention capacity than other polymers, which can improve the survival rate of vegetation in severely arid areas [19,52].

3.4. Direct Shear Test

3.4.1. Shear Strength Characteristics of Silt

Figure 15 shows the shear strength–displacement relation curves of pure silt with water contents of 2%, 4%, 8%, 16%, and 22%. The peak strength and residual strength of the silt specimens gradually increased with increasing vertical stress.

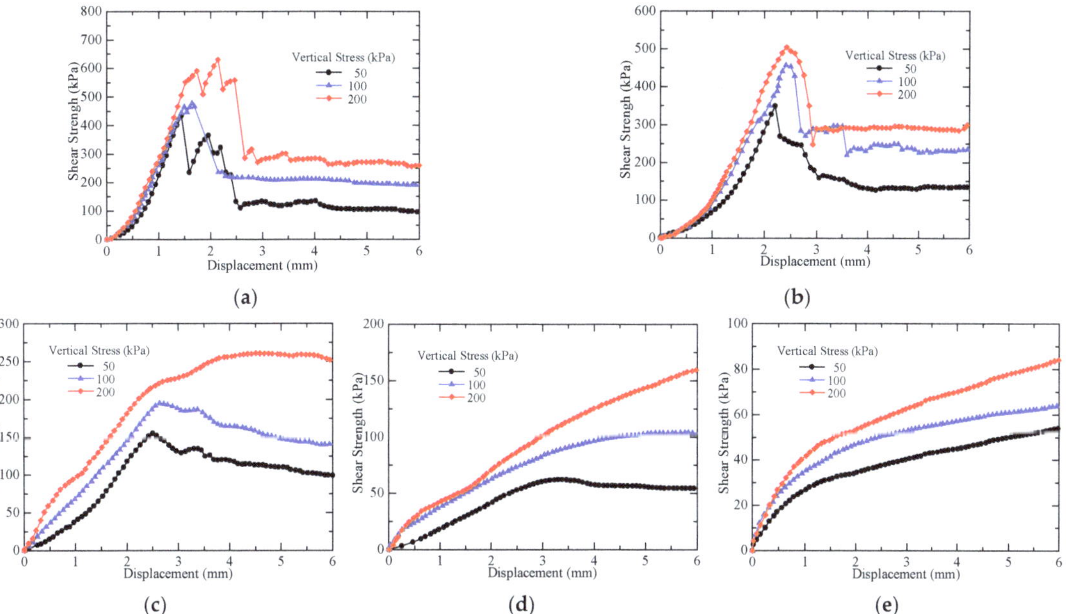

Figure 15. Shear strength of the silt samples with different water contents (m_{bp}/m_s = 0%): (**a**) w = 2%; (**b**) w = 4%; (**c**) w = 8%; (**d**) w = 16%; (**e**) w = 22%.

When the water content was 2% and 4%, the shear strength–displacement relation curves of pure silt showed a stress-softening phenomenon, and the peak strength was higher than the residual strength. When the water content was 16% and 22%, the shear strength–displacement relation curves of pure silt showed a stress-hardening phenomenon. When the water content was 8%, and the vertical stress was 50 kPa and 100 kPa, the shear strength–displacement relation curves of pure silt showed a stress-softening phenomenon, and when the vertical stress was 200 kPa, the shear strength–displacement relation curve displayed a stress-hardening phenomenon. The results show that with the increase in vertical pressure, the soil began to show a stress-hardening phenomenon [53,54].

Figure 15c,d show the stress softening into stress hardening of pure silt with water content increasing from 8% to 16%.

3.4.2. Strength Characteristics of Xanthan Gum-Treated Silt

Figure 16 shows the shear strength–displacement relation curves of the XG-silt samples with different water content. The peak strength and residual strength of the treated silt were found to gradually increase with the increase in vertical stress.

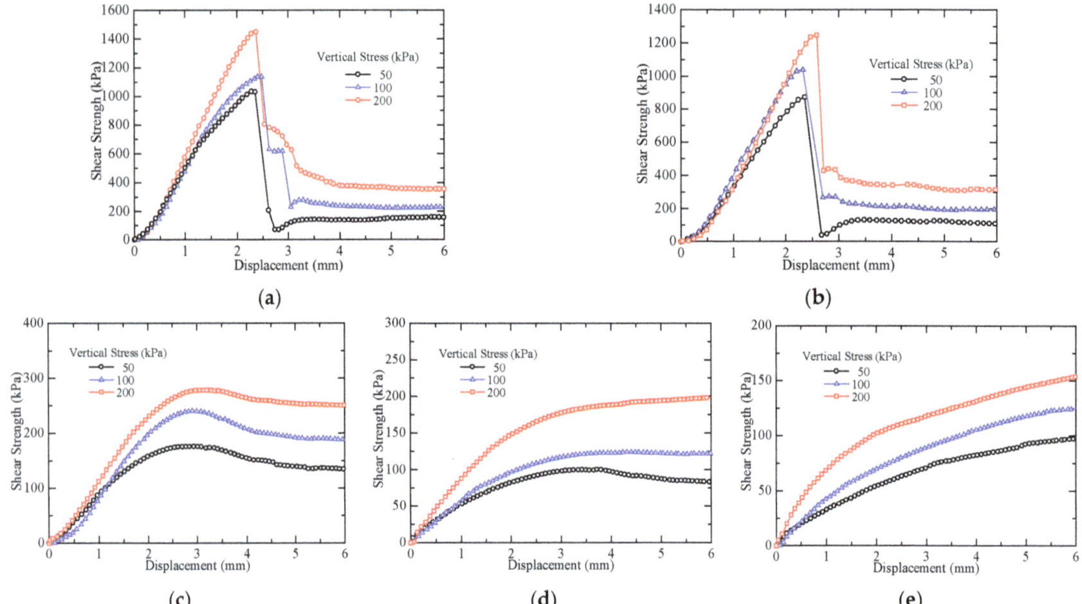

Figure 16. Shear strength of xanthan gum-treated silt with different water contents (m_{bp}/m_s = 2%): (**a**) w = 2%; (**b**) w = 4%; (**c**) w = 8%; (**d**) w = 16%; (**e**) w = 22%.

When the water content was 2%, 4%, and 8%, the shear strength–displacement relation curves of the treated silt showed the stress-softening phenomenon. When the water content was 22%, the shear strength–displacement relation curves of the treated silt showed a stress-hardening phenomenon. When the water content was 16%, and the vertical pressure was 50 kPa, the shear strength–displacement relation curve of pure silt showed a stress-softening phenomenon, and when the vertical stress was 100 kPa and 200 kPa, shear strength–displacement relation curves showed a stress-hardening phenomenon.

Figure 16c,d show the stress softening into stress hardening with water content increasing from 8% to 16% after the addition of XG.

3.4.3. Relationship between Strength and Water Content

Figure 17 shows the shear strength–displacement relation curves of the pure silt and XG-silt specimens with different water contents under vertical stress of 100 kPa. The strength of the silt specimens gradually increased with decreasing water content, as did the strength of the XG-silt specimens.

3.4.4. Variation Rules of Shear Strength Parameters

Figure 18 shows the relationship curves of water content, cohesion c, and internal friction angle ϕ of silt. Table 3 summarizes the parameters of the direct shear strength tests before and after silt treatment during the wetting process. Table 3 and Figure 18 show that at low water content (2–4%), the c values of XG-silt increased by more than a factor of 2.3, and ϕ increased by more than a factor of 1.5. However, when the water content was lower than 8%, the sample strength was enhanced both before and after treatment, whereas c and ϕ only slightly increased. At high water contents (16% and 22%), c increased by factors of 1.4 and 1.8 times, respectively, and ϕ increased by factors of 1.4 and 1.9. The experimental results are consistent with those of Soldo et al. [35,36] and Chang et al. [55]. Additionally, the test results showed that the addition of XG can greatly improve the strength and stability of silt during the wetting process.

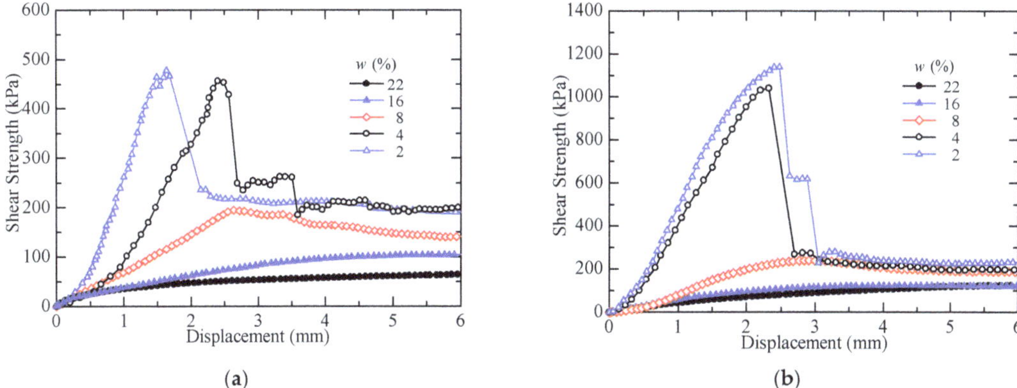

Figure 17. Wetting strength of the silt specimens with different water contents under a vertical stress of 100 kPa: (**a**) m_{bp}/m_s = 0%; (**b**) m_{bp}/m_s = 2%.

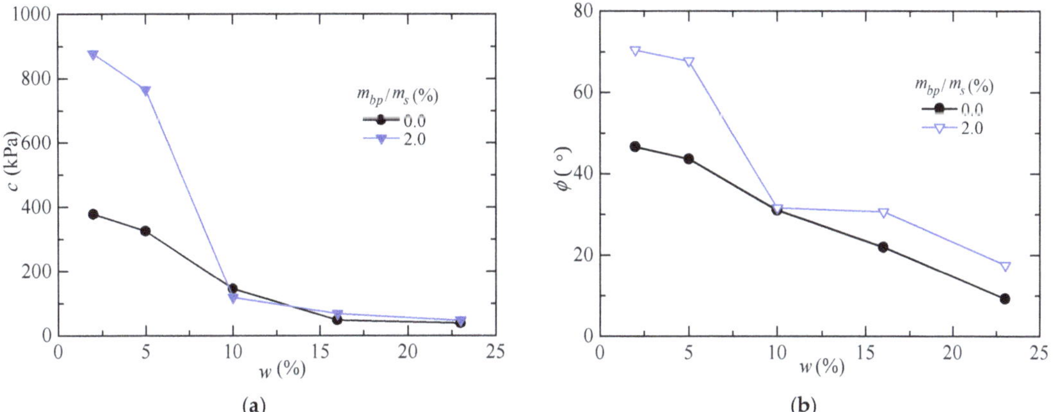

Figure 18. Comparison of shear strength parameters before and after treatment: (**a**) *w*-*c* relation curve; (**b**) *w*-*ϕ* relation curve.

3.4.5. Discussion on Strength Mechanisms

According to the mercury intrusion test result, the pore size distribution of the silt specimen is bimodal. Therefore, in the shear process, the internal stress of the silt specimen is not uniform. By comparing Figures 15a and 16a, it is obvious that the pore structure in the silt greatly affects the shear stress of the soil and thus further affects the strength and stability of the soil in the range of low water content. According to SEM pictures and mercury intrusion test results, with the increase in XG content, the pores of silt are gradually filled with XG cement. Compared with Figures 15 and 16, it can be concluded that the stress-hardening phenomenon of the silt treated by XG is more obvious than silt when the water content is 2%, 4%, and 8% because the pores of the silt treated by XG are supported by XG, which expands after absorbing moisture. Therefore, the water content of stress softening into stress hardening increased from 8% to 16%. The results indicate that XG can improve the strength and stability of soil after improving the micropore structure of the soil.

Table 3. Strength properties of the XG-silt specimens under different conditions (m_{bp}/m_s).

w (%)	σ_v (kPa)	m_{bp}/m_s = 0%			m_{bp}/m_s = 2%			Increment of c	Increment of ϕ
		τ_f (kPa)	c (kPa)	ϕ (°)	τ_f (kPa)	c (kPa)	ϕ (°)		
2	50	434.4	378.15	46.55	1033	878.50	70.50	2.32	1.515
	100	478.5			1141				
	200	591			1450				
4	50	349	325.45	43.57	870.7	766.70	67.82	2.36	1.557
	100	456.6			1040				
	200	503.7			1248				
8	50	155.2	147.3	31.05	155.2	165.20	31.65	1.22	1.019
	100	240.8			256.9				
	200	256.6			278.5				
16	50	62.11	48.09	21.91	100.5	68.15	30.71	1.42	1.402
	100	97.45			123.6				
	200	125.5			188.3				
22	50	45.03	38.58	9.19	59.03	69.63	17.61	1.81	1.916
	100	57.2			70.94				
	200	70.1			93.42				

The mercury intrusion test results indicate that the pore size distribution of the silt specimen is bimodal. The dispersed pore distribution of the specimens results in an unstable structure. The silt specimen is easily disturbed by external loads; thus, the strength of the silt specimen is small. After adding XG, the silt pore size decreased, the pore size distribution changed into a unimodal pore size distribution, and the pore structure stability improved. The water retention capacity and shear strength of the silt, therefore, significantly increased. The SEM test results showed that the cohesion and internal friction angle of the silt specimens under different water content was significantly higher because XG blocks the pores of soil and produces a hydrogel that can bond silt particles together [39,55–57].

4. Conclusions

A series of microscopic tests, water retention tests, and shear tests were carried out to explore the effect and mechanism of the strength characteristics of silt treated with xanthan gum. The main conclusions are as follows.

The drying and wetting curves of the specimens all shifted to higher water retention capacity with increasing xanthan gum content, and the hysteretic circle area between the drying and wetting curves also gradually increased. Microscopic analysis showed that the pores of the specimens gradually filled with increasing xanthan gum content, and xanthan gum itself has strong water-retaining properties. The water retention capacity of the treated silt therefore increased.

The peak strength and residual strength of the specimens gradually increase with increasing vertical stress. The shear strength, cohesion, and internal friction angle of the pure and treated silt specimens significantly increased with decreasing water content. During the wetting process, the strength of the pure and treated silt specimens all significantly weakened with increasing water content, but the strength of the treated silt was notably higher than that without treatment. The microscopic analysis showed that xanthan gum fills the uneven pores in silt during the wetting process, which makes the pore distribution more uniform and the pore structure more stable. Xanthan gum forms hydrogel bonded with the silt particles, and the cementation between the soil particles was notable, thus significantly increasing the strength of the silt treated by xanthan gum.

As an eco-friendly and efficient material, xanthan gum can improve the strength and stability of silt after rainfall infiltration. Therefore, xanthan gum can be used to improve the engineering properties of soil during many projects, such as deep foundation treatment, slope treatment, highway subgrade treatment, etc. The shear strength prediction model based on experimental data needs to be further studied.

Author Contributions: Conceptualization, J.Z. (Junran Zhangand), Z.M. and T.J.; methodology, J.Z. (Junran Zhangand), Z.M. and J.Z. (Jindi Zhao); data curation, J.Z. (Junran Zhangand), J.Z. (Jindi Zhao), X.Z. and Z.M.; writing—original draft preparation, Z.M.; writing—review and editing, J.Z. and S.W.; visualization, Z.M.; supervision, J.Z. (Junran Zhangand); project administration, J.Z. (Junran Zhangand) and T.J.; funding acquisition, J.Z. (Junran Zhangand) and T.J. All authors have read and agreed to the published version of the manuscript.

Funding: This study is financially supported by the National Natural Science Foundation of China (Grant No. 41602295), the Foundation for University Key Teacher by the Ministry of Education of Henan Province (Grant No. 2020GGJS-094), the Key Scientific Research Projects of Colleges and Universities in Henan Province (Grant No. 21A410002), and the Doctoral Student Innovation Foundation of NCWU.

Conflicts of Interest: The authors declare no conflict of interest. The funders had no role in the design of the study; in the collection, analyses, or interpretation of data; in the writing of the manuscript, or in the decision to publish the results.

References

1. Wang, Y.; Zhang, H.; Zhang, Z. Experimental Study on Mechanics and Water Stability of High Liquid Limit Soil Stabilized by Compound Stabilizer: A Sustainable Construction Perspective. *Sustainability* **2021**, *13*, 5681. [CrossRef]
2. Kim, Y.; Worrell, E. CO_2 Emission Trends in the Cement Industry: An International Comparison. *Mitig. Adapt. Strat. Glob. Change* **2002**, *7*, 115–133. [CrossRef]
3. Thangaraj, R.; Thenmozhi, R. Sustainable concrete using high volume fly ash from thermal power plants. *Ecol. Environ. Ment Conservat.* **2013**, *19*, 461–466.
4. Alonso, S.; Palomo, A. Alkaline activation of metakaolin and calcium hydroxide mixtures: Influence of temperature, activator concentration and solids ratio. *Mater. Lett.* **2001**, *47*, 55–62. [CrossRef]
5. Krivenko, P.V.; Kovalchuk, G.Y. Directed synthesis of alkaline aluminosilicate minerals in a geocement matrix. *J. Mater. Sci.* **2007**, *42*, 2944–2952. [CrossRef]
6. Sofi, M.; van Deventer, J.S.J.; Mendis, P.A.; Lukey, G.C. Bond performance of reinforcing bars in inorganic polymer concrete (IPC). *J. Mater. Sci.* **2007**, *42*, 3107–3116. [CrossRef]
7. Lee, S.; Chung, M.; Park, H.M.; Song, K.-I.; Chang, I. Xanthan Gum Biopolymer as Soil-Stabilization Binder for Road Construction Using Local Soil in Sri Lanka. *J. Mater. Civ. Eng.* **2019**, *31*, 06019012. [CrossRef]
8. Chang, I.; Jeon, M.; Cho, G.-C. Application of Microbial Biopolymers as an Alternative Construction Binder for Earth Buildings in Underdeveloped Countries. *Int. J. Polym. Sci.* **2015**, *2015*, 326745. [CrossRef]
9. Georgees, R.N.; Hassan, R.A.; Evans, R.P. A potential use of a hydrophilic polymeric material to enhance durability properties of pavement materials. *Constr. Build. Mater.* **2017**, *148*, 686–695. [CrossRef]
10. Fatehi, H.; Abtahi, S.M.; Hashemolhosseini, H.; Hejazi, S.M. A novel study on using protein based biopolymers in soil strengthening. *Constr. Build. Mater.* **2018**, *167*, 813–821. [CrossRef]
11. Liu, J.; Ali, A.; Su, J.; Wu, Z.; Zhang, R.; Xiong, R. Simultaneous removal of calcium, fluoride, nickel, and nitrate using microbial induced calcium precipitation in a biological immobilization reactor. *J. Hazard. Mater.* **2021**, *416*, 125776. [CrossRef] [PubMed]
12. Hosseinpour, Z.; Najafpour-Darzi, G.; Latifi, N.; Morowvat, M.; Manahiloh, K.N. Synthesis of a biopolymer via a novel strain of Pantoea as a soil stabilizer. *Transp. Geotech.* **2020**, *26*, 100425. [CrossRef]
13. Choi, S.-G.; Chang, I.; Lee, M.; Lee, J.-H.; Han, J.-T.; Kwon, T.-H. Review on geotechnical engineering properties of sands treated by microbially induced calcium carbonate precipitation (MICP) and biopolymers. *Constr. Build. Mater.* **2020**, *246*, 118415. [CrossRef]
14. Bai, B.; Nie, Q.; Zhang, Y.; Wang, X.; Hu, W. Cotransport of heavy metals and SiO2 particles at different temperatures by seepage. *J. Hydrol.* **2021**, *597*, 125771. [CrossRef]
15. Van de Velde, K.; Kiekens, P. Biopolymers: Overview of several properties and consequences on their applications. *Polym. Test.* **2002**, *21*, 433–442. [CrossRef]
16. Lora, J.H.; Glasser, W.G. Recent Industrial Applications of Lignin: A Sustainable Alternative to Nonrenewable Materials. *J. Polym. Environ.* **2002**, *10*, 39–48. [CrossRef]

17. Ortsi, W.J.; Roa-Espinosa, A.; Sojka, R.E.; Glenn, G.M.; Imams, S.H.; Erlacher, K.; Pedersen, J.S. Use of Synthetic Polymers and Biopolymers for Soil Stabilization in Agricultural, Construction, and Military Applications. *J. Mater. Civ. Eng.* **2007**, *19*, 58–66. [CrossRef]
18. Chang, I.; Prasidhi, A.K.; Im, J.; Cho, G.-C. Soil strengthening using thermo-gelation biopolymers. *Constr. Build. Mater.* **2015**, *77*, 430–438. [CrossRef]
19. Chang, I.; Prasidhi, A.K.; Im, J.; Shin, H.-D.; Cho, G.-C. Soil treatment using microbial biopolymers for anti-desertification purposes. *Geoderma* **2015**, *253–254*, 39–47. [CrossRef]
20. Im, J.; Tran, A.T.; Chang, I.; Cho, G.-C. Dynamic properties of gel-type biopolymer-treated sands evaluated by Resonant Column (RC) Tests. *Géoméch. Eng.* **2017**, *12*, 815–830. [CrossRef]
21. Ayeldeen, M.; Negm, A.; El-Sawwaf, M.; Kitazume, M. Enhancing mechanical behaviors of collapsible soil using two biopolymers. *J. Rock Mech. Geotech. Eng.* **2017**, *9*, 329–339. [CrossRef]
22. Dehghan, H.; Tabarsa, A.; Latifi, N.; Bagheri, Y. Use of xanthan and guar gums in soil strengthening. *Clean Technol. Environ. Policy* **2019**, *21*, 155–165. [CrossRef]
23. Ghasemzadeh, H.; Modiri, F. Application of novel Persian gum hydrocolloid in soil stabilization. *Carbohydr. Polym.* **2020**, *246*, 116639. [CrossRef] [PubMed]
24. Chang, I.; Lee, M.; Tran, A.T.P.; Lee, S.; Kwon, Y.-M.; Im, J.; Cho, G.-C. Review on biopolymer-based soil treatment (BPST) technology in geotechnical engineering practices. *Transp. Geotech.* **2020**, *24*, 100385. [CrossRef]
25. Wallingford, L.; Labuza, T.P. Evaluation of the Water Binding Properties of Food Hydrocolloids by Physical/Chemical Methods and in a Low Fat Meat Emulsion. *J. Food Sci.* **1983**, *48*, 1–5. [CrossRef]
26. Sánchez, V.; Bartholomai, G.; Pilosof, A. Rheological properties of food gums as related to their water binding capacity and to soy protein interaction. *LWT Food Sci. Technol.* **1995**, *28*, 380–385. [CrossRef]
27. Khachatoorian, R.; Petrisor, I.G.; Kwan, C.-C.; Yen, T.F. Biopolymer plugging effect: Laboratory-pressurized pumping flow studies. *J. Pet. Sci. Eng.* **2003**, *38*, 13–21. [CrossRef]
28. Zhou, C.; So, P.; Chen, X. A water retention model considering biopolymer-soil interactions. *J. Hydrol.* **2020**, *586*, 124874. [CrossRef]
29. Bouazza, A.; Gates, W.; Ranjith, P. Hydraulic conductivity of biopolymer-treated silty sand. *Geotechnique* **2009**, *59*, 71–72. [CrossRef]
30. Rosenzweig, R.; Shavit, U.; Furman, A. Water Retention Curves of Biofilm-Affected Soils using Xanthan as an Analogue. *Soil Sci. Soc. Am. J.* **2012**, *76*, 61–69. [CrossRef]
31. Qureshi, M.U.; Chang, I.; Al-Sadarani, K. Strength and durability characteristics of biopolymer-treated desert sand. *Géoméch. Eng.* **2017**, *12*, 785–801. [CrossRef]
32. Cabalar, A.F.; Awraheem, M.H.; Khalaf, M.M. Geotechnical Properties of a Low-Plasticity Clay with Biopolymer. *J. Mater. Civ. Eng.* **2018**, *30*, 04018170. [CrossRef]
33. Chang, I.; Im, J.; Prasidhi, A.K.; Cho, G.-C. Effects of Xanthan gum biopolymer on soil strengthening. *Constr. Build. Mater.* **2015**, *74*, 65–72. [CrossRef]
34. Latifi, N.; Horpibulsuk, S.; Meehan, C.L.; Majid, M.Z.A.; Tahir, M.M.; Mohamad, E.T. Improvement of Problematic Soils with Biopolymer—An Environmentally Friendly Soil Stabilizer. *J. Mater. Civ. Eng.* **2016**, *29*, 10. [CrossRef]
35. Soldo, A.; Miletić, M.; Auad, M.L. Biopolymers as a sustainable solution for the enhancement of soil mechanical properties. *Sci. Rep.* **2020**, *10*, 267. [CrossRef]
36. Soldo, A.; Miletić, M. Study on Shear Strength of Xanthan Gum-Amended Soil. *Sustainability* **2019**, *11*, 6142. [CrossRef]
37. Sujatha, E.R.; Atchaya, S.; Sivasaran, A.; Keerdthe, R.S. Enhancing the geotechnical properties of soil using xanthan gum—An eco-friendly alternative to traditional stabilizers. *Bull. Eng. Geol. Environ.* **2021**, *80*, 1157–1167. [CrossRef]
38. Chang, I.; Im, J.; Lee, S.-W.; Cho, G.-C. Strength durability of gellan gum biopolymer-treated Korean sand with cyclic wetting and drying. *Constr. Build. Mater.* **2017**, *143*, 210–221. [CrossRef]
39. Lee, S.; Chang, I.; Chung, M.-K.; Kim, Y.; Kee, J. Geotechnical shear behavior of Xanthan Gum biopolymer treated sand from direct shear testing. *Géoméch. Eng.* **2017**, *12*, 831–847. [CrossRef]
40. Zhu, Z.-D.; Liu, S.-Y.; Shao, G.-H.; Hao, J.-X. Research on silts and silts treated with stabilizers by triaxial shear tests. *Rock Soil Mech.* **2005**, *12*, 1967–1971. [CrossRef]
41. Zhang, J.-W.; Kang, F.-X.; Bian, H.-L.; Yu, H. Experiments on unconfined compressive strength of lignin modified silt in Yel-low river flood area under freezing-thawing cycles. *Rock Soil Mech.* **2020**, *S2*, 1–6. [CrossRef]
42. Ni, J.; Li, S.-S.; Ma, L.; Geng, X.-Y. Performance of soils enhanced with eco-friendly biopolymers in unconfined compression strength tests and fatigue loading tests. *Constr. Build. Mater.* **2020**, *263*, 120039. [CrossRef]
43. Latifi, N.; Rashid, A.S.A.; Siddiqua, S.; Majid, M.Z.A. Strength measurement and textural characteristics of tropical residual soil stabilised with liquid polymer. *Measurement* **2016**, *91*, 46–54. [CrossRef]
44. Chang, I.; Im, J.; Cho, G.-C. Introduction of Microbial Biopolymers in Soil Treatment for Future Environmentally-Friendly and Sustainable Geotechnical Engineering. *Sustainability* **2016**, *8*, 251. [CrossRef]
45. Kodikara, J.; Barbour, S.; Fredlund, D. Changes in clay structure and behaviour due to wetting and drying. In Proceedings of the 8th Australian-New Zealand Conference on Geomechanics, Hobart, TAS, Australia, 15–17 February 1999.

46. Lee, J.; Fratta, D.; Akin, I. Shear strength and stiffness behavior of fine-grained soils at different surface hydration conditions. *Can. Geotech. J.* **2021**, *99*, 1–12. [CrossRef]
47. Bai, B.; Wang, Y.; Rao, D.-Y.; Bai, F. The effective thermal conductivity of unsaturated porous media deduced by pore-scale SPH simulation. *Front. Earth Sci.* **2022**. [CrossRef]
48. Bai, B.; Xu, T.; Nie, Q.; Li, P. Temperature-driven migration of heavy metal Pb2+ along with moisture movement in unsaturated soils. *Int. J. Heat Mass Transf.* **2020**, *153*, 119573. [CrossRef]
49. Bai, B.; Zhou, R.; Cai, G.; Hu, W.; Yang, G. Coupled thermo-hydro-mechanical mechanism in view of the soil particle rearrangement of granular thermodynamics. *Comput. Geotech.* **2021**, *137*, 104272. [CrossRef]
50. Sun, D.A.; Zhang, J.R.; Lu, H.B. Soil-water characteristic curve of Nanyang expansive soil in full suction range. *Rock Soil Mech.* **2013**, *34*, 1839–1846. [CrossRef]
51. Narjary, B.; Aggarwal, P.; Singh, A.; Chakraborty, D.; Singh, R. Water availability in different soils in relation to hydrogel application. *Geoderma* **2012**, *187–188*, 94–101. [CrossRef]
52. Tran, T.P.A.; Chang, I.; Cho, G.-C. Soil water retention and vegetation survivability improvement using microbial biopolymers in drylands. *Geomech. Eng.* **2019**, *17*, 475–483. [CrossRef]
53. Wang, Y.-X.; Shao, S.-J.; Wang, Z.; Liu, A.-G. Experimental study on the unloading characteristics of coarse aggregate under true triaxial shear loading. *J. Rock Mech. Eng.* **2020**, *39*, 1503–1512. [CrossRef]
54. Bai, B.; Yang, G.-C.; Li, T.; Yang, G.-S. A thermodynamic constitutive model with temperature effect based on particle rearrangement for geomaterials. *Mech. Mater.* **2019**, *139*, 103180. [CrossRef]
55. Chang, I.; Cho, G.-C. Shear strength behavior and parameters of microbial gellan gum-treated soils: From sand to clay. *Acta Geotech.* **2019**, *14*, 361–375. [CrossRef]
56. Chen, C.; Wu, L.; Perdjon, M.; Huang, X.; Peng, Y. The drying effect on xanthan gum biopolymer treated sandy soil shear strength. *Constr. Build. Mater.* **2019**, *197*, 271–279. [CrossRef]
57. Chang, I.; Kwon, Y.-M.; Im, J.; Cho, G.-C. Soil consistency and interparticle characteristics of xanthan gum biopolymer–containing soils with pore-fluid variation. *Can. Geotech. J.* **2018**, *56*, 1206–1213. [CrossRef]

Article

Long-Term Performance of the Water Infiltration and Stability of Fill Side Slope against Wetting in Expressways

Yuedong Wu, Xiangyu Zhou and Jian Liu *

Key Laboratory of Ministry of Education for Geomechanics and Embankment Engineering, Hohai University, Nanjing 210098, China; hhuwyd@163.com (Y.W.); zxy1152021469@hhu.edu.cn (X.Z.)
* Correspondence: 20170053@hhu.edu.cn; Tel.: +86-1836-296-0323

Abstract: Different settlements and instabilities of unsaturated subgrade subjected to wetting have been paid increasing attention in the southeast coastal areas of China. However, the treatments are costly when they are used in engineering. In addition, the long-term performances of the treatments are unclear. Based on seepage theory for unsaturated soils, a novel subgrade using a capillary barrier was proposed in this study to reduce the different settlements and stabilities. Compared with previous studies, a capillary barrier was merely applied in the landfill. The long-term performance and feasibility of a capillary barrier applied in a tilted subgrade slope is worthy of study, particularly in humid climates. Using Geo-Studio, the feasibility was verified by comparing a conventional subgrade with a subgrade using a capillary barrier in southeast coastal areas in terms of pore-water pressure, water content, settlement, and the safety factor. The numerical results showed that the subgrade using a capillary barrier could provide significant improvements in the performance of reducing the impact of pore-water pressure distribution it suffered from, so as to lead to smaller different settlements. The vertical settlement of the pavement using a capillary barrier over a 1 year period was 1 cm. Compared with a conventional subgrade, the settlement fell by 94%, and the safely factor increased by 15% for the subgrade using the capillary barrier.

Keywords: unsaturated subgrade; capillary barrier; distress control of wetting

Citation: Wu, Y.; Zhou, X.; Liu, J. Long-Term Performance of the Water Infiltration and Stability of Fill Side Slope against Wetting in Expressways. *Appl. Sci.* **2022**, *12*, 5809. https://doi.org/10.3390/app12125809

Academic Editor: Bing Bai

Received: 2 April 2022
Accepted: 4 June 2022
Published: 7 June 2022

1. Introduction

With the development of social economy and the requirements of modern transportation, expressways have become a focus during city construction [1]. Different from buildings, strict requirements are considered for the foundation of expressways such as the evaluation index of the expressway (i.e., fast, security, economy, and comfort). However, many problems still exist after an expressway's construction. Firstly, the safety of the expressway is affected directly by the instability of the subgrade. Secondly, if different settlement occurs on the expressway pavement, the driving speed is reduced. Moreover, the impact force caused by slowing down the driving speed is further exacerbated by the extent of the unevenness of the expressway pavement. Therefore, a vicious circle is formed, which further affects driving safety and causes huge economic losses [2]. These problems have seriously affected the use of expressways, especially in areas in southeast coastal provinces where the economy is developing rapidly. They are located in the region of a continental monsoon climate zone with relatively high rainfall in China. Therefore, in these areas, the subgrades have been greatly affected after completion by atmospheric conditions that include rainfall and evaporation. The instability and different settlement become more prominent under the influence of wet conditions [3].

For expressways, the foundation soil is located above the groundwater table isoline and is in an unsaturated state. In the unsaturated subgrade, the volumetric moisture content increases with rainfall. Therefore, the strength of the foundation soil is minimized, and it leads to instability and different settlement of the subgrade. In order to prevent instability and reduce different settlements after completion, a long–short pile composite foundation [4,5] is adopted. This enables piles of different lengths and stiffnesses to carry more loads with a decrease in the proportion of the load on the soil in the subgrade. Moreover, it makes up for the decrease in the soil strength caused by wetting deformation. However, the pile foundation greatly increases the cost of an expressway's construction, and the long-term performance of the pile is still unclear. The composite foundation using foamed cement banking (FCB) as a replacement is another effective treatment measure [6]. FCB is a light-weight material with low density, high strength, and low permeability. By minimizing its weight and permeability, the instability and settlement problems are effectively delayed due to the wetting deformation of the expressway pavement. However, FCB is still in the research stage. The influencing factors of its strength and permeability are still unclear. Wang [7] used a soil stabilizer to reinforce the silty clay for the subgrade. He studied the pavement's performance with a mixture composed of the best proportion, which provided the basis and reference for similar engineering applications. Apart from using a soil stabilizer, Zhao [8] proposed that the recycled fine soil of construction waste with a diameter under 5 mm could be used in expressway subgrade. The soil was validated as a good subgrade material through basic experiments. However, the existing improvements were all treatment methods based on materials without considering the wetting distress on the long-term performance. The construction cost has increased because of these methods, and long-term performance under wetting and drying cycles is unclear. Hence, we propose effective measures to block the infiltrating rainwater in the subgrade through a capillary barrier.

A capillary barrier has been widely used in landfill cover systems, because of its advantages of long service life, ease of construction, and environment conservation [9,10]. It consists of fine layers over coarse layers, which are composed of different properties. Soil properties are very important in engineering. Different soil particles play an important role in the subgrade soil, which affect soil erodibility and the infiltration rate of water. Arunrat et al. [11] pointed out the variations in soil properties and explained how soil erodibility is affected. Wang et al. [12] showed that the different properties and the different poultry compost amendments to the soils resulted in distinct runoff, sediment yield, and soil erodibility values. With the different permeability coefficients of these two soil layers under unsaturated conditions, the fine-grained layer, used as a buffer, can store and transfer infiltrating water effectively [13]. Ross [14] and Stormont [15] derived the evaluation index of capillary barriers through the theory of the permeability coefficients of two soils. The evaluation index is the distance downslope, called the diversion length. It has been verified that a capillary barrier can effectively block the infiltrating rainwater within the diversion length [16,17]. Rahardjo [18] proposed that a capillary barrier could be used to protect the slope. The waterproof effect of a capillary barrier was verified through field tests and through the same way the application of a capillary barrier to the slope was studied and found to be feasible. Wu et al. [19] proposed that a capillary barrier could be applied as a cover barrier to protect the expressway subgrade from wetting distress. Previous studies focused on the effect and mechanism of a capillary barrier applied to landfills in dry climates. However, the performance of a capillary barrier in humid climates is still unclear. In other words, the performance of a capillary barrier applied to a slope shoulder is still unclear in humid climates in China's southeast coastal areas. Meanwhile, the environment of expressways in the southeast coastal areas of China is generally humid compared with the relatively dry landfill cover system. Due to the presence of these peculiarities, it is necessary to verify the rationality of the application of the capillary barrier on the expressway.

This paper firstly explains the mechanism of moisture migration in the unsaturated subgrade and proposes effective measures to block the infiltrating rainwater in the subgrade through capillary barrier. Then, the anti-seepage effect of the capillary barrier is introduced. Comparing the numerical simulation results of the subgrade with or without a capillary barrier, the feasibility of applying a capillary barrier to the subgrade was verified. Finally, the results were evaluated and focused on the role of the capillary barrier in the anti-seepage performance by simulating their changes in moisture content, pore-water pressure, settlement, and safety factor of the subgrade under rainfall conditions in the southeast areas of China.

2. Reinforcement Mechanism of Subgrades with a Capillary Barrier

In the southeast coastal areas of China, the groundwater table isoline is generally below the original ground, and the subgrade is generally in an unsaturated state. Therefore, the soil strength is directly influenced by the distribution of matric suction in the subgrade soil. The matric suction is a manifestation of the energy produced by the water in the soil due to the capillary effect. Numerically, it is equal to the pore-air pressure minus the pore-water pressure ($u_a - u_w$). Based on the M-C model, Fredlund and Rahardjo [20] proposed a formula for the shear strength of unsaturated soil:

$$\tau_u = c' + (\sigma - u_a)\tan(\phi') + (u_a - u_w)\tan\left(\phi^b\right) \quad (1)$$

where τ_u is the shear strength of unsaturated soil; c' is effective cohesion; u_a is the pore air pressure; u_w is the pore-water pressure; ϕ' is the effective internal friction angle; ϕ^b is the shear strength increase rate with the change in matric suction. Figure 1 shows the traditional subgrade structure. After the expressway is completed, the pavement is an asphalt layer with good waterproofness. However, the fill slope is generally exposed in the atmosphere. When it rains, moisture flows into the subgrade from the slope shoulder. Because the subgrade is under unsaturated conditions, the soil suction in the subgrade is large, and the suction at the shoulder slope is small. Moisture gradually inflows into the subgrade soil along the suction gradient line. It is worth noting that the almost horizontal water flow is unsaturated flow. Therefore, the moisture content of the subgrade soil continues to increase, and the suction continues to decrease. According to formula (1), it can be seen that the reduction in shear strength is directly influenced by the reduction in suction, which may lead to the shear deformation and instability of the subgrade. Figure 2a shows the soil water characteristic curve of the subgrade soil. The water-entry value was approximately 6 kPa and the air-entry value was approximately 28 kPa. The difference between the water-entry value and the air-entry value of the subgrade soil was approximately 22 kPa. However, the standard design value of the vehicle load on the expressway was only 10.5 kPa [21]. Therefore, the strength change caused by the wetting subgrade should not be ignored.

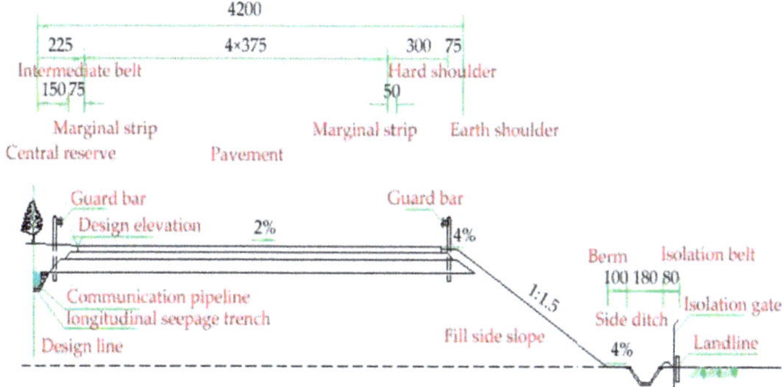

Figure 1. Schematic diagram of a standard subgrade (unit: mm).

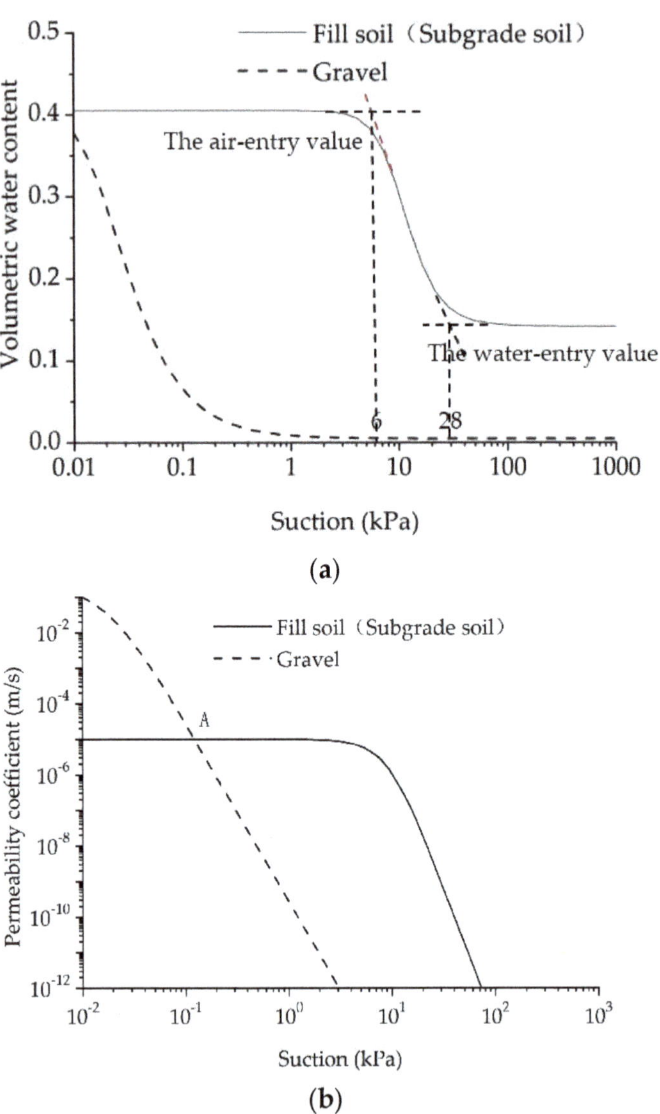

Figure 2. (**a**) Soil water characteristic curves and (**b**) permeability function of soils used in the subgrade.

In order to avoid the decrease in soil strength due to the presence of rainfall, this paper proposes to add a capillary barrier on the fill slope to protect the subgrade. The fine-grained soil in the capillary barrier was filled with the soil near the expressway. Gravel or sand was used as the coarse-grained soil. The fill slope soil from the surface to bottom was the fill near the expressway, gravel, and subgrade soil. Figure 2 shows the soil water characteristic curves and permeability coefficient curves of the two soils. The unsaturated permeability coefficients were estimated using the Mualem model [22].

$$k = k_s S_e^{0.5} \left(\int_0^{S_e} \frac{dS_e}{\psi} / \int_0^1 \frac{dS_e}{\psi} \right)^2 \quad (2)$$

where k_s is equal to the saturated permeability coefficient; S_e is equal to the effective saturation; ψ is equal to the soil suction. The slope in the soil water characteristic curve of the coarse-grained soil was steeper because of its large pores. Thereby, its water-holding capacity was worse, and the permeability coefficient curve was also steeper. The volumetric water content and permeability coefficient have a nonlinear distribution under different suction ranges. The suction where the permeability coefficients of two soils are equal is the critical failure suction, as shown in Figure 2b, which is the suction at point A.

With the water content continuously increasing when it rains, soil suction decreases. Rainwater enters the capillary barrier along the slope shoulder and inflows into the fill due to the fact of gravity. When soil suction is higher than the critical failure suction, the permeability coefficient of the fill is higher than that of the gravel. At this time, water accumulates only at the interface between the gravel and the fill. It does not breakthrough into the gravel layer. Because the capillary barrier is tilted, the moisture accumulated at the interface is drained laterally along the interface. When soil suction is less than the critical failure suction, the capillary barrier fails. At the interface, the diversion length is the distance from the point where the suction is equivalent to the critical failure suction to the surface of slope. It was evidently shown that the capillary barrier is effective within the diversion length. Walter [23], Tami [24], and Aubertin [25] pointed out that the diversion length of the capillary barrier was influenced by the slope angle, thickness of the fine-grained soil, the properties of the fine-grained and coarse-grained soils, layer thickness, rainfall condition, etc. They all focused on the landfill cover system. Compared with the landfill, the slope length of the subgrade is relatively shorter, and the blocking effect of rainwater is better.

3. Calculation Model of the Subgrade with a Capillary Barrier

It is difficult to study wetting deformation using an analytical solution. For one thing, the soil water characteristic curve and the permeability coefficient equation of soils are both nonlinear. For another, the strength constitutive relationships of soils are complicated. To quantitatively analyze the deformation and stability of the subgrade protected by a capillary barrier during the rainfall, Geo-Studio 2018 R2 software (version 9.1.1.16749) was used to carry out a numerical simulation study in this paper. Geo-Studio is an overall analysis tool for a set of geological structure model software. The test sections in this paper included SLOPE/W (slope stability analysis module), SEEP/W (groundwater seepage analysis module), and SIGMA/W (rock and soil stress and deformation analysis module).

3.1. Mesh and Model

Figure 3 shows the model of the subgrade with a capillary barrier. The figure on the horizontal axis means the distance from the road's central line. The figure on the vertical axis means the height of the pavement to the bedrock. A subgrade section was selected in Guangzhou. The height of the foundation was 20 m. The height of the subgrade was 6 m. The slope of the shoulder was 1:1.5. Due to the symmetry of the subgrade, the simulation only took half of the subgrade for the research. This paper studied two subgrade models: (1) a conventional subgrade; (2) a subgrade using a capillary barrier. The conventional subgrade was composed of a single soil layer, while the subgrade protected by a capillary barrier added a capillary barrier on the slope shoulder. The subgrade soil was used as fill with a thickness of 0.3 m. The thickness of the gravel was 0.2 m.

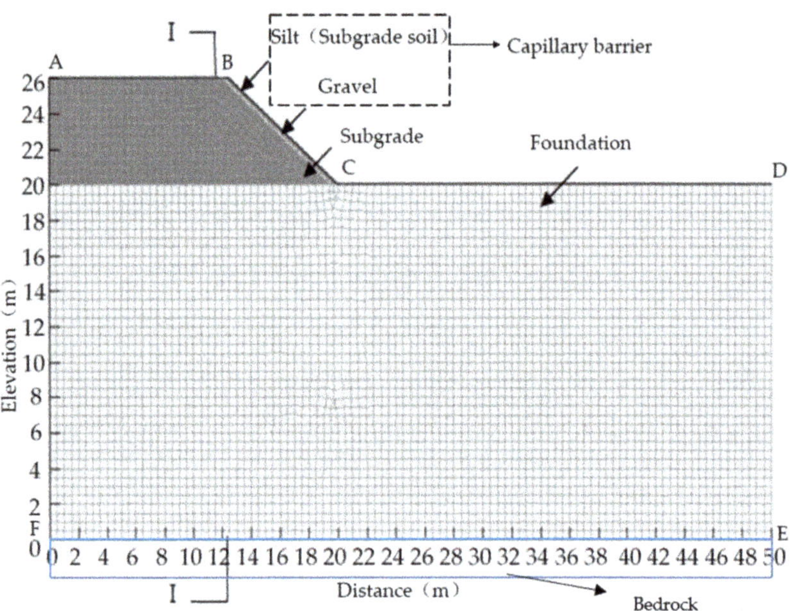

Figure 3. Mesh of the subgrade with a capillary barrier (unit: m).

3.2. Material Properties

The physical properties of soils for the subgrade with a capillary barrier are shown in Tables 1 and 2. Based on the data in Figure 4, the coefficient of uniformity of the gravel was 14 and that of the fill soil was 6.4. The coefficient of the curvature of the gravel was 2.8 and that of the fill soil was 1.2. According to the Standard for Classification of Engineering Soils [26], the fill soil was classified as the silty clay of well gradation, and the gravel was well graded. The soil involved in this paper included foundation soil, subgrade soil, and gravel. In view of the similar moisture migration properties of the foundation and the subgrade, the identical hydraulic characteristic parameters were used for the subgrade and the foundation to facilitate the study. In the numerical simulation, the soil water characteristic curve of the subgrade soil was based on the test data from Jun Luo [3]. The soil water characteristic curve of the gravel was based on the test data used by Morris et al. [27]. The saturated permeability coefficient of the soil was measured by the constant head method. However, the unsaturated permeability coefficient equation was predicted on the basis of the soil water characteristics using the van Genuchten functions as shown in Figure 2. The ideal elastoplastic M-C model was adopted as the mechanical model for the deformation, and the strength parameters of the foundation soil and subgrade soil were determined by consolidated drained triaxial tests. The results of the triaxial shear test are shown in Table 1. The gravel was only 0.2 m thick due to the fact of its small thickness. In order to ensure the calculations were convenient, the strength characteristics were set to elastic materials and the data, listed in Table 1, were selected based on the elastic modulus test by Yuedong Wu et al. [28].

Table 1. The physical properties of various soils for the subgrade with a capillary barrier.

Type of Soil Layer	Maximum Dry Density (g/cm^3)	Optimum Moisture Content (%)	Liquid Limit (%)	Plastic Limit (%)	Plasticity Index
Foundation	1.675	15.03	82	23	11
Subgrade	1.675	15.03	82	23	11
Gravel	2.168				

Table 2. Strength properties of the various soils for the subgrade with a capillary barrier.

Type of Soil Layer	Bulk Density (kN/m^3)	Elastic Modulus (MPa)	Poisson's Ratio	c' (kPa)	φ' (°)	φ^b (°)
Foundation	18.3	20	0.35	1	23	11
Subgrade	18.3	20	0.35	5	26	13
Gravel	18.3	40	0.35	-	-	-

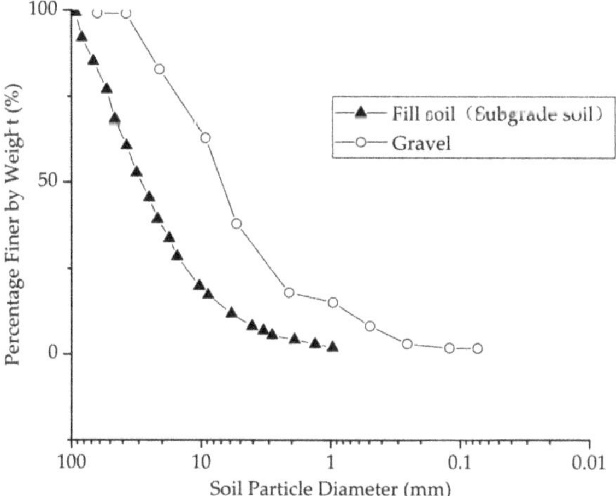

Figure 4. Grain size distribution of the soils.

3.3. Simulation Steps and Boundary Conditions

In order to simulate the impact of rainfall on the expressway, it was necessary to eliminate other factors that cause settlement. This paper simulated the following four steps.

The first step was to perform an analysis on the stress and deformation (SIGMA/W) of the subgrade. The design load on the pavement (AB) was set to complete the load and geo-stress balance. According to the expressway design code, the standard design load of the pavement was 10.5 kPa. The left and right boundaries (i.e., AF and DE) were set to the displacement boundary condition, restraining in the horizontal direction. The bottom boundary (EF) was set to the displacement boundary conditions, restraining in the horizontal and vertical directions.

The second step was to conduct a water seepage analysis (SEEP/W) on the subgrade and design the flow boundary conditions on the shoulder slope (BC) and the ground (CD). The annual precipitation in Guangdong, eastern Guangxi, Fujian, Jiangxi, and most of Zhejiang along the southeast coast of China is 1500–2000 mm. The middle and lower reaches of the Yangtze River are 1000–1600 mm. The Huaihe River, Qinling Mountains, and the Liaodong Peninsula have an annual precipitation of 800–1000 mm. Moreover, the seasonal distribution is uneven, with summer accounting for approximately 50% of the annual rainfall. Particularly, in this study, we focused on the long-term performance of the subgrade. The subgrade soil can store water. Unsaturated drainage still occurs from the topsoil when it does not rain. Therefore, the most extreme rainfall intensity was selected, which was 1000 mm in the summer and the rainfall on the rainfall surface was set to 1.27×10^{-7} m/s. In order to fully study the law of subgrade infiltration, the duration was chosen as 365 days, which considered the humid climate in the eastern coastal area. Due to the large stiffness and small deformation, we assumed that the impermeable bedrock was the bottom of the groundwater. Considering the surface runoff, the set flow boundary was a flow boundary condition that allowed for correction. The bottom boundary condition (EF) was the pressure head boundary condition. Assuming that the initial water level was 3 m below the ground, the pressure head of EF was set to 17 m.

The third step was to perform stress and deformation analysis (SIGMA/W) on the subgrade at the corresponding time. Using the stress conditions in step 1 and the pore-water pressure results at different moments in step 2 as suction conditions, the deformation response in the wetting conditions was calculated. The boundary conditions were the same as in Step 1 and Step 2. The results obtained were calculated with the load and geo-stress balance as the starting point.

In the fourth step, the slope stability analysis (SLOPE/W) of the subgrade was performed. The initial stress condition was the result of step 3. The initial pore-water pressure distribution condition was the result of step 2. The boundary conditions were the same as in step 1.

4. Numerical Simulation Results

4.1. The Law of Moisture Migration in the Subgrade

When the conventional subgrade was subjected to rainfall, the rainwater flowed from the slope shoulder of the expressway to the inside of the subgrade soil, as shown in Figure 5a. This led to a decrease in the pore-water pressure of the soil near the shoulder. But far away from the expressway shoulder, the distribution of the pore-water pressure was almost unaffected. This was mainly due to the effect of gravity, which limits the scope of moisture migration. The influence range of the rainfall in the conventional subgrade was approximately 3 m.

Compared with the conventional subgrade, the subgrade using a capillary barrier was far less affected by rain. In the subgrade with a capillary barrier, water was mainly diverted laterally from the fine-grained soil of the capillary barrier. The distribution of the pore-water pressure in the subgrade and the foundation was almost horizontal, but the flow was concentrated at the bottom of the expressway's slope shoulder. In order to more effectively minimize the impact of rainfall on the pore-water pressure in the foundation, it is recommended that a drainage ditch connecting the fine-grained soil layer be constructed at the base of the slope during the site's construction to facilitate drainage.

It is worth noting that the impermeability of the pavement is very important for the subgrade using a capillary barrier. If there are many cracks in the expressway pavement, rainwater will infiltrate into the subgrade along the cracks, which will lead to a decrease in the pore-water pressure of the subgrade. Under this condition, the suction in the subgrade soil is first reduced to a critical failure suction, and the capillary barrier loses its blocking effect.

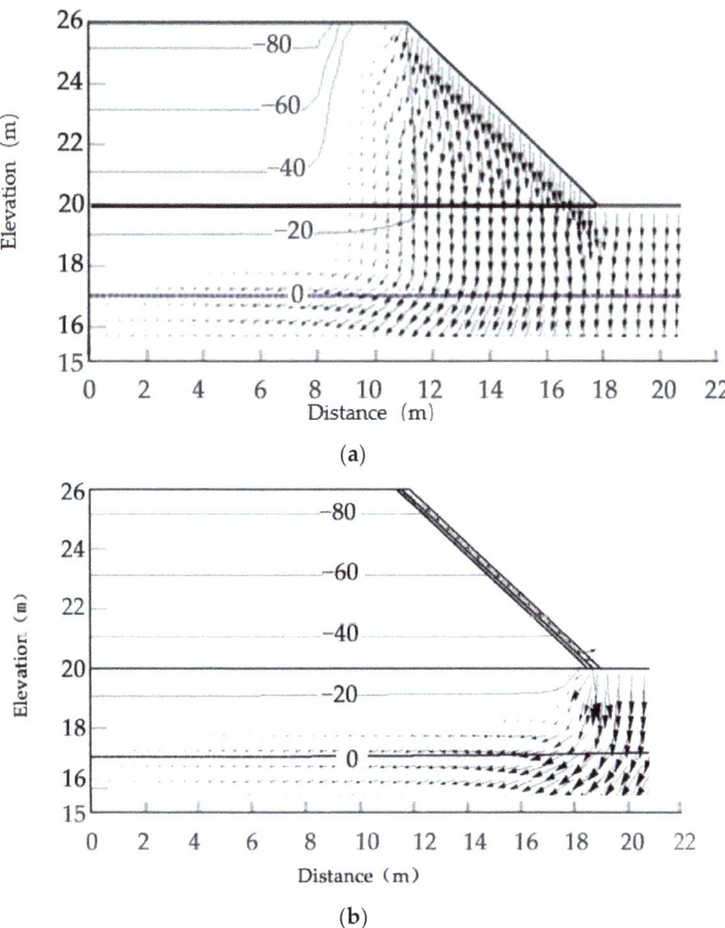

Figure 5. Pore-water pressure distribution in (**a**) traditional subgrade; (**b**) the subgrade with a capillary barrier at the end of the year.

Due to a deeper-seated slide, which often happens on the top of the slope, the pore-water pressure at the interface I-I was taken as the abscissa, and the elevation was taken as the ordinate (the elevation of CD was 0) to quantitatively study the impact of rainwater on the pore-water pressure, which is plotted as Figure 6. Under the action of rainwater, the soil in the middle and upper layers of the conventional subgrade was greatly affected. In the upper soil, the largest pore-water pressure was −31 kPa; in the middle soil, the pore-water pressure was maintained at approximately −25 kPa; the bottom soil was almost unaffected by rain and almost coincided with the hydrostatic pressure distribution. This distribution curve was not the same as the wetting law of a single-layer of soil [29], which is mainly due to the different intrusion surfaces of rainwater. The lateral slope in the subgrade was the rainwater immersion surface.

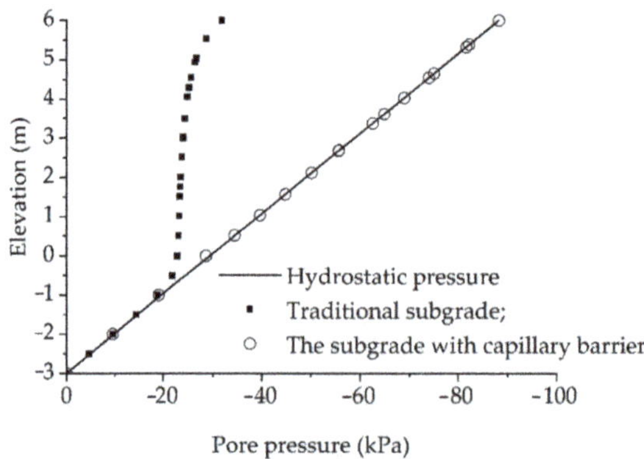

Figure 6. Pore-water pressure profiles at section I-I.

The subgrade using a capillary barrier was less affected by rain, and its pore-water pressure distribution almost coincided with the hydrostatic pressure distribution line. It can be seen that the blocking effect of the capillary barrier on the subgrade was more obvious. The capillary barrier used as a landfill cover system has been proven to be unsuitable for humid climates [16,17], but as a subgrade protection layer, the capillary barrier is suitable for humid climates. This is mainly due to the short length of the tilted slope in the subgrade, which is completely within the diversion length. In the middle and upper soils, the pore-water pressure in the subgrade protected by a capillary barrier is lower than that of the conventional subgrade. The difference between the two subgrades shows a nonlinear distribution, and the maximum was 57 kPa. Compared with the traditional capillary barrier, the pore-water pressure was reduced by 180%. It can be seen that the impact of the rainfall on the strength of subgrade cannot be ignored. It is worth noting that the value of ϕ^b in Formula (1) was generally different from that of ϕ'. Therefore, the contribution of the pore-water pressure to the shear strength was also different from that of the additional stress (σ).

Figure 7 shows the change trend in the volumetric moisture content with elevation. The volumetric moisture content in the subgrade (elevation 0–6 m) was basically constant, which was mainly because the pore-water pressure in the subgrade was basically greater than the inflow value of the fill. In this case, a very small change in the water content caused a huge change in the pore-water pressure. The water-holding capacity of conventional subgrade was basically maintained at 0.17, while the water content of the subgrade with capillary barrier was at 0.14. Compared with the conventional subgrade, the moisture content of the subgrade protected by the capillary barrier was reduced by 18%.

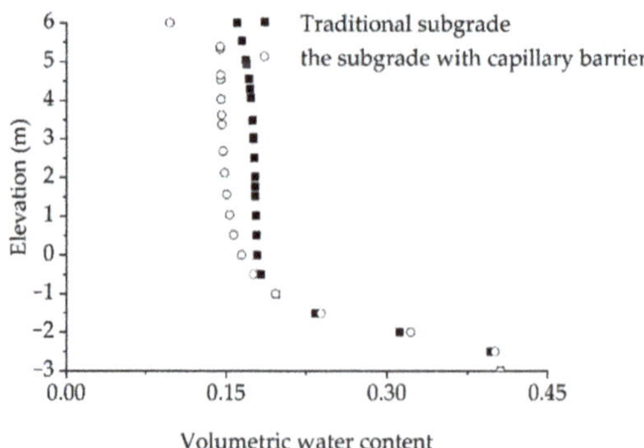

Figure 7. Volumetric water content profiles at section I-I.

4.2. Settlement and Deformation Law

Changes in the pore-water pressure caused changes in the shear strength. Figure 8 shows the deformation vector arrows of the subgrade protected by a capillary barrier. The subgrade with a capillary barrier was basically not affected by rainfall, while settlement still occurred. This was mainly because the slight change in the pore-water pressure at the bottom of the expressway shoulder slope caused the strength in the uplift area of the slip surface to decrease, which led to the shear deformation of the subgrade.

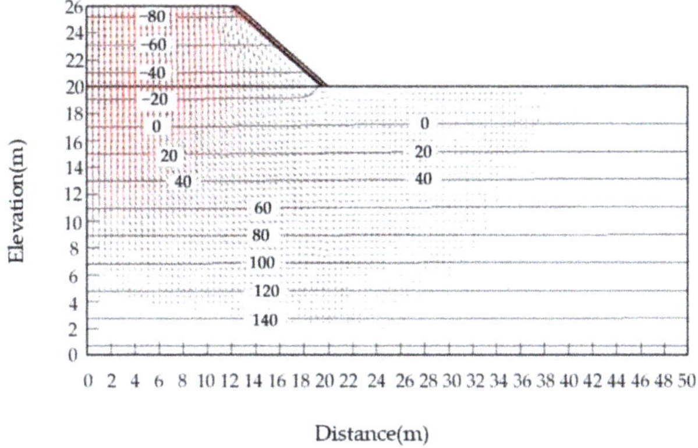

Figure 8. Deformation vector arrows of the subgrade with a capillary barrier.

Figure 9 shows the change trend in the expressway's pavement AB settlement with horizontal displacement. In the nonaffected area, the settlement of the conventional subgrade was relatively small, which was only 1 cm. While in the affected area, the settlement was large, which was up to 16 cm. Large different settlements occurred at the shoulder and in the subgrade. The different settlement was up to 15 cm. However, the subgrade using a capillary barrier had a ground settlement of approximately 1 cm in the affected zone and the noninfluenced zone. The different settlement was also relatively small. Surprisingly, even if the pore-water pressure in the subgrade did not change, the subgrade still had a certain amount of settlement. This was mainly caused by the shear deformation of the foundation.

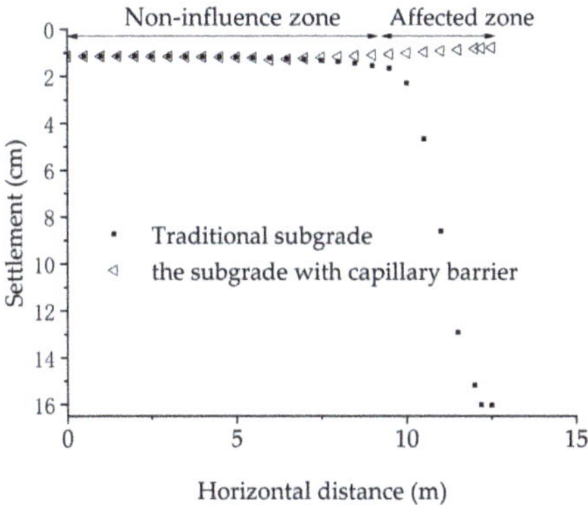

Figure 9. Comparison of the settlement in the different subgrades.

4.3. Variation Law of Safety Factor

The safety factor can be used to reflect the possibility of slope instability to a certain extent. As the subgrade and foundation are influenced by rainfall, the pore-water pressure in the traditional subgrade decreased, which reduced the overall safety factor of the subgrade.

The safety factor was reduced from 1.35 to 1.15 in one year as shown in Figure 10, and the decrease in the safety factor with time was nonlinear. The safety factor of the subgrade protected by the capillary barrier was almost unchanged. This was mainly due to the better blocking effect of the capillary barrier. In one year, the safety factor of the subgrade using a capillary barrier increased by 15% compared to the traditional subgrade.

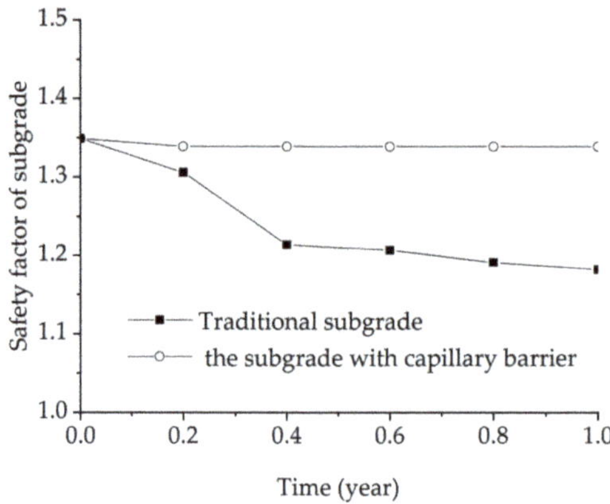

Figure 10. Comparison of the safety factor in the different subgrades.

5. Conclusions

In this study, we first proposed effective measures to block the infiltrating rainwater in the subgrade through a capillary barrier. Secondly, using the numerical simulation method, we conducted studies on the mechanism and stability of the capillary barrier used in the subgrade to control distress against wetting. The subgrade with a capillary barrier could effectively use the different permeability coefficients of the different types of soil to achieve an anti-seepage effect, which minimized the deformation of the subgrade and increased stability. Using the numerical simulation method, this paper analyzed the pore-water pressure distribution, settlement, and safety factor of the conventional subgrade and the subgrade protected by the capillary barrier. Through a comparison of the numerical results of the two subgrades, the following conclusions were drawn:

(1) The pore-water pressure of the subgrade protected by the capillary barrier was hardly affected by rainfall. Compared with the conventional subgrade, the pore-water pressure was reduced by 180%, and the pore-water pressure distribution was close to the hydrostatic pressure distribution. The moisture content of the conventional subgrade was effectively reduced by approximately 18% by using a capillary barrier;

(2) The settlement of the subgrade using a capillary barrier was uniform and small, and is only 1 cm. Compared with the conventional subgrade, the settlement was reduced by 94%. The safety factor of the subgrade using a capillary barrier was almost unchanged. Compared with the conventional subgrade, the safety factor increased by 15%;

(3) In humid climates, it is feasible to use subgrades with a capillary barrier to reduce the deformation and stability problems caused by humidification.

6. Recommendations for Future Study

This was a preliminary study that merely focused on the feasibility of a capillary barrier for the subgrade. The parameters of soil were selected from the literature to study the feasibility. In the future, specific subgrade soils will be used to conduct experiments for more in-depth research. Furthermore, the capillary barrier was used as the cover layer for natural materials without manual treatments, which can effectively block the infiltrating rainwater. It can be applied in many areas including slope stability. It can prevent rainwater infiltration so as to improve the slope's stability. However, it cannot be applied in humid climates due to the fact of its poor water storage capacity. In the future, more studies should be carried out on improving its water storage capacity by using new materials with high water storage capacity and methods so as to be applies more in engineering.

Author Contributions: The formal analysis, formula reasoning, and the performance of the experiment, as well as the original draft preparation were completed by X.Z.; Review and editing were completed by Y.W. and J.L. All authors have read and agreed to the published version of the manuscript.

Funding: This research was funded by the Fundamental Research Funds for the Central Universities (grant numbers: B210202034, B200204036, and B200204032).

Institutional Review Board Statement: Not applicable.

Informed Consent Statement: Not applicable.

Data Availability Statement: Data sharing not applicable. No new data were created or analyzed in this study.

Conflicts of Interest: The authors declare no conflict of interest.

References

1. Ni, X.; Ma, Y.; Gao, B.; Xue, S.; Bai, P.; Huo, X.; Bian, X. *2013–2017 China Highway Industry Investment Analysis and Prospect Forecast Report*; CIC: Beijing, China, 2012.
2. Zheng, S. Study on the Settlement of Bridge Embankment Expressway Opened Traffic in Soft Subsoil. Master's Thesis, Hohai University, Nanjing, China, 2010.
3. Luo, J. The Research on Moisture Migration In Silt Embankment and Differential Settlement Formation Mechanism for Widened Road. Master's Thesis, Changsha University of Science & Technology, Changsha, China, 2006.
4. Gong, X. *Theory of Composite Foundations and Engineering Applications*; China Architecture and Building Press: Beijing, China, 2002; pp. 76–79.
5. Liu, J. Application and Research of Long-Short-Pile Composite Foundation on Soft Foundation of Motorway. Master's Thesis, Hohai University, Nanjing, China, 2010.
6. Wu, Y.; Liu, J. Research on foamed cement banking for non-uniform differential settlement in used super highway. In *Architecture, Civil and Environmental Engineering*; ACEE 2011; Hohai University: Nanjing, China, 2011; Volume 6, pp. 5334–5337.
7. Wang, D. Experimental Study on Pavement Performance of the Fine-Grained Soil Stabilized by Soil Stabilizer. Master's Thesis, Zhengzhou University, Zhengzhou, China, 2010.
8. Zhao, D. Research on properties of expressway subgrade using the recycled fine soil of construction waste. *World Build. Mater.* **2010**, *34*, 42–44.
9. Hauser, V.L.; Weand, B.L.; Gill, M.D. *Alternative Landfill Covers*; Prepared for Air Force Center for Environmental Excellence, Technology Transfer Division; Brooks AFB: San Antonio, TX, USA, 2001.
10. Benson, C.H.; Khire, M.V. *Earthen Covers for Semi-Arid and Arid Climates*; Geotechnical Special Publication: Reston, VA, USA, 1995.
11. Arunrat, N.; Sereenonchai, S.; Kongsurakan, P.; Hatano, R. Assessing Soil Organic Carbon, Soil Nutrients and Soil Erodibility under Terraced Paddy Fields and Upland Rice in Northern Thailand. *Agronomy* **2022**, *12*, 537. [CrossRef]
12. Wang, G.; Fang, Q.; Wu, B.; Yang, H.; Xu, Z. Relationship between soil erodibility and modeled infiltration rate in different soils. *J. Hydrol.* **2015**, *528*, 408–418. [CrossRef]
13. Rahardjo, H.; Krisdani, H.; Leong, E. Application of unsaturated soil mechanics in capillary barrier system. In *Third Asian Conference on Unsaturated Soils*; Science Press: Alexandria, NSW, Australia, 2005; pp. 127–137.
14. Benjamin, R. The diversion capacity of capillary barriers. *Water Resour. Res.* **1990**, *26*, 2652–2659.
15. Stormont, J.C. The effect of constant anisotropy on capillary barrier performance. *Water Resour. Res.* **1995**, *31*, 783–785. [CrossRef]
16. Stormont, J.C. The effectiveness of two capillary barriers on a 10% slope. *Geotech. Geol. Eng.* **1996**, *14*, 243–267. [CrossRef]
17. Qian, X.; Koerner, R.M.; Gray, D.H. *Geotechnical Aspects of Landfill Design and Construction*; Prentice Hall: Hoboken, NJ, USA, 2002; pp. 562–567.
18. Rahardjo, H.; Hua, C.J.; Leong, E.C.; Santoso, V.A. Performance of an instrumented slope under a capillary barrier system. In Proceedings of the Fifth International Conference of Unsaturated Soils, Barcelona, Spain, 6–8 September 2010; pp. 1279–1284.
19. Wu, Y.; Diao, H.; Liu, J.; Chen, R.; Shi, X. A Embankment Construction and Method to Prevent Wetting Distress Control for Highway. China Patent 201310408929.3, 30 August 2016.
20. Fredlund, D.G.; Rahardjo, H. *Soil Mechanics for Unsaturated Soils*; John Wiley & Sons: New York, NY, USA, 1993; pp. 267–269.
21. *JTG D60-2004*; General Code for Design of Highway Bridges and Culverts. China Communications Press: Beijing, China, 2004; pp. 53–55.
22. Mualem, Y. A new model for predicting the hydraulic conductivity of unsaturated porous media. *Water Resour.* **1976**, *12*, 513–522. [CrossRef]

23. Walter, M.T.; Kim, J.S.; Steenhuis, T.S.; Parlange, J.Y.; Heilig, A.; Braddock, R.D.; Selker, J.S.; Boll, J. Funneled flow mechanisms in a sloping layered soil-Laboratory investigation. *Water Resour. Res.* **2000**, *36*, 841–849. [CrossRef]
24. Tami, D.; Rahardjo, H.; Leong, E.C.; Fredlund, D.G. Design and laboratory verification of a physical model of sloping capillary barrier. *Can. Geotech. J.* **2004**, *41*, 814–830. [CrossRef]
25. Aubertin, M.; Cifuentes, E.; Apithy, S.A.; Bussière, B.; Molson, J.; Chapuis, R.P. Analyses of water diversion along inclined covers with capillary barrier effects. *Can. Geotech. J.* **2009**, *46*, 1146–1164. [CrossRef]
26. *GB/T 50145-2007*; Standard for Classification of Engineering Soils. Ministry of Construction of the People's Republic of China: Beijing, China, 2007.
27. Morris Carl, E.; Stormont John, C. Parametric study of unsaturated drainage layers in a capillary barrier. *J. Geotech. Geoenviron. Eng.* **1999**, *125*, 1057–1065. [CrossRef]
28. Wu, Y.; Wang, W.; Liu, J.; Ji, K. Test research on compaction effect of expressway embankment with sand-gravel-cobble mixture. *Rock Soil Mech.* **2012**, *33*, 211–216.
29. Zhan, L.T.; Ng, C.W. Analytical analysis of rainfall infiltration mechanism in unsaturated soils. *Int. J. Geomech.* **2004**, *4*, 273–284. [CrossRef]

Article

An Experimental Study on the Migration of Pb in the Groundwater Table Fluctuation Zone

Jihong Qu [1,2,*], Tiangang Yan [1,2], Yifeng Zhang [1,2], Yuepeng Li [1,2], Ran Tian [1,2], Wei Guo [1,2] and Jueyan Jiang [1,2]

1 College of Geosciences and Engineering, North China University of Water Resources and Electric Power, Zhengzhou 450046, China; ytg@stu.ncwu.edu.cn (T.Y.); yyds1231110@163.com (Y.Z.); liyuepeng@ncwu.edu.cn (Y.L.); tianran0227@163.com (R.T.); gwmengyeah@163.com (W.G.); jiangjueyan@stu.ncwu.edu.cn (J.J.)

2 Collaborative Innovation Center for Efficient Utilization of Water Resources, Zhengzhou 450046, China

* Correspondence: qujihong@ncwu.edu.cn; Tel.: +86-133-9370-9096

Abstract: As a result of fluctuations in the shallow groundwater table, hydrodynamic conditions change alongside environmental conditions and hydrogeochemical processes to affect pollutant migration. The study aimed to investigate the migration, adsorption, and desorption characteristics of Pb on fine, medium, and coarse sand in the water table fluctuation zone by using several laboratory methods, including the kinetic aspects of Pb^{2+} adsorption/desorption and water table fluctuation experiments. The results showed that the adsorption and desorption curves fit the Elovich equation well at a correlation coefficient above 0.9. In the adsorption and desorption kinetic experiments for fine, medium, and coarse sand collected and from the floodplain, the maximum adsorption capacity of Pb^{2+} was 2367 $mg \cdot kg^{-1}$, 1848 $mg \cdot kg^{-1}$, and 1544 $mg \cdot kg^{-1}$, respectively. The maximum desorption capacity of Pb^{2+} was 29.18 $mg \cdot kg^{-1}$, 62.38 $mg \cdot kg^{-1}$, and 81.60 $mg \cdot kg^{-1}$, respectively. In environments with pH greater than 4, the adsorption capacity was proportional to the pH, but the desorption capacity decreased as the pH increased in water. As the water table varied, the lowest pH occurred in the polluted medium we set initially. When the distance between the pollutants and sample solution grew further, pH increased, and the Pb^{2+} concentration decreased in the sample solution. In the column experiment of water table fluctuations on coarse sand, Pb^{2+} migrated nearly 5 cm upward from the original pollutant and migrated less than 10 cm downward from that. In our experiments on medium and fine sand, the upward and downward migration distances were <5 cm. The groundwater table fluctuations, pH variation, and Pb concentration currently influence the migration of Pb.

Keywords: kinetic adsorption and desorption; groundwater table fluctuations; Pb; migration; experimental study

Citation: Qu, J.; Yan, T.; Zhang, Y.; Li, Y.; Tian, R.; Guo, W.; Jiang, J. An Experimental Study on the Migration of Pb in the Groundwater Table Fluctuation Zone. *Appl. Sci.* **2022**, *12*, 3870. https://doi.org/10.3390/app12083870

Academic Editor: Bing Bai

Received: 1 March 2022
Accepted: 10 April 2022
Published: 12 April 2022

1. Introduction

As an essential water source, groundwater plays a significant role in supplying urban and rural residents with water. The water table fluctuates under natural and human factors, such as rainfall, evaporation, exploitation [1], and recharge [2]. The area between the highest and lowest groundwater table is referred to as the groundwater table fluctuation zone [3,4]. Water table fluctuations allow both saturated and unsaturated soil to alternate within the environment. These fluctuations cause significant changes in the biochemical characteristics of that zone [5], including the adsorption, desorption, and dilution of pollutants, which correspondingly alter the groundwater environment. The influence of water table fluctuations on pollutant migration and transformation has attracted the attention of scholars.

Current studies mainly focus on solute migration and transformation, such as nitrogen [6], organic matter [7], heavy metals [8–10], and groundwater quality [11] under water table fluctuations. Liu et al. studied the nitrate change law under water table rise using a sandbox model experiment, finding that the water environment gradually changed into a state of relative reduction, and the lateral flow of the experiment was conducive to the migration of nitrate [12]. Liu et al. studied the variation in nitrate concentration for two water table fluctuation conditions (the water table stayed constant and varied by 20 cm/10 days) [13]. Tian et al. adopted an experimental simulation involving three kinds of groundwater tables and different surface runoff velocities to study nitrate change laws on solute migrates to the soil surface. Their results showed that the solute transport process shares an essential relationship with surface runoff velocity and the groundwater table [14]. Wang et al. simulated and verified the regularity of soil salt migration under water table fluctuations at different groundwater depths using a laboratory column experiment with homogenic medium. They assumed that the capillary pressure and temperature field variation caused by water table fluctuations significantly influence the migration of organic pollutants [15]. Wang et al. used the TMVOC model (a numerical simulator for three-phase non-isothermal flow of water, soil gas, and a multi-component mixture of volatile organic chemicals) to simulate how benzene, toluene, ethylbenzene, and o-xylene (BTEX) migrate in areas caused by steam extraction under natural attenuation and groundwater table fluctuations [16]. Oostorm et al. used a two-dimensional sandbox to study the distribution of pollutants migrating from their source into soil and groundwater under water table fluctuations. They observed that pollutants migrate with the infiltrated water flow and dissolve when the water table rises, increasing the pollutant concentration in the sample solution [17]. Bustos Medina et al. studied iron hydroxide blockage in wells and its effect on groundwater table fluctuations. This blockage affects indexes such as the water table, pH, EC (electrical conductivity), and DO (dissolved oxygen). By adopting hydrogeochemical simulations, Medina et al. determined the minerals' reaction mode, such as iron ions and manganese ions in the aquifer [18]. Li et al. explored the nitrogen transport law for fluctuations in different water tables using laboratory-based column experiments and numerical simulation. Their results showed that groundwater table fluctuations influence the variation of dissolved oxygen in the solution and decrease the nitrate and ammonium concentrations, which is conducive to removing ammonium [19]. Liu et al. employed a numerical model to predict changes in Beijing's groundwater table when the South–North Water Diversion Project was open. They also analyzed changes to the vadose zone and the impact of solid waste on the groundwater environment [20]. On account of the South–North Water Diversion Project, Cao et al. analyzed chemical quality predictions in the Baoding Plain for when the groundwater table rises using hydrogeochemical simulations [21].

Heavy metals (Pb) pose potential risks to the soil ecosystem and human health [22–24]. China and other nations have focused on soil pollution prevention and control plans and the study of Pb migration characteristics [25,26]. Domestic and foreign scholars have studied the adsorption, desorption, and migration of heavy metal pollutants such as Cu, Cr, and Zn in soil [27–30]. Nonetheless, there are few reports on the migration laws of heavy metals under groundwater table fluctuations. Therefore, in this article, technical methods, including adsorption and desorption tests, water table fluctuation experiments, etc., were used to analyze the distribution characteristics of Pb in the vadose and saturated zones under different typical media. Our methods help research the migration of Pb in the groundwater table fluctuation zone and provide theoretical support for heavy metal pollution treatments for soil and groundwater, remediation, and protection.

2. Materials and Methods

2.1. Sample Collection and Process

The sample medium used in our laboratory was collected from the floodplain of the Yellow River in Mengjin District, Luoyang City, Henan Province. The sampling location is scoured by the Yellow River all the year round. As a result, the soil medium is relatively clean, and the background concentration of Pb is small, which has little influence on the laboratory experiment. Laboratory experiments need to separate sand into coarse medium and fine medium. The sample sand was collected and processed using the Technical

Specification for Soil Environmental Monitoring's quarter method [31]. Then, the sample sand was dried and crushed in the laboratory. Finally, the sample was divided into three typical media, including coarse, medium, and fine sand in the pH range of 8.5–9.3 with an organic matter content range of 0.241–1.070 $g{\cdot}kg^{-1}$. The basic physical and chemical properties of the soil are shown in Table 1.

Table 1. Physical and chemical properties of soil.

Sample	Particle Size (mm)	pH	Organic Matter Content ($g{\cdot}kg^{-1}$)
Coarse sand	0.5–1.0	8.5	0.241
Medium sand	0.25–0.5	8.8	0.587
Fine sand	0.125–0.25	9.3	1.070

2.2. Experimental Equipment

The following devices were employed for the adsorption and desorption kinetic aspects of Pb^{2+}: a digital water bath oscillator, high-speed desktop centrifuge, and microwave.

The experimental devices for water table fluctuations include a 90 cm long column, water tank with 25 L volume, peristaltic pump with 100 rpm speed, piezometer tube, and soil solution sampler. We set eight openings for sampling in every column at heights of 20, 30, 35, 40, 45, 50, 55, and 60 cm. The experimental devices are shown in Figure 1.

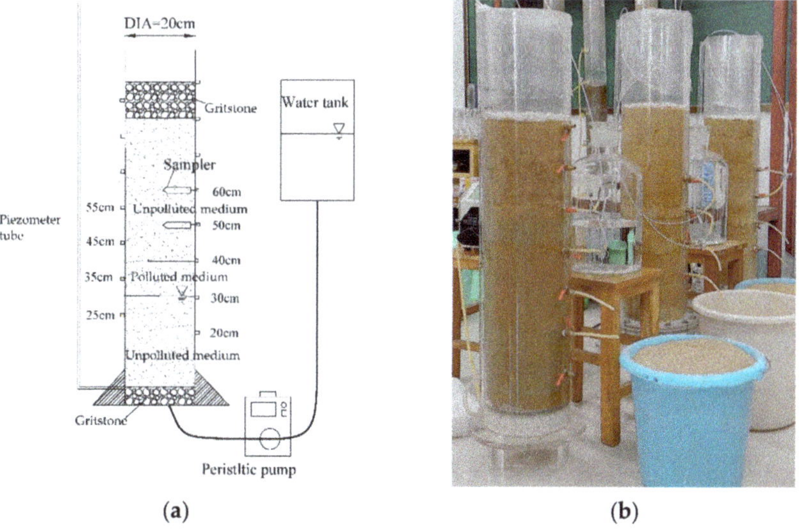

Figure 1. Experimental devices for the water table fluctuations. (**a**) The schematic diagram. (**b**) The field experiment diagram.

2.3. Experimental Methods

2.3.1. Adsorption and Desorption Experimental Methods of Pb

In this work, we completed the adsorption and desorption experiments, including the adsorption kinetics experiment, the desorption kinetics experiment, and the Pb adsorption and desorption quantities due to pH variation, in the laboratory. The experimental principle for adsorption was as follows: For the experiment, 50 mL of a 200 $mg{\cdot}L^{-1}$ $Pb(NO_3)_2$ solution was prepared mixing sample sand (the sample sand was divided into coarse, medium, and fine sand). After mingling, the supernatant liquids, which were samples obtained in 5, 10, 15, 20, 30, 60, 90, 120, 180, 240, 360, 480, 720, and 1440 min, passed through the oscillator for centrifugal filtration. The experimental principle for desorption was as follows. The desorption process resembles the first step of the adsorption process, i.e., the pollutant was mixed with the sand medium. After shaking this mixture for 24 h, the filtered samples were passed through centrifugal filtration. Then, 50 mL of a 0.01 mg·L^{-1} $NaNO_3$ solution appended the filtered samples. The desorption time was the same as the adsorption time. The supernatant liquids were left behind by the oscillator via centrifugal filtration. The adsorption experiment at pH range of 4 to 6 and desorption experiment at pH range of 4 to 9 were as follows: Combining 50 mL of 500 mg·L^{-1} $Pb(NO_3)_2$ solution with 10 g of sample sand, the pH of the background solution was adjusted to target values, such as pH 4, 5, 6, with either HCl or NaOH. When reaching the equilibrium (1440 min), the supernatant liquids and residue for adsorption were obtained through centrifugal filtration. The residue was obtained at the end of the experiment on the adsorption due to pH variation. The solutions at pH 4, 5, 6, 7, 8, 9 were added to the residue. Then the mixture was shaken using oscillator (200 rpm·min^{-1}), the supernatant liquids for desorption were obtained through centrifugal filtration. The following equations reveal the calculation formulae for the adsorption and desorption capacities [32]:

$$Q_{\text{ads}} = \frac{V(C_1 - C_2)}{M} \tag{1}$$

$$Q_{\text{des}} = \frac{VC_3}{M} \tag{2}$$

where Q_{ads} is the adsorption capacity for the medium (mg·kg^{-1}), Q_{des} is the desorption capacity for the medium (mg·kg^{-1}), V is the supernatant volume (mL), C_1 is the original concentration for adsorption aspect (mg·L^{-1}), C_2 is the supernatant concentration for adsorption aspect (mg·L^{-1}), M is the mass of the sand sample (g), C_3 is the supernatant concentration for desorption aspect (mg·L^{-1}).

2.3.2. Experimental Methods for Water Table Fluctuations

The columns were filled with sand collected and processed from the field, including coarse, medium, and fine sand. The column was filled with gritstone around the 5 cm length, unpolluted medium around the 30 cm length, polluted medium around the 10 cm length, unpolluted medium around the 40 cm length, and gritstone around the 10 cm length, respectively, from bottom to top. A $Pb(NO_3)_2$ solution was mixed with the sample sand to produce 2000 mg·kg^{-1} contaminated mixture. After loading, the deionized water at pH 7 flowed up from the base of the column, whose pressure tube was used to measure the water table. The peristaltic pump controlled the water table variation, rising or falling. The initial water table was set in the column at the beginning of the experiment at about 20 cm high. Then, the water table was adjusted to increase 10 cm per day until it had continuously risen to 60 cm by adjusting the peristaltic pump. The same method was applied to the water table decrease, which decreased 10 cm per day until it reached a height of 20 cm. This process represented the completion of one water table fluctuation cycle. In the column for water table fluctuations, two cycles were continuously conducted. Figure 2 shows how the water table varies with time. In order to further understand the migration process of Pb in the medium under water table fluctuations, the Pb^{2+} concentration adsorbed in the medium was measured by digestion treatment after the experiment. The sampling medium was obtained from the 20–60 cm height at an interval of 5 cm.

2.3.3. Experimental Test Instrument

According to the groundwater quality standard (GB/T 14848-2017) and other standards, the Pb concentrations for the sample solution and sample sand were detected by a flame atomic adsorption spectrophotometer (FAAS, Manufacturer: Beijing Ruili Analytical Instrument Co. Ltd., Beijing, China). The pH of the solution was detected using the glass electrode method (Micro600, Manufacturer: Palintest Co. Ltd., Newcastle, UK). The medium sample was digested under microwave irradiation. The Pb contents were determined by FAAS.

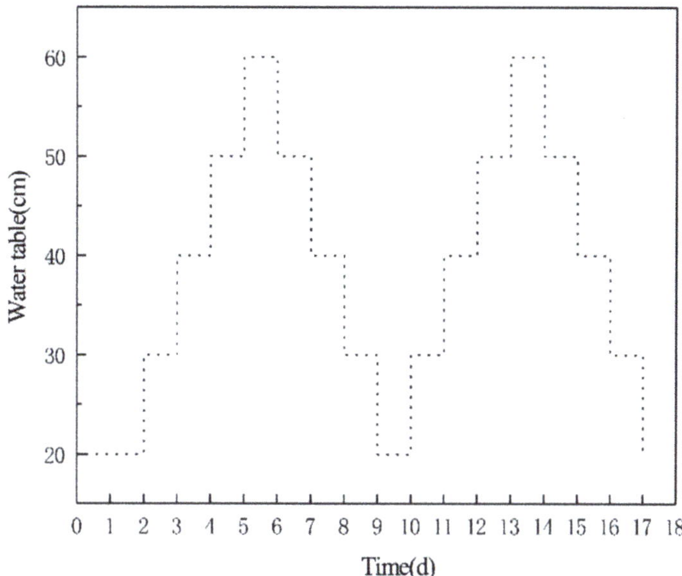

Figure 2. Water table fluctuations process.

3. Results and Discussion

3.1. The Curve of Adsorption and Desorption Kinetics

Figure 3 shows the curves of adsorption and desorption kinetics of Pb. The adsorption and desorption capacities of Pb^{2+} in the three sand media were significantly dissimilar. The adsorption and desorption quantity reached equilibrium in 1440 min. The maximum adsorption capacity of fine sand, medium sand, and coarse sand were 2366.6 $mg \cdot kg^{-1}$, 1847.6 $mg \cdot kg^{-1}$, and 1543.8 $mg \cdot kg^{-1}$, respectively. The maximum desorption capacity of coarse sand, medium sand, and fine sand were 81.6 $mg \cdot kg^{-1}$, 62.38 $mg \cdot kg^{-1}$, and 29.18 $mg \cdot kg^{-1}$, respectively. Compared to the hyperbolic diffusion model, pseudo-second-order model and Weber–Morris model, the data fit well to the Elovich model against various time ranges because the minimum R^2 value was 0.90. The fitting parameters are shown in Table 2. Based on the adsorption kinetics experiment data, Pb^{2+} in the contaminated solution was adsorbed rapidly onto the sampling sand within 240 min. The adsorption quantity gradually was stable from 240 to 1440 min in the experiment; this implies several points that adsorbed Pb^{2+} quickly on the medium surface at the initial stage. The effective points that adsorb Pb^{2+} gradually decreased with increasing reaction time, which gradually weakened the adsorption capacity until it reached equilibrium. The experiment's result aligned with the study of Ren, L. [33]. The same conditions occurred in the desorption kinetics experiment.

Figure 4 illustrates the effects of pH on the adsorption and desorption of sand medium. The adsorption capacity of Pb^{2+} in three sand media gradually increased at pH range of 4 to 6. The desorption capacity gradually decreased with increasing pH and the desorption quantity steadily approached equilibrium when the pH was between 8 and 9. Fine sand's adsorption and desorption capacities showed almost no change at different pH levels compared with medium and coarse sand. When the pH increased, the competitive adsorption sites of hydrogen ions in the medium decreased, and heavy metals mainly existed in the combined state of hydroxide or carbonic acid. This state is not conducive to their migration in the medium and increases adsorption capacity. These experimental conclusions are consistent with the literature [30,34].

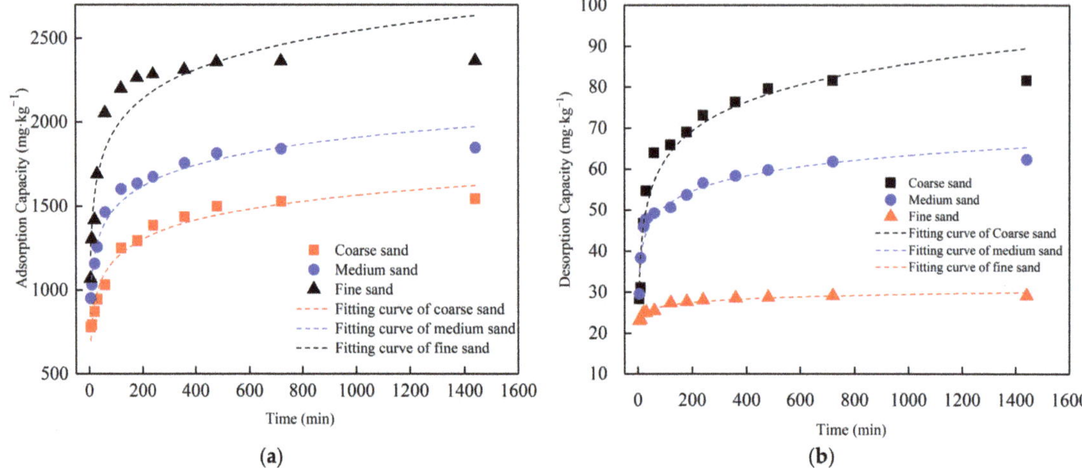

Figure 3. Curves of adsorption and desorption kinetics of Pb in different media: (**a**) the adsorption quantity variation of Pb based on adsorption kinetics; (**b**) the desorption quantity variation of Pb based on desorption kinetics.

Table 2. Kinetic parameters for adsorption and desorption of Pb in different media.

Category	Media	Elovich Equation			Hyperbolic Diffusion Equation			Pseudo-Second-Order Equation		Weber–Morris Equation		
		a_1	b_1	R^2	a_2	b_2	R^2	k_1	R^2	k_2	c	R^2
adsorption	Coarse sand	437.20	163.15	0.96	0.5513	0.0159	0.81	0.00005	0.72	24.52	851.0	0.81
	Medium sand	667.04	179.76	0.97	0.6171	0.0140	0.74	0.00006	0.83	25.81	1139.9	0.74
	Fine sand	799.14	252.62	0.90	0.6326	0.0143	0.60	0.00005	0.95	33.89	1497.1	0.60
desorption	Coarse sand	15.18	10.21	0.94	0.5208	0.0176	0.69	0.0008	0.96	1.43	42.50	0.69
	Medium sand	26.14	5.39	0.94	0.6465	0.0124	0.72	0.0022	0.81	0.77	40.33	0.72
	Fine sand	21.00	1.24	0.97	0.8298	0.0062	0.76	0.0165	0.55	0.18	24.21	0.76

Note: The data were fitted to the Elovich equation: $Q = a_1 + b_1 \ln t$, the hyperbolic diffusion equation: $Q/Q_{max} = a_2 + b_2\, t^{1/2}$, the pseudo-second-order equation: $Q = k_1\, Q_{max}^2\, t/(1 + k_1\, Q_{max}\, t)$ and the Weber–Morris equation: $Q = k_2\, t^{1/2} + c$, respectively, where Q is the adsorption/desorption capacity; t is time; a_1 and a_2 refer to constant associated with maximum adsorption/desorption amount for the Elovich equation and the pseudo-second-order equation, respectively, b_1 and b_2 refer to adsorption/desorption rate coefficient, k_1 and k_2 refer to adsorption/desorption rate coefficient for pseudo-second-order model and Weber–Morris model, respectively, c is constants related to the medium for Weber–Morris model, Q_{max} refers the equilibrium adsorption/desorption capacity for coarse sand, medium sand and fine medium.

3.2. pH Variation of Sample Solution on Water Table Fluctuation Zone

In Section 3.1, pH is one of the major factors affecting the adsorption and desorption capacities of Pb. Figure 5 illustrates pH variation at different sampling port heights in the water table fluctuation experiment on three typical media. The average pH variation at different sampling port heights is shown in Figure 6. The vertical direction of the column shows the distribution properties of pH in the sample solution. Firstly, pH decreased and then increased from the top to the bottom. The lowest pH often occurred at the height of 35 cm (pH = 4.7), where the location of original pollutants was consistent. The average pH of the three typical media followed the sequence of fine sand > medium sand > coarse sand. The aforementioned changes may be attributed to the increasing groundwater table since Pb^{2+} was not only adsorbed by the adsorption sites on the surface of soil particles, but also possibly combined with OH^- ions to obtain $Pb(OH)_2$ precipitation. Consequently, the concentration of OH^- ions decreased, so that the pH decreased in the solution, which resulted in a higher pH the further the distance. As the water table rose, the Pb^{2+} in the solution was absorbed to saturation by the medium. A pH range from 4.5 to 7.5 was detected in the sample solution by coarse sand, at pH range from 5.1 to 7.8 by medium sand, and a pH range from 7.1 to 7.7 by fine sand. A comparison of the average pH values for the three typical media revealed that pH at different grain diameters generally followed the order of: coarse sand > medium sand > fine sand.

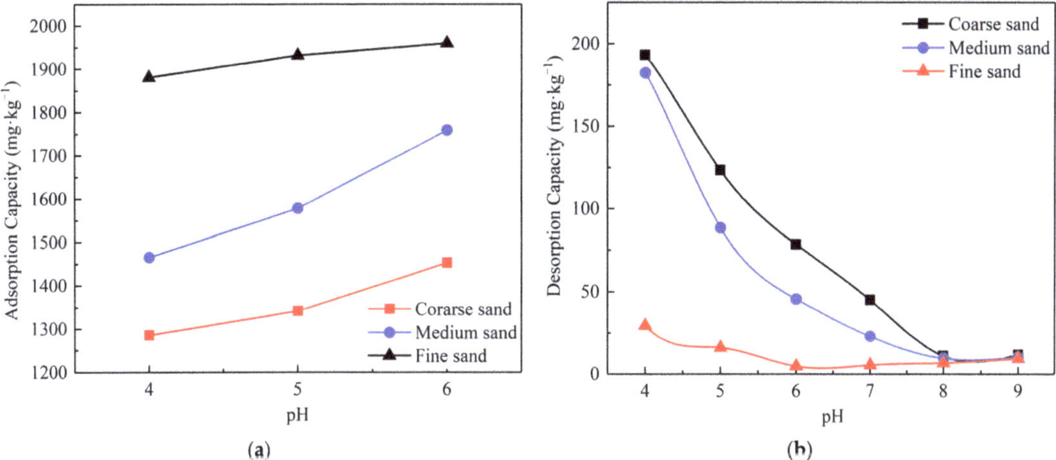

Figure 4. Effects of pH on adsorption (**a**) and desorption (**b**) on different media, including coarse, medium, and fine sand.

3.3. Migration Law of Pb Due to Water Table Fluctuations

Due to the strong adsorption capacity of fine sand, the migration ability of Pb in fine sand is weak. The Pb of each sample solution in the columns did not reach the instrument's detection limit (the instrument's detection limit (TAS-990A) is 0.01 mg/L). Therefore, we only analyzed the migration law of Pb in coarse and medium sand in this article. During groundwater table fluctuations, the variation in Pb concentration at different sample solutions is shown in Figure 7a (coarse sand) and Figure 7b (medium sand). The results of how Pb concentration is altered with water table fluctuations are shown in Table 3 and can be used to study the migration of Pb when the water table rises and falls in the experiment. Our analysis is as follows. (1) In the sample solution, when the height was 30 cm high in coarse sand, the water table height increased from 20 to 40 cm, and Pb^{2+} concentration increased in the range of 8.29–42.42% in the early-stage relative to the initial concentration on the first day. Then, with the water table fluctuations, Pb^{2+} concentration decreased in the range of 37.12–94.34%. This concentration declined by 9.54% on average within 8 days. (2) At the height of 35 cm for sample solutions in coarse sand, Pb^{2+} concentration decreased in the range of 4.01–97.47% with water table fluctuations. This concentration decreased by 6.68% on average within 14 days. (3) At the height of 40 cm for the sample solution in coarse sand, the concentration declined by 5.54% on average within 12 days. (4) At the height of 45 cm for the sample solution, Pb^{2+} was absorbed by coarse sand after 4 days. (5) At the heights of 30 and 40 cm for the sample solution in medium sand, Pb^{2+} was only detected on the first day. Afterward, Pb^{2+} was completely absorbed by the medium. (6) At the 30 and 40 cm heights for the sample solution in medium sand, Pb^{2+} concentration decreased in the range of 24.26–100%. This concentration decreased by 7.57% on average until the water table fluctuated at between 50 and 60 cm.

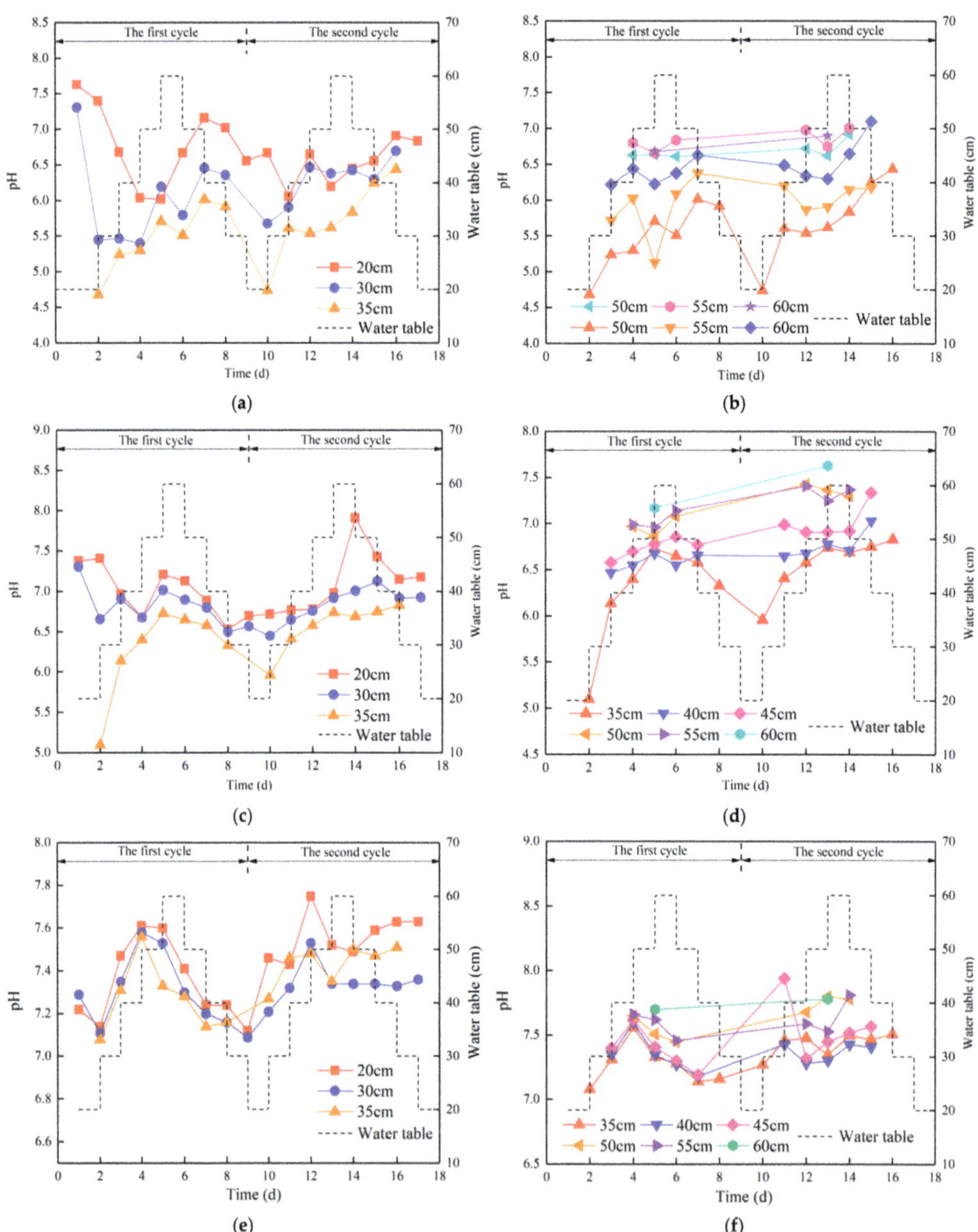

Figure 5. pH variation at different heights of sampling locations in the first and second cycles: (**a**) pH variation of coarse sand at sampling positions with heights of 20, 30, and 35 cm; (**b**) pH variation of coarse sand at sampling positions with heights of 35, 40, 45, 50, 55, and 60 cm; (**c**) pH variation of medium sand at sampling positions with heights of 20, 30, and 35 cm; (**d**) pH variation of medium sand at sampling positions with heights of 35, 40, 45, 50, 55, and 60 cm; (**e**) pH variation of fine sand at sampling positions with heights of 20, 30, and 35 cm; (**f**) pH variation of fine sand at sampling positions with heights of 35, 40, 45, 50, 55, and 60 cm.

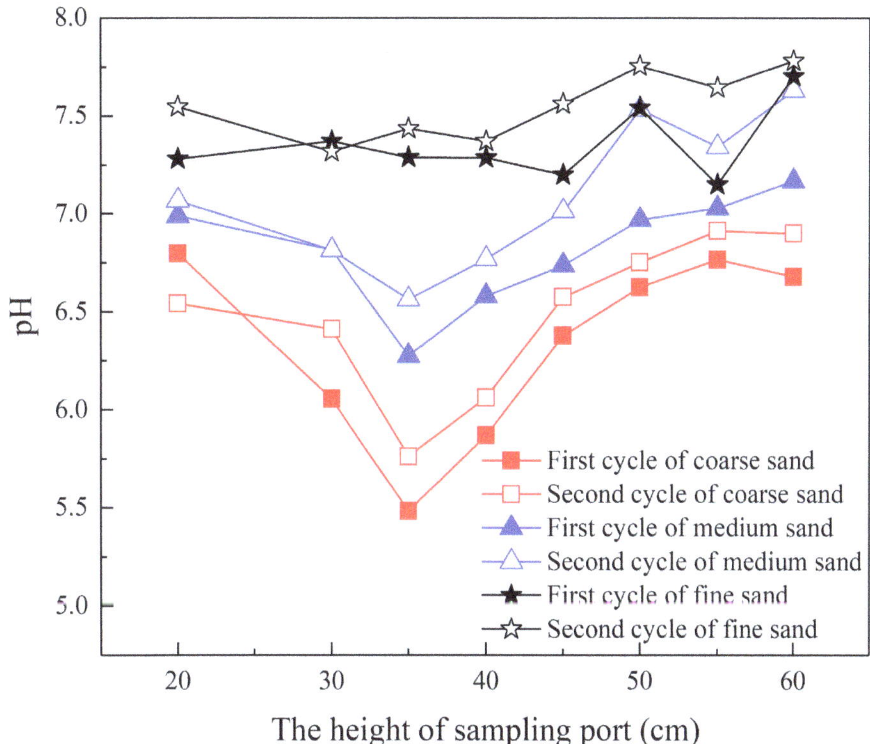

Figure 6. Average pH at different heights of sampling locations.

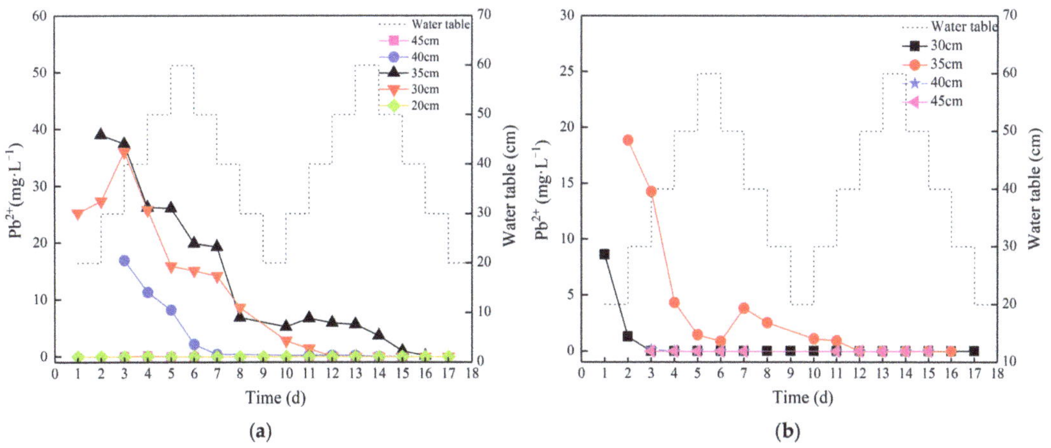

Figure 7. The Pb^{2+} concentrations vary in water table fluctuations. (**a**) The Pb^{2+} concentrations vary in columns filled with coarse sand; (**b**) the Pb^{2+} concentrations vary in columns filled with medium sand.

As seen in the analysis in Figure 5 of the pH and the water table, the closer the sample solution was to the pollutant in the rising water table, the lower the pH became, the greater the activity of Pb became, and the more desorption quantities resulted from polluted sand.

Therefore, at the height of 35 cm for the sample solution, Pb^{2+} concentration reached its maximum compared with the others. When the sample solutions were further away, the maximum concentration of Pb^{2+} was smaller. In a column of coarse sand, when the water table rose to 30 cm, the desorption quantity of Pb^{2+} reached its maximum, and the Pb^{2+} concentration in the solution reached its maximum. Throughout the whole experiment, the maximum concentration of Pb^{2+} at the 20, 30, 35, 40, and 45 cm sample solutions were 0.072, 36.061, 38.973, 16.941, and 0.042 mg·L^{-1}, respectively. Figure 7b shows that the adsorption capacity of medium sand is greater than that of coarse sand; therefore, the migration capacity of Pb^{2+} in medium sand becomes weak compared with that in coarse sand. Based on Figure 7b, Pb^{2+} is only detected at the heights of 30, 35, and 40 cm. In these sample solutions, the highest concentrations are 8.619, 18.862, and 0.164 mg·L^{-1}, respectively.

Table 3. Variation range of concentration of Pb in coarse and medium sand sampling ports with water table fluctuations.

Media	How Did the Water Table Fluctuate	Concentration Variation of Pb in the First Cycle (Compared with the Initial Concentration)				Concentration Variation of Pb in the Second Cycle (Compared with the Initial Concentration)			
		Sampling Port with Height of 40 cm	**Sampling Port with Height of 35 cm**	**Sampling Port with Height of 40 cm**	**Sampling Port with Height of 45 cm**	**Sampling Port with Height of 30 cm**	**Sampling Port with Height of 35 cm**	**Sampling Port with Height of 40 cm**	**Sampling Port with Height of 45 cm**
Coarse sand	20→30 cm	8.29%	-	-	-	-	-	-	-
	30→40 cm	42.42%	−4.01%	-	-	−88.82%	−86.67%	-	-
	40→50 cm	2.00%	−32.66%	−33.14%	285.71%	−94.34%	−82.83%	−98.91%	-
	50→60 cm	−37.12%	−33.04%	−51.56%	−100.00%	-	−84.93%	−98.48%	-
	60→50 cm	−40.23%	−49.12%	−87.05%	-	-	−85.60%	−98.82%	-
	50→40 cm	−43.94%	−50.60%	−97.42%	-	-	−90.68%	−99.68%	-
	40→30 cm	−65.96%	−82.47%	-	-	-	−97.47%	-	-
	30→20 cm	-	-	-	-	-	-	-	-
Medium sand	20→30 cm	−85.09%	-	-	-	-	-	-	-
	30→40 cm	-	−24.26%	-	-	-	−94.05%	-	-
	40→50 cm	-	−76.93%	−100.00%	-	-	−94.94%	-	-
	50→60 cm	-	−92.21%	-	-	-	−100.00%	-	-
	60→50 cm	-	−95.21%	-	-	-	-	-	-
	50→40 cm	-	−79.59%	-	-	-	-	-	-
	40→30 cm	-	−86.50%	-	-	-	-	-	-

Note: During the whole experiment, the concentrations in the solution sample at each sampling port did not reach the detection limit for columns filled with fine sand because of the strong adsorption capacities of fine and medium sand; therefore, they are not listed in this table. The "-" in the table indicates that the water sample was not obtained due to water table fluctuation limitations.

The experimental results of groundwater fluctuation showed that the underground water table rose to a height of 30 cm in coarse and medium sand after coming into contact with pollutants. The capillary band rose when the moisture content of the medium increased at the height of 30 cm. Then, Pb^{2+} in the medium dissolved into water under the effect of desorption. Pb^{2+} in the solution was still detectable since the water table fluctuated rapidly and Pb^{2+} in the solution had not been fully adsorbed by the medium. Reasons for why the concentration of Pb^{2+} decreased as the water table rose are as follows. (1) Due to the alkaline soil, the solution's pH increased with the rising water table, resulting in OH^- and Pb^{2+} combining in the water to form precipitation. (2) The adsorption quantity of the unpolluted medium was stronger than that of the pollution medium. The rising of the water table brought about Pb^{2+} adsorption one more time. (3) The adsorption and desorption of Pb is affected by hydrodynamic conditions. The change in hydrodynamic conditions caused Pb^{2+} to be desorbed and Pb^{2+} adsorbed by the medium. These two reactions were mutual until the adsorption equilibrium was reached.

3.4. Migration Characteristics of Pb in Soils with a Fluctuating Water Table

Figure 8 shows how the contents of Pb adsorbed by the sand sample varied at different heights. The maximum Pb content occurred at the height of 35 cm in three columns. Trace amounts of Pb were detected at the height range of 25–45 cm. The analyses in Figures 5 and 8 show that the maximum Pb content occurred where the minimum pH was detected. Therefore, the presence of heavy metal Pb decreases pH as a consequence of acidic soil. Zhai L et al. determined that precipitates and complexes may solidify the soil, causing grievous pollution to soils and crops [35].

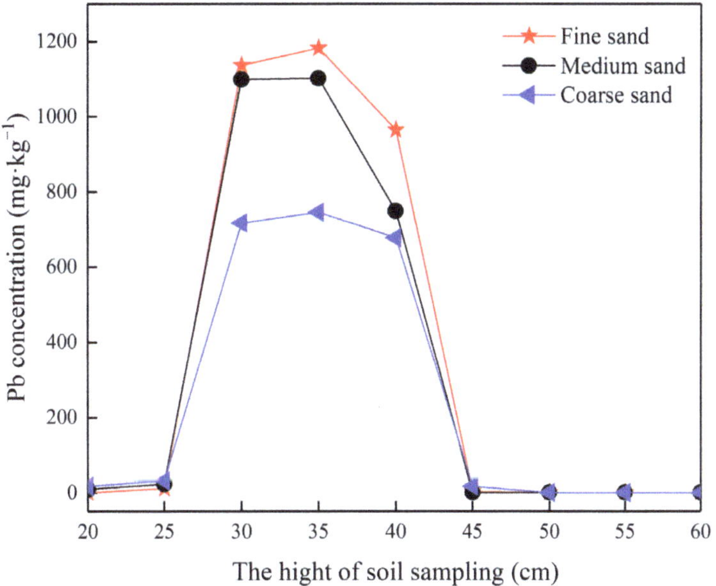

Figure 8. The variation of Pb concentration in the medium at different heights after experiments.

As the water table fluctuated, Pb^{2+} migrated upward or downward in the medium, but the migration distance of Pb^{2+} was more positively affected by the adsorption. The migration distance of Pb^{2+} in coarse sand was further than that in the medium sand or fine sand. The water table varied in coarse sand, the upward migration distance was only 5 cm high, and the downward migration distance was <10 cm high. In medium and fine sand, the upward and downward migration distance height was <5 cm.

4. Conclusions

- In the adsorption and desorption kinetic experiments for fine, medium, and coarse sand, the maximum adsorption capacity of Pb^{2+} was 2367 mg·kg^{-1}, 1848 mg·kg^{-1}, and 1544 mg·kg^{-1}, respectively. The maximum desorption capacity of Pb^{2+} was 29.18 mg·kg^{-1}, 62.38 mg·kg^{-1}, and 81.60 mg·kg^{-1}, respectively. For adsorption and adsorption experiments at different pH levels, the adsorption capacity of Pb^{2+} gradually increased and the desorption capacity gradually decreased with increasing pH in the sample solution at pH 4.0 and above. In these experiments, the desorption capacity held steady with an increase in pH from 8.0 to 9.0;
- In the water table fluctuation experiments, the pH detected in the sample solution varied with the water table fluctuation. The location was further away from the origin pollution we set, resulting in decreased pH in the sample solution. The minimum pH was consistent with the original pollution. At the same sample port, this universally followed the order of the medium's average pH: fine sand > medium sand > coarse sand. The Pb^{2+} concentration in the sample solution varies with time and water table fluctuations. Pb concentration in the medium results in Pb^{2+} forming desorption products in the solution that adsorb the experimental medium surfaces once again;

- The migration of Pb^{2+} is affected by pH in the solution, water table fluctuations, and the adsorption capacity of the medium. Water table fluctuations affect the desorption of Pb in the polluted medium. Consequently, the Pb^{2+} concentration in the solution changes. Then, the pH variation results in the adsorption and desorption capacities of Pb^{2+} changing as well;
- The effects of Pb's migration on groundwater table fluctuation zones are due to various factors. In this article, we only considered the water table, pH, and adsorption and desorption capacities in our analysis, which are the deficiencies of this work. Other meaningful factors will be added for future studies, such as dissolved oxygen, redox potential, etc.

Author Contributions: Conceptualization, J.Q.; methodology, T.Y.; validation, Y.Z.; formal analysis, Y.L.; investigation, R.T.; resources, W.G.; data curation, J.J.; writing—original draft preparation, J.Q.; writing—review and editing, J.Q. and T.Y.; funding acquisition, J.Q. All authors have read and agreed to the published version of the manuscript.

Funding: This research was funded by the Key R&D and Promotion Projects in Henan Province (202102310012), the Doctoral Research Fund of North China University of Water Resources and Electric Power (40651), and the Key Project of Science and Technology Research of Henan Education Department (14A170006).

Institutional Review Board Statement: Not applicable.

Informed Consent Statement: Not applicable.

Data Availability Statement: Data are contained within the article.

Conflicts of Interest: The authors declare no conflict of interest.

References

1. Li, W.; Wang, L.; Yang, H.; Zheng, Y.; Cao, W.; Liu, K. The groundwater overexploitation status and countermeasure suggestions of the North China Plain. *China Water Resour.* **2020**, *13*, 26–30.
2. Zhang, C.; Duan, Q.; Yeh, P.J.F.; Pan, Y.; Gong, H.; Gong, W.; Di, Z.; Lei, X.; Liao, W.; Huang, Z.; et al. The Effectiveness of the South-to-North Water Diversion Middle Route Project on Water Delivery and Groundwater Recovery in North China Plain. *Water Resour. Res.* **2020**, *56*, 10. [CrossRef]
3. Liu, X.; Zuo, R.; Wang, J.; He, Z.; Li, Q. Advances in researches on ammonia, nitrite and nitrate on migration and transformation in the groundwater level fluctuation zone. *Hydrogeol. Eng. Geol.* **2021**, *48*, 27–36.
4. Mu, E.; Ou, Y.; Dong, S.; Yang, J. Summary of research progress on groundwater level management. *Ground Water* **2019**, *41*, 33–34.
5. Rezanezhad, F.; Couture, R.M.; Kovac, R.; O Connell, D.; Van Cappellen, P. Water table fluctuations and soil biogeochemistry: An experimental approach using an automated soil column system. *J. Hydrol.* **2014**, *509*, 245–256. [CrossRef]
6. Zhang, D.; Cui, R.; Fu, B.; Yang, Y.; Wang, P.; Mao, Y.; Chen, A.; Lei, B. Shallow groundwater table fluctuations affect bacterial communities and nitrogen functional genes along the soil profile in a vegetable field. *Appl. Soil Ecol.* **2020**, *146*, 103368. [CrossRef]
7. Liu, Y.; Ding, A.; Liu, B.; Liang, X.; Li, S.; Zhang, L.; Yin, H. A review of the petroleum hydrocarbon contamination transformation performance in the zone of intermittent saturation. *Sci. Technol. Eng.* **2018**, *18*, 172–178.
8. Wu, Z.; Liu, G.; Qian, J.; Wei, Y.; Yue, Q. Experimental Study on the Vertical Migration of Pb from the Saturated Zone to Unsaturated Zone. *Environ. Sci. Technol.* **2019**, *42*, 36–41.
9. Rezaei, A.; Hassani, H.; Hassani, S.; Jabbari, N.; Fard Mousavi, S.B.; Rezaei, A. Evaluation of Groundwater Quality and Heavy Metal Pollution Indices in Bazman Basin, Southeastern Iran. *Groundw. Sustain. Dev.* **2019**, *9*, 100245. [CrossRef]
10. Rezaei, A.; Hassani, H.; Tziritis, E.; Fard Mousavi, S.B.; Jabbari, N. Hydrochemical characterization and evaluation of groundwater quality in Dalgan basin, SE Iran. *Groundw. Sustain. Dev.* **2020**, *10*, 100353. [CrossRef]
11. Yu, X. Groundwater Level Numerical Simulation Prediction and Its Environmental Influence Evaluation after the South-to-North Water Transferring Project. Master's Thesis, Jilin University, Changchun, China, 2004.
12. Liu, X.; Zuo, R.; Meng, L.; Li, P.; Li, Z.; He, Z.; Li, Q.; Wang, J. Study on the variation law of nitrate pollution during the rise of groundwater level. *China Environ. Sci.* **2021**, *41*, 232–238.
13. Liu, J.; Zhu, X.; Li, S.; Kang, L.; Ma, M.; Du, L. Effects of groundwater fluctuation on nitrate nitrogen transport after nitrogen application in cropland soil. *Chin. J. Eco-Agric.* **2021**, *29*, 154–162.
14. Tian, K.; Zhang, G.; Zheng, F. Chemical transport from soil into surface runoff under different ground-water tables. *J. Northwest A&F Univ.* **2009**, *37*, 193–200.
15. Wang, H.; Gao, W.; Ma, X. Research on the influence of groundwater level fluctuating on soil salin movement. *Sci. Technol. Innov. Her.* **2013**, *34*, 13–14.

16. Wang, Y.; Chen, L.; Yang, Y.; Li, J.; Tang, J.; Bai, S.; Feng, Y. Numerical simulation of BTEX migration in groundwater table fluctuation zone based on TMVOC. *Res. Environ. Sci.* **2020**, *33*, 634–642.
17. Oostrom, M.; Dane, J.H.; Wietsma, T.W. A review of multidimensional, multifluid, intermediate-scale experiments: Flow Behavior, Saturation Imaging, and Tracer Detection and Quantification. *Vadose Zone J.* **2007**, *6*, 610–637. [CrossRef]
18. Bustos Medina, D.A.; van den Berg, G.A.; van Breukelen, B.M.; Juhasz-Holterman, M.; Stuyfzand, P.J. Iron-hydroxide clogging of public supply wells receiving artificial recharge: Near-well and in-well hydrological and hydrochemical observations. *Hydrogeol. J.* **2013**, *21*, 1393–1412. [CrossRef]
19. Li, X.; Yang, T.; Bai, S.; Xi, B.; Zhu, X.; Yuan, Z.; Wei, Y.; Li, W. The effects of groundwater table fluctuation on nitrogen migration in aeration zone. *J. Agro-Environ. Sci.* **2013**, *32*, 2443–2450.
20. Liu, Y.; Sun, Y.; Yin, K. Prediction of groundwater environment in Beijing after water entering the capital by the South-north Water Diversion. *Hydrogeol. Eng. Geol.* **2005**, *05*, 93–96.
21. Cao, W.; Yang, H.; Gao, Y.; Nan, T.; Wang, Z.; Xu, S. Prediction of groundwater quality evolution in the Baoding Plain of the SNWDP benefited regions. *J. Hydraul. Eng.* **2020**, *51*, 924–935.
22. Cheng, X. Research progress of lead pollution on soil. *Ground Water* **2011**, *33*, 65–68.
23. Lee, J.; Son, Y.; Pratheeshkumar, P.; Shi, X. Oxidative stress and metal carcinogenesis. *Free Radic. Biol. Med.* **2012**, *53*, 742–757. [CrossRef] [PubMed]
24. Rezaei, A.; Hassani, H.; Fard Mousavi, S.B.; Jabbari, N. Evaluation of Heavy Metals Concentration in Jajarm Bauxite Deposit in Northeast of Iran Using Environmental Pollution Indices. *Malays. J. Geosci.* **2019**, *3*, 12–20. [CrossRef]
25. Shi, X.; Li, L.; Zhang, T. Water Pollution Control Action Plan, a Realistic and Pragmatic Plan—An Interpretation of Water Pollution Control Action Plan. *Environ. Prot. Sci.* **2015**, *41*, 1–3.
26. Ge, F.; Yun, J.; Xu, K.; Zhang, M.; Xu, L.; Li, Y.; Zhang, A. Progress of research on environmental criteria for lead in soil. *J. Ecol. Rural Environ.* **2019**, *35*, 1103–1110.
27. Singh, O.V.; Labana, S.; Pandey, G.; Budhiraja, R.; Jain, R.K. Phytoremediation: An overview of metallic ion decontamination from soil. *Appl. Microbiol. Biotechnol.* **2003**, *61*, 405–412. [CrossRef]
28. Singh, S.P.; Ma, L.Q.; Harris, W.G. Heavy Metal Interactions with Phosphatic Clay: Sorption and Desorption Behavior. *J. Environ. Qual.* **2001**, *30*, 1961–1968. [CrossRef]
29. Jacques, D.; Šimůnek, J.; Mallants, D.; van Genuchten, M.T. Modelling coupled water flow, solute transport and geochemical reactions affecting heavy metal migration in a podzol soil. *Geoderma* **2008**, *145*, 449–461. [CrossRef]
30. Zhang, R.; Yao, S.; Ao, Y. Vertical transport rules of Cu, Cd, Pb in the surface of soil. *J. Shanghai Jiaotong Univ. (Agric. Sci.)* **2013**, *31*, 67–71.
31. China National Environmental Monitoring Centre; Nanjing Environmental Monitoring Center Station. Technical Specification for Soil Environmental Monitoring (HJ/T 166-2004). Available online: http://www.mee.gov.cn/image20010518/5406.pdf (accessed on 9 December 2004).
32. Azouzi, R.; Charef, A.; Hamzaoui, A.H. Assessment of effect of pH, temperature and organic matter on zinc mobility in a hydromorphic soil. *Environ. Earth Sci.* **2015**, *74*, 2967–2980. [CrossRef]
33. Ren, L. Adsorption-Desorption Characteristics Research of Pb, Cr, Ni in Sediment and Soil. Master's Thesis, Jilin Agricultural University, Changchun, China, 2017.
34. Zheng, S. Studies on the Transformation and Transport of Heavy Metals in Typical Chinese Agricultural Soi. Master's Thesis, Zhejiang University, Hangzhou, China, 2010.
35. Zhai, L.; Chen, T.; Liao, X.; Yan, X.; Wang, L.; Xie, H. Pollution of agricultural soils resulting from a tailing spill at a Pb-Zn mine:A case study in HuanJiang Guangxi Province. *Acta Sci. Circumstantiae* **2008**, *6*, 1206–1211.

Article

Experimental and Simulation Research on the Process of Nitrogen Migration and Transformation in the Fluctuation Zone of Groundwater Level

Yepeng Li [1,2,*], Liuyue Wang [1,2], Xun Zou [1,2], Jihong Qu [1,2] and Gang Bai [1,2]

[1] College of Geosciences and Engineering, North China University of Water Resources and Electric Power, Zhengzhou 450046, China; z20201020259@stu.ncwu.edu.cn (L.W.); x201810207166@stu.ncwu.edu.cn (X.Z.); qujihong@ncwu.edu.cn (J.Q.); x20201020232@stu.ncwu.edu.cn (G.B.)

[2] Collaborative Innovation Center for Efficient Utilization of Water Resources, Zhengzhou 450046, China

* Correspondence: liyuepeng@ncwu.edu.cn; Tel.: +86-139-3716-5752

Abstract: The fluctuation of groundwater causes a change in the groundwater environment and then affects the migration and transformation of pollutants. To study the influence of water level fluctuations on nitrogen migration and transformation, physical experiments on the nitrogen migration and transformation process in the groundwater level fluctuation zone were carried out. A numerical model of nitrogen migration in the Vadose zone and the saturated zone was constructed by using the software HydrUS-1D. The correlation coefficient and the root mean square error of the model show that the model fits well. The numerical model is used to predict nitrogen migration and transformation in different water level fluctuation scenarios. The results show that, compared with the fluctuating physical experiment scenario, when the fluctuation range of the water level increases by 5 cm, the fluctuation range of the nitrogen concentration in the coarse sand, medium sand and fine sand media increases by 37.52%, 31.40% and 21.14%, respectively. Additionally, when the fluctuation range of the water level decreases by 5 cm, the fluctuation range of the nitrogen concentration in the coarse sand, medium sand and fine sand media decreases by 36.74%, 14.70% and 9.39%, respectively. The fluctuation of nitrogen concentration varies most significantly with the amplitude of water level fluctuations in coarse sand; the change in water level has the most significant impact on the flux of nitrate nitrogen and has little effect on the change in nitrite nitrogen and ammonium nitrogen, and the difference in fine sand is the most obvious, followed by medium sand, and the difference in coarse sand is not great.

Keywords: groundwater level fluctuation zone; nitrogen; migration and transformation; HYDRUS-1D model

Citation: Li, Y.; Wang, L.; Zou, X.; Qu, J.; Bai, G. Experimental and Simulation Research on the Process of Nitrogen Migration and Transformation in the Fluctuation Zone of Groundwater Level. *Appl. Sci.* **2022**, *12*, 3742. https://doi.org/10.3390/app12083742

Academic Editor: Bing Bai

Received: 1 March 2022
Accepted: 5 April 2022
Published: 8 April 2022

1. Introduction

Due to industry [1,2], agriculture [3,4], life [5], aquaculture [6], atmospheric deposition [7] and other factors, the groundwater in many countries and regions around the world are affected by nitrogen pollution, including the United States [8], China [9], the United Kingdom [10], South Korea [11] and other countries with severe pollution. In 2005, 34.1% of 1139 groundwater samples in northern China failed to meet World Health Organization (WHO) criteria. [12]. Currently, China has become one of the countries with the highest nitrogen fertilizer used in the world [13]. Nitrate from nitrogen may rapidly leach into groundwater, affecting the ecology and human health [14]. Wild et al. concluded that even if nitrate input is significantly reduced in the future, it will take decades to significantly reduce nitrate concentrations in porous aquifers through denitrification [15]. In 2000, the Water Framework Directive to bring water bodies to good chemical and ecological status by 2027 was issued [16]. Therefore, how to scientifically understand the nitrogen migration and transformation patterns under fluctuating groundwater level conditions has become one of the hot issues in the field of environmental research.

During groundwater level fluctuations, soluble substances in the aquifer medium are gradually dissolved into the groundwater, thus changing the chemical composition of the groundwater [17]. Guo Huaming et al. [18] studied the influencing factors of arsenic enrichment in groundwater, and the results showed that a high pH was unfavorable to the adsorption of arsenic in the form of anions by the aquifer medium. In their study, Sorensen J et al. [19] found that chemical and biological contaminants near the surface were transported to groundwater with minimal attenuation. Water level fluctuations also have a significant effect on nitrogen. Hefting M et al. [20] selected 13 riparian sites to analyze nitrogen cycling processes and confirmed a direct positive correlation between denitrification and elevated water table levels. Heather L. Welch et al. [21] analyzed and found that nitrate-N is weakened by denitrification during downward transport using redox sensitivity metrics at the water table without a location in the vertical direction. Jurado A et al. [22] concluded that the accumulation of N_2O in groundwater is also mainly due to denitrification and, to a lesser extent, nitrification. Yang L.P. et al. [23] conducted indoor soil column experiments to study the effect of pH on nitrogen transport and transformation processes and showed that pH 6.5 was the most efficient for the removal of "tri-nitrogen" in the optimal pH range for adsorption nitrification and denitrification. The authors in [24] analyzed the effect of water level fluctuation on nitrogen transformation by simulating aerobic, anoxic and anaerobic zones, and the results showed that denitrification played the greatest role in the anaerobic zone.

Groundwater level fluctuations significantly affect the migration and transformation of groundwater pollutants [25–29]. Scholars at home and abroad have conducted studies on groundwater level fluctuations. Most of their research focuses on the migration and transformation processes of soil salinity, iron, manganese, arsenic, iodin, benzene and other characteristic components when subjected to water level fluctuations, etc. [18,30,31]. Davis et al. [32] studied the changes of organic matter and oxygen at the fluctuation of water level, and the results showed that BTEX and oxygen concentration showed a relationship between this and that near the water level, and the main reason for oxygen reduction was due to microbial degradation. Kamon et al. [33] conducted an experimental and numerical study of the migration of LNAPL with water level fluctuations and showed that the entry and displacement pressures were greater for the air-water system than for the LNAPL-water system. Xiaoxi Xie et al. [34] analyzed the hysteresis relationship between saturation and the capillary pressure in a medium under the water table fluctuation conditions during alternating drying and wetting processes, and the results demonstrated that when the initial water saturation of the drying process is similar, the greater the initial water saturation of the wetting process, the degree of hysteresis gradually decreases. Xiang Li et al. [35] analyzed and verified that groundwater level fluctuations affected the physicochemical properties of soil-water bodies and further affected the movement of nitrate in soil solids.

Research on the mechanisms of solute migration and transformation in the fluctuation zone of groundwater level, mainly through technical methods, such as field investigation, indoor experiments, theoretical analysis and numerical simulation, was conducted to predict the temporal and spatial distribution characteristics of solute pollution. Chen et al. [36] found through field experiments, that the depth of shallow groundwater has more influence on the nitrogen concentration in shallow groundwater than other factors. Zhang Dan et al. [27] conducted on-site monitoring of plots at different altitudes for a year. They found that the fluctuation of shallow groundwater levels significantly affected the soil profile and the nitrogen concentration of shallow groundwater. Farnsworth et al. [37] established an indoor soil column experiment, using a set of 1.3 m quartz sand columns inoculated with microorganisms, and changing the water level in the sand column every 30~50 h to simulate the periodic production wells of production wells and groundwater caused by the cessation of mining and the influence of the level fluctuation on the oxidation of manganese—the content of manganese increases as the water level drops, and decreases as the water levels rise. Yang Yang [38] studied the influence of the groundwater level rise and fall on the migration of cadmium, and argued that the rise and fall in the water level mainly affect the transport of cadmium ions through convection. Zhang Xuejing et al. [39] used inverse distance weighting (IDW) interpolation and the water chemistry Piper graphic method to analyze the response relationship between the groundwater chemistry characteristics and the depth of the water level in the Ejina Oasis after the ecological water transport (2001–2017). Cao Wengeng et al. [40] used the potential distribution-multi-point complexation model (CD-MUSIC) to predict the evolution of groundwater chemical composition and hydrochemical types in the Baoding Plain of the South-to-North Water Transfer Project under the condition of groundwater level rebound. Arash Tafteh and Ali Reza Sepaskhah [41] successfully simulated the leaching of water and nitrate from two crops in the field with high accuracy using HYDRUS-1D. Mo Xiaoyu et al. [42] used HYDRUS-1D to simulate the changes in nitrogen leaching under different rainfall intensities, analyzed the influencing factors, and found that high intensity would reduce the nitrogen utilization rate.

In summary, the impact of water level fluctuations on the migration and transformation of pollutants has attracted the attention of scholars. Still, the study of nitrogen migration and transformation under different media water level fluctuations is not systematic enough. In this paper, the indoor soil column is used to simulate water level fluctuations to study the temporal and spatial distribution of nitrate nitrogen, nitrite nitrogen and ammonium nitrogen in three typical soil media subjected to water level fluctuations. According to physical experimental conditions, a numerical model of nitrogen migration and transformation in the fluctuation zone of the groundwater level was established to predict the spatial distribution and temporal change of nitrogen pollutants in the fluctuation zone of the groundwater level under different scenarios. This paper provides a scientific basis for the treatment, restoration and protection of groundwater nitrogen pollution.

2. Materials and Methods

2.1. Experimental Design

2.1.1. Experimental Materials

The experimental soil samples were collected on the floodplain of the Yellow River in Mengjin District, Luoyang City, Henan Province (Figure 1). The groundwater level in this area fluctuates frequently. Sample collection and processing were carried out by the requirements of the "Technical Specifications for Soil Environmental Monitoring" (HJ/T 166-2004). Sampling was performed using the quarter method, followed by drying, crushing and sieving, delineating three media, namely coarse sand, medium sand and fine sand. The pH value of the sand was between 8.5 and 9.3, and the organic matter content was between 0.241 and 1.070 $g{\cdot}kg^{-1}$. The basic physical and chemical properties of the soil are shown in Table 1. The water used in the laboratory is ultrapure water made by the German Millipore ultrapure water machine.

2.1.2. Experimental Device

The experiment designed three plexiglass columns with the following exact specifications: the inner diameter was 20 cm, the height was 110 cm, the top was open, and the opening of the glass column was covered with plastic film; the bottom was connected to a pressure-measuring tube and a Markov flask through a three-way valve. Between the Markov flask and the cylinder, a peristaltic pump was equipped to control the water level to simulate the process of groundwater fluctuation (rising–falling); the organic glass column was equipped with four sampling ports from top to bottom, at 15, 25, 35, and 45 cm away from the bottom of the column; adjacent points were separated by 10 cm. The sampling points were equipped with a Rhizon solution sampler (inner diameter: 1 mm); we covered the top and bottom of each column with 5 cm of quartz sand (diameter 2~3 mm), to ensure that the water level rose and fell uniformly. The experimental device is shown in Figure 2.

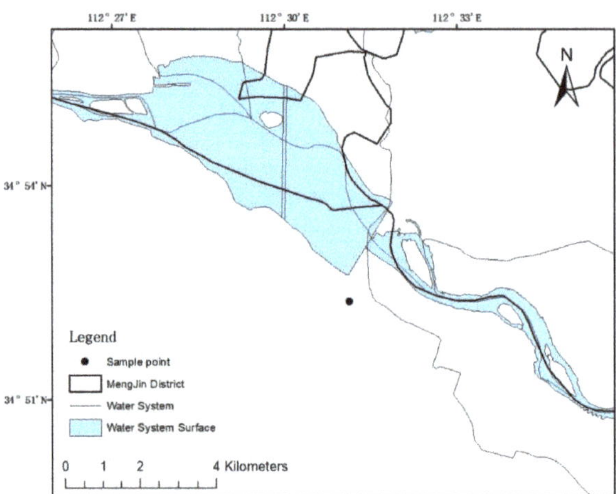

Figure 1. Location of sampling points.

Table 1. Soil particle composition and its parameters.

Serial Number	Sample	Particle Size (mm)	pH	Organic Matter Content ($g \cdot kg^{-1}$)
1	Coarse sand	0.5–1.0	8.5	0.241
2	Medium sand	0.25–0.5	8.8	0.587
3	Fine sand	0.125–0.25	9.3	1.070

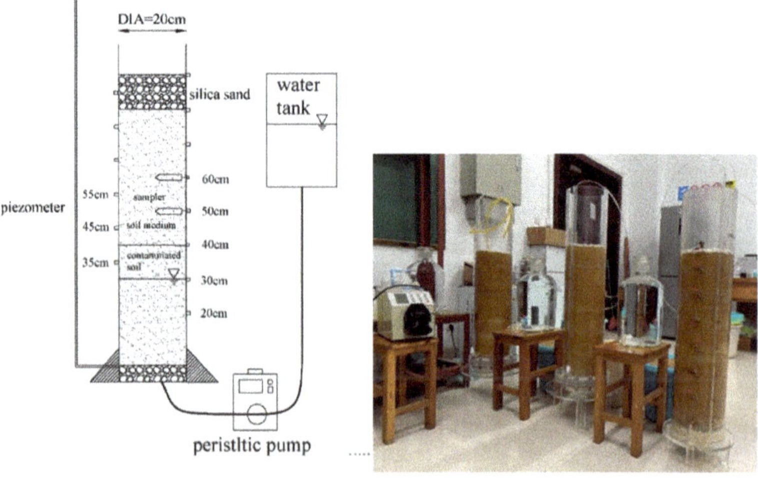

Figure 2. Schematic diagram of the experimental device.

2.1.3. Experimental Method

First, the experiments were conducted using soil from different media to fill the columns, filled with distilled water and stabilized at a 30 cm scale at the initial water level. Then, potassium nitrate and ammonium chloride solutions were injected from the top of the column. Three days later, we took out the water sample, and adjusted the water level using a peristaltic pump, raising its height by 10 cm each time until the water level reached 50 cm; after that, we decreased the water level by 10 cm each time, until the water level dropped to 10 cm. Finally, we raised the water level again, by 10 cm each time, until it reached 30 cm. The water levels during the experimental period were 30, 40, 50, 40, 30, 20, 10, 20, and 30 cm, respectively. Each water level was maintained for 3 d, and the change in the water level over time is shown in Figure 3, with the experiment using 24 d as a complete change period. The investigation was carried out for two cycles.

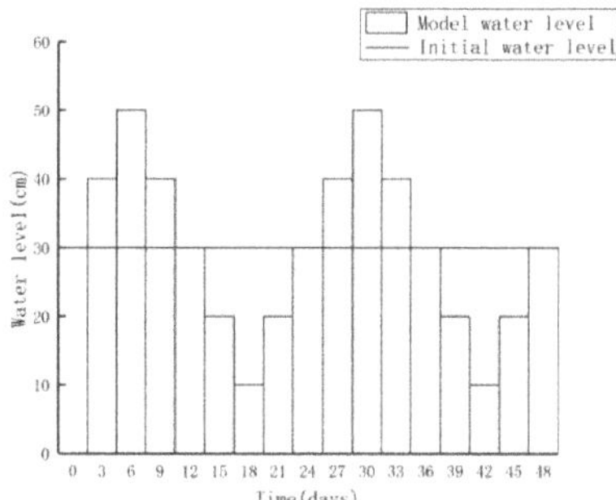

Figure 3. Time variation of water level.

Before each water level change, water samples were collected at four sampling ports using a diaphragm vacuum pump and a soil solution sampler to measure the concentrations of three forms of ammonium nitrogen, nitrate nitrogen and nitrite nitrogen.

2.2. Numerical Model

HYDRUS-1D is widely used in saturated-unsaturated zones of water, heat and solute transport to study the process of nitrogen migration and the transformation in the fluctuation zone at the groundwater level. In this paper, based on indoor physical experiments, a numerical model of nitrogen migration and change in the Vadose zone–saturated zone coupled with water transport and solute transport, was constructed using HYDRUS-1D.

2.2.1. Mathematical Model

- Water Movement Model.

The mathematical model of water movement in the water level fluctuation zone can be expressed as the Richards equation, as follows:

$$C(h)\frac{\partial \theta}{\partial t} = \frac{\partial}{\partial z}\left[K(h)[\frac{\partial h}{\partial t} - \cos(\alpha)]\right] - S(z,t) \tag{1}$$

where $C(h)$ is the water capacity (cm^{-1}); $K(h)$ is the hydraulic conductivity (cm/d); h is the negative pressure (cm); z represents the position coordinates in the parallel water flow direction (cm); t is the time (d); α is the angle between the water flow direction and the vertical (°); θ is the volumetric water content (cm^3/cm^3); and $S(z,t)$ is the water absorption strength of plant roots (cm^3/cm^3·d^{-1}).

- Solute transport model in the Vadose zone.

The solute transport model only considers the behavioral characteristics of convection, diffusion, adsorption, degradation, etc., and uses the traditional convection–diffusion equation to describe the transport process. The equation is expressed as:

$$\frac{\partial}{\partial t}(\theta C) = \frac{\partial}{\partial z}(\theta D_L \frac{\partial C}{\partial z} - vC) - \rho_b \frac{\partial(\rho_s K_L C)}{\partial t} - C_0 \exp(-kt) \tag{2}$$

where C is the solute concentration in the liquid phase (mg/L); D_L is the longitudinal dispersion coefficient (cm/d); v is the Darcy flow velocity (cm/d); ρ_b is the soil bulk density (mg/cm); ρ_s is the soil bulk density (mg/cm^3); K_L is the adsorption distribution coefficient (cm^3/g); $C_0 \exp(-kt)$ is the source-sink term ($cm^3/cm^3 \cdot d^{-1}$); the others are the same as above.

- Nitrogen migration and transformation model.

The transport model of nitrogen in the soil varies according to the different forms of nitrogen. The transport and transformation process of NH_4^+ is mainly subjected to adsorption and nitrification. The equation is as follows:

$$\begin{cases} \frac{\partial(\theta c_1)}{\partial t} = \frac{\partial}{\partial z}\left(\theta D_L \frac{\partial c_1}{\partial z}\right) - \frac{\partial}{\partial z}(V c_1) - \rho_b \frac{\partial(\rho_s k_e c_1)}{\partial t} - k_1 \theta c_1 & \\ c_1(Z,0) = c_{10}(Z) & 0 \le Z \le L, t = 0 \\ -\left(\theta D_L \frac{\partial c_1}{\partial z} - q c_1\right) = \varepsilon(t) c_0 & Z = 0, t = 0 \\ c_1(L,t) = c_1 L & Z = L, t > 0 \end{cases} \quad (3)$$

Meanwhile NO_2^- and NO_3^- are mainly affected by nitrification and denitrification, and the equation is as follows:

$$\begin{cases} \frac{\partial(\theta c_i)}{\partial t} = \frac{\partial}{\partial z}\left(\theta D_L \frac{\partial c_i}{\partial z}\right) - \frac{\partial}{\partial z}(v_i c_i) - k_i \theta c_i & \\ c_i(Z,0) = c_{i0}(Z) & 0 \le Z \le L, t = 0 \\ c_i(0,t) = c_{i0}(t) & Z = 0, t > 0 \\ c_i(L,t) = c_i L & Z = L, t > 0 \end{cases} \quad (4)$$

where θ is the soil volume water content (cm^3/cm^3); c_1 is the soil solution NH_4^+ concentration (mg/L); D_L is the longitudinal diffusion coefficient (cm/d); k_e is the adsorption distribution coefficient (cm^3/g) for NH_4^+ in soil; k_1, k_2 are the denitrification rate constants (d^{-1}) for NH_4^+ and NO_2^-, respectively; $c_{10}(t)$ is the soil NH_4^+ initial concentration (mg/L); c_0 is the inlet solution NH_4^+ concentration (mg/L); c_{1L} is the diving NH_4^+ concentration (mg/L); c_2,c_2 are the NO_2^- and NO_3^- concentration (mg/L), respectively; k_3 is the denitrification rate constant (d^{-1}); $c_{20}(z)$, $c_{30}(z)$ are the soil NO_2^- and NO_3^- initial concentration (mg/L), respectively; and c_{2L},c_{3L} are the diving NO_2^- and NO_3^- concentration (mg/L), respectively. No nitrogen input through the water.

2.2.2. Initial Conditions and Boundary Conditions

- Initial conditions.

At the initial moment, the water level was set to 30 cm, and the initial concentration of pollutants is shown in Table 2.

Table 2. Initial concentration of pollutants.

Intake	Coarse Sand			Medium Sand			Fine Sand		
	Nitrite Nitrogen	Nitrate	Ammonium Nitrogen	Nitrite Nitrogen	Nitrate	Ammonium Nitrogen	Nitrite Nitrogen	Nitrate	Ammonium Nitrogen
15 cm	0.19	74.85	6.17	0.37	35.81	1.97	0.70	0.55	0.08
25 cm	0.15	77.14	6.97	0.27	36.47	3.07	0.84	2.20	0.09
35 cm	0.37	47.56	7.03	0.36	78.39	11.87	0.44	12.61	5.14
45 cm	0.33	45.25	8.23	0.25	84.41	12.61	0.49	76.77	9.76

- Boundary conditions.

1. Water transport and boundary conditions.

According to the experimental model, the upper boundary was in direct contact with the atmosphere and was set as the atmospheric boundary. The lower boundary was set as the variable head boundary due to the rise and fall of the water level.

2. Solute transport boundary.

According to the model, the upper boundary condition was the pollutant concentration boundary, and the lower boundary condition was the zero concentration gradient boundary. In the model setting, the primary considerations were adsorption and desorption, as well as nitrification and denitrification.

2.2.3. Model Parameters

The parameters of the numerical model mainly included soil hydraulic parameters and solute transport parameters. The initial values of the soil hydraulic parameters were determined according to the soil medium hydraulic parameter database of the Hydrus-1D software. The initial values of solute transport parameters were based on the measured results of this physical experiment, and the practical value was determined. The inverse solution module of the Hydrus-1D software was used for inversion to obtain the final parameters of the numerical model (Tables 3 and 4).

Table 3. Soil and water characteristic parameters.

Parameter	θr Residual Volume Water Content	θs Saturated Volume of Water Content	α/cm^{-1} Soil Moisture Characteristic Parameters	n Soil Moisture Characteristic Index	Ks/(cm·d^{-1}) Saturated Hydraulic Conductivity
Coarse sand	0.045	0.43	0.1450	2.68	712.8
Medium sand	0.051	0.42	0.1045	2.08	550.0
Fine sand	0.057	0.41	0.1240	2.28	350.2

Table 4. Solute transport parameters.

Medium	Solute	ρ/mg·cm^{-3} Bulk Density	$Disp$/cm Longitudinal Diffusion	D/cm^2·d^{-1} Solute Diffusion Coefficient	Kd/cm^3·mg^{-1} Adsorption Coefficient
Coarse sand	Nitrite Nitrogen	1600	1.568	1.29085	0.000256
	Nitrate			1.69085	0.001584
	Ammonium Nitrogen			1.69085	0.008779
Medium sand	Nitrite Nitrogen	1800	1.233	1.29085	0.000103
	Nitrate			1.69085	0.000413
	Ammonium Nitrogen			1.69085	0.008895
Fine sand	Nitrite Nitrogen	1800	1.116	1.29085	0.000767
	Nitrate			1.69085	0.003094
	Ammonium Nitrogen			1.69085	0.008976

2.2.4. Calibration and Evaluation of the Model

The reliability was verified and analyzed by inputting the solute transport parameters and soil hydraulic parameters through the HYDRUS-1D software, and the simulation's accuracy was evaluated by the coefficient of determination R^2, and the root mean square error ($RMSE$). The closer the coefficient of determination R^2 was to 1, the closer the root mean square error ($RMSE$) was to 0, which means that the model simulation results and the measured results of nitrogen had higher fitting accuracy. The calculation formula is:

$$R^2 = \frac{\sum_{i=1}^{n} (\hat{y}_i - \overline{y})^2}{\sum_{i=1}^{n} (y_i - \overline{y})^2} \tag{5}$$

$$RMSE = \sqrt{\frac{1}{N} \sum_{i=1}^{N} (S_i - M_i)^2} \tag{6}$$

where S_i and M_i are simulated and measured values, respectively; N is the number of samples.

According to the simulation results of the numerical model of nitrogen migration and transformation in the water level fluctuation zone, the simulated values of ammonia nitrogen, nitrate nitrogen and nitrite nitrogen are shown in Figures 4–6, and the model fitting results are shown in Table 5. The correlation coefficients are mostly above 0.8.

The root mean square error (*RMSE*) is small. Except for the significant simulation error of individual pollutants, the simulation results of the numerical model fit well with the measured results, indicating that the numerical model can better reflect the nitrogen transfer and transformation process under the fluctuation of the groundwater level.

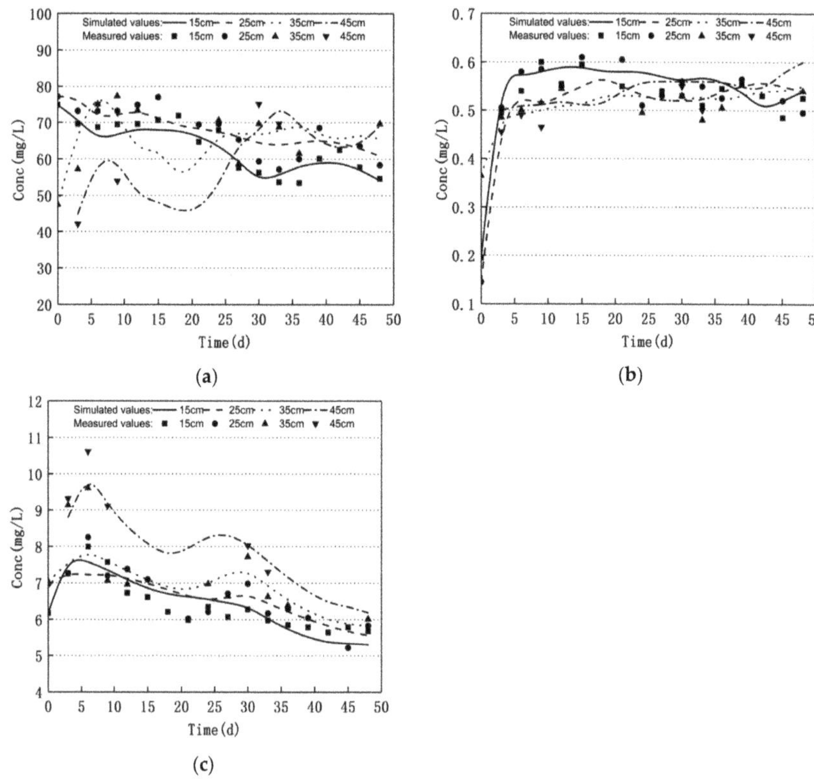

Figure 4. The process of nitrogen change in coarse sand ((**a**) nitrate nitrogen; (**b**) nitrite nitrogen; (**c**) ammonium nitrogen).

Table 5. Model fitting effect.

Soil Media	Pollutants	Decisive Factor R^2	*RMSE*
Coarse sand	Nitrite nitrogen	0.71284	0.0883
	Nitrate nitrogen	0.70572	5.4257
	Ammonia nitrogen	0.82099	0.4863
Medium sand	Nitrite nitrogen	0.74610	0.0486
	Nitrate nitrogen	0.80656	9.1317
	Ammonia nitrogen	0.98810	0.5115
Fine sand	Nitrite nitrogen	0.87036	0.0309
	Nitrate nitrogen	0.98398	2.7137
	Ammonia nitrogen	0.97552	0.3860

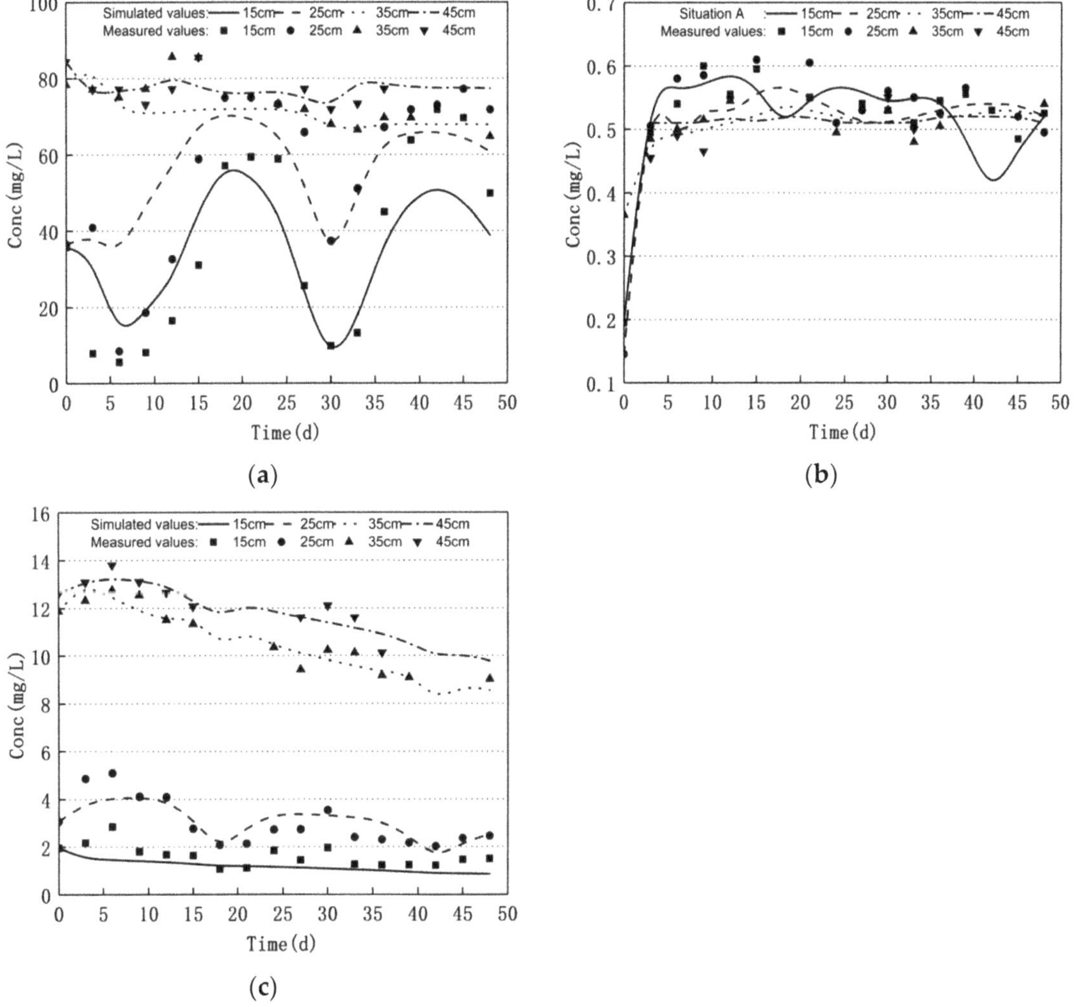

Figure 5. The process of nitrogen change in medium sand ((**a**) nitrate nitrogen; (**b**) nitrite nitrogen; (**c**) ammonium nitrogen).

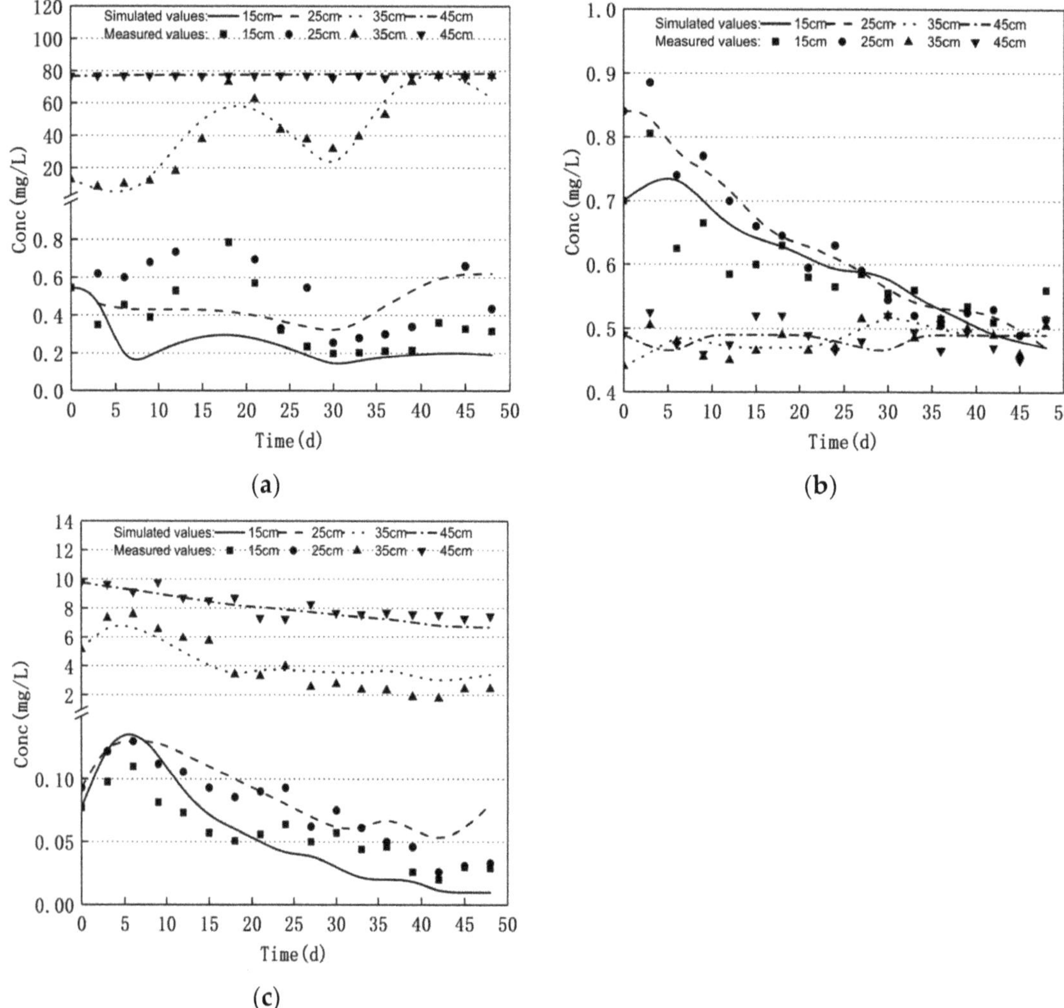

Figure 6. The process of nitrogen change in fine sand ((**a**) nitrate nitrogen; (**b**) nitrite nitrogen; (**c**) ammonium nitrogen).

3. Results and Discussion

3.1. Nitrogen Changes and Model Validation

The process curves of the measured and simulated results of ammonia nitrogen, nitrate nitrogen and nitrite nitrogen in the coarse sandy soil column under the condition of water level fluctuation are shown in Figure 4.

In the 15 cm and 25 cm sampling ports in the coarse sand column, NO_3^--N showed a decreasing trend in the rising stage of the water level, with a decrease of 8.55% on average, the maximum decrease of 14.94% in the second cycle of the second rising stage of the water level, and the minimum decrease of 2.27% in the first rising stage of the second cycle of the water level. The concentration of NO_3^--N showed an increasing trend in the declining water level stage, with an average increase of 9.10%, a maximum increase of 15.47% and a minimum increase of 4.62%, and the second fluctuation cycle was more obvious than the first fluctuation cycle of nitrate heel fluctuation. The average increase of NH_4^+-N concentration was 9.16% at the stage of the water level rise, and the average decrease of NH_4^+-N concentration was 14.355% at the stage of water level fall, and the fluctuation change was more obvious in the first cycle. The concentration of NO_2^--N was much smaller than that of the NO_3^--N and NH_4^+-N concentrations, and NO_2^--N showed fluctuating changes and eventually stabilized.

The process curves of the measured and simulated results of ammonia nitrogen, nitrate nitrogen and nitrite nitrogen in the medium sand soil column under the water level fluctuation conditions are shown in Figure 5.

In the medium sand column, the trend of NO_3^--N was similar to that of the coarse sand soil column, with insignificant changes in concentrations at the 35 cm and 45 cm sampling ports, and obvious fluctuation trends at the 15 cm and 25 cm sampling ports. The concentration of NO_3^--N at the 15 cm sampling port decreased by 780.30% on average, especially during the first water level rise in the first cycle. The concentration of NO_2^--N was fluctuating at the beginning and stabilized later.

The process curves of the measured and simulated results of ammonia nitrogen, nitrate nitrogen and nitrite nitrogen in the fine sandy soil column under the condition of water level fluctuation are shown in Figure 6.

In the fine sand soil column, the trend of the NO_3^--N concentration changes at the 15 cm, 25 cm and 35 cm sampling ports were basically the same, and the decreases at 15 cm, 25 cm and 35 cm were 25.93%, 68.05% and 19.19, respectively, during the water level rise stage, and the increases were at 15 cm, 25 cm and 35 cm during the water level fall stage. The trends of NH_4^+-N concentrations at the 15 cm, 25 cm and 35 cm sampling ports were basically the same, and the concentrations at the 15 cm sampling port did not change much, and the increases at the water level rising stage were 32.44%, 23.39% and 27.03% for 15 cm, 25 cm and 35 cm, respectively, and the decreasing water level for the NO_2^--N concentration was low and stabilized after fluctuating changes.

Figures 4–6 show the actual measurement process curve of ammonia nitrogen, nitrate nitrogen and nitrite nitrogen under the conditions of water level fluctuation. When the water level rises, the dissolved oxygen content decreases, and the NH_4^+-N concentration should fall. Still, denitrifying bacteria become active and dominant under hypoxic conditions, promoting the increase in the NH_4^+-N concentration and the significant decrease in the NO_3^--N concentration. When the water level drops, the dissolved oxygen content increases and nitrification plays a central role. The concentration of NO_3^--N increases significantly. However, due to the strong adsorption of the soil, the concentration of NH_4^+-N in the free water of the soil solution decreases [43,44]. Nitrite nitrogen, in the three media, fluctuated and eventually stabilized, but its concentration was much smaller than the concentrations of NO_3^--N and NH_4^+-N. Before the water level rises, when the dissolved oxygen is sufficient, the reproduction rate of nitrifying bacteria is slower than that of nitrosating bacteria; at this time, the nitrosation reaction dominates, causing the accumulation of NO_2^--N, resulting in an increased concentration of NO_2^--N in the soil solution during the rising water level stage. During the falling phase of the water level, nitrification dominates, a lot of H^+ is produced in the solution, and the soil solution becomes weakly acidic, which strengthens the conversion of nitrite to nitrate. NO_2^--N no longer accumulates, and the concentration gradually decreases [29,45,46]. In the initial stage, the increase in NO_2^--N satisfies coarse sand, then medium sand, then fine sand. The more significant the particle size is, the more particle surfaces there are that microbial flocs can come into contact with, and more microbes can participate in the nitrification reaction. With a stronger microbial nitrification ability, it is understood that the nitrification ability of microbial flocs will increase with the increase in particle size [47].

3.2. Scenario Simulation

3.2.1. Increasing Water Level Fluctuation

Scenario 1 is set to: keep the initial pollutant concentration unchanged, and increase the water level fluctuation range. That is, the initial water level is set to 30 cm, and the height is raised by 15 cm at a time, and then rises twice until the water level reaches 60 cm; then it is dropped by 15 cm each time and drops four times until the water level drops to 0 cm; finally, the water level is raised again twice, by 15 cm each time, until it reaches 30 cm, which is the initial water level. The comparison chart of water level changes is shown in Figure 7, recorded as situation A.

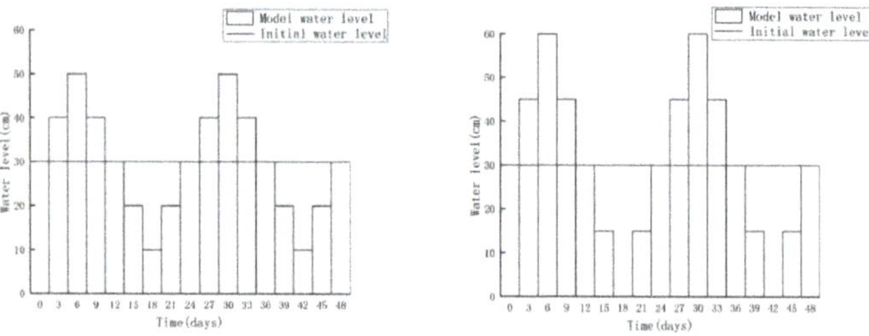

Figure 7. Comparison of water level changes.

Synchronously with the previous experiment scenario of groundwater level fluctuation, nitrogen migration and transformation (henceforth referred to as the fluctuation experiment), the simulation period is set to two by increasing the amplitude of the water level fluctuation, and the range of solute concentration in the water body increases. The model is built and run according to the scenario, and the simulation result is shown in Figures 8–10.

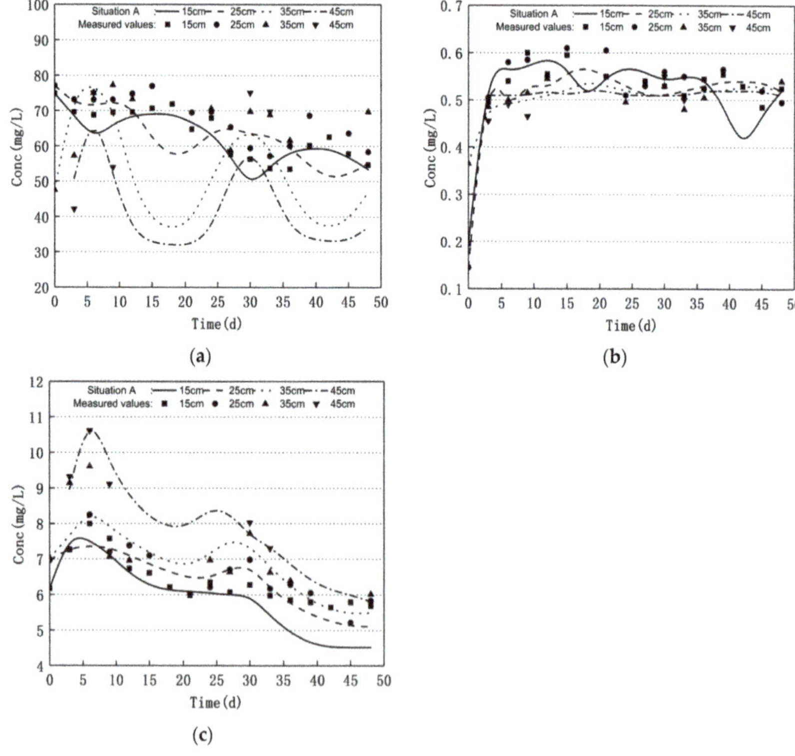

Figure 8. The measured results of coarse sand and the process of dynamic change of nitrogen in situation A ((**a**) nitrate nitrogen; (**b**) nitrite nitrogen; (**c**) ammonium nitrogen).

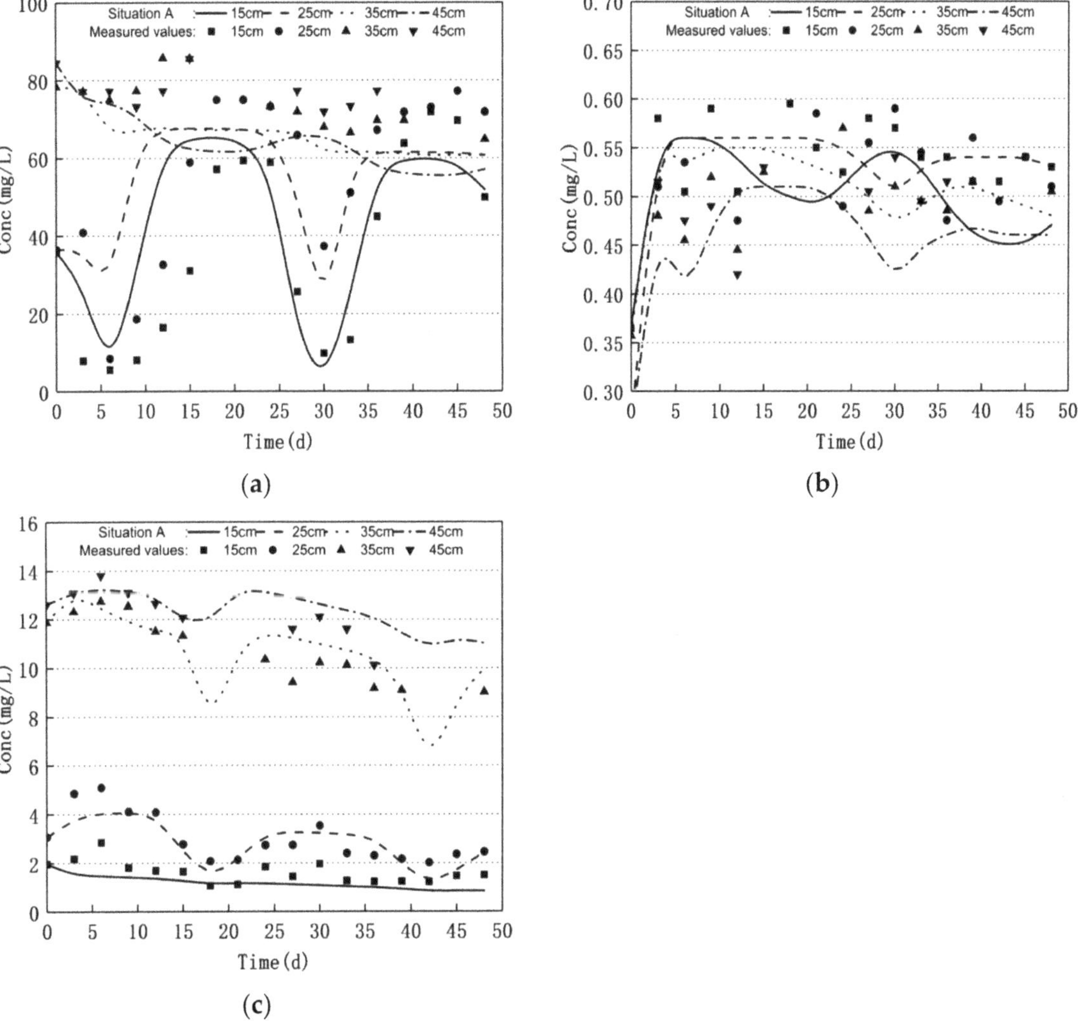

Figure 9. The measured results of medium sand and the dynamic process of nitrogen in situation A ((**a**) nitrate nitrogen; (**b**) nitrite nitrogen; (**c**) ammonium nitrogen).

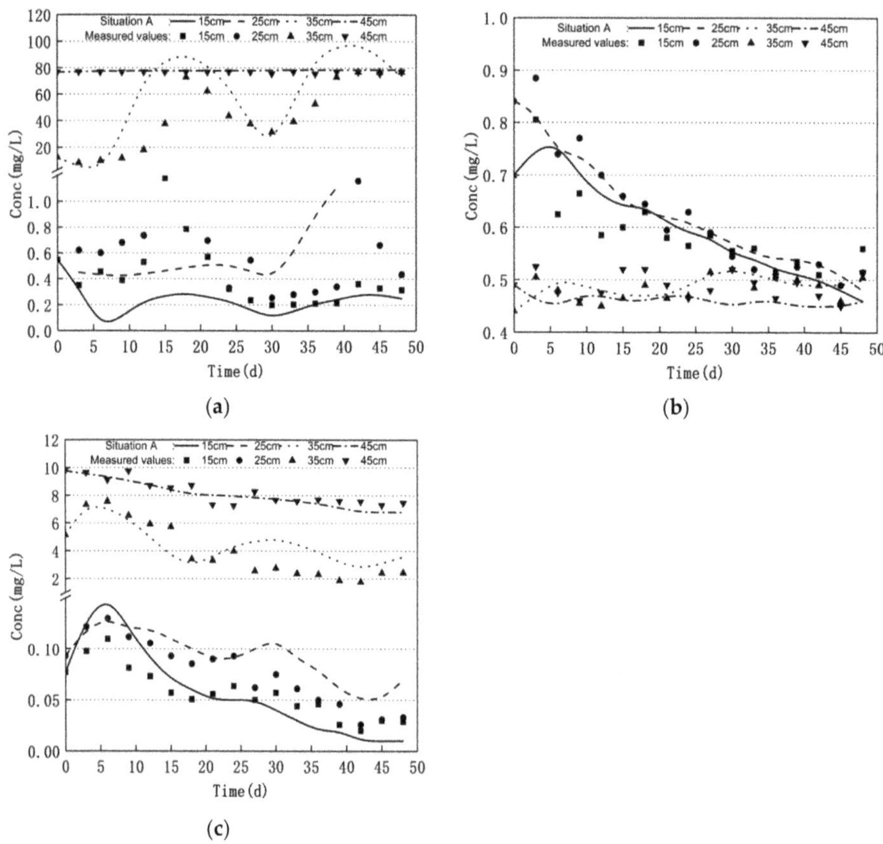

Figure 10. The measured results of fine sand and the dynamic process of nitrogen in situation A ((**a**) nitrate nitrogen; (**b**) nitrite nitrogen; (**c**) ammonium nitrogen).

The comparison of the standard deviation of nitrogen concentration obtained by the three media simulations is shown in Table 6.

Table 6. Scenario A model standard deviation comparison.

	Observation Hole	Coarse Sand		Medium Sand		Fine Sand	
		Fluctuation Experiment	Situation A	Fluctuation Experiment	Situation A	Fluctuation Experiment	Situation A
Nitrite nitrogen	15 cm	0.0249	0.0488	0.0348	0.0384	0.0838	0.0896
	25 cm	0.0197	0.0262	0.0141	0.0168	0.1037	0.0924
	35 cm	0.0159	0.0135	0.0261	0.0263	0.0166	0.0159
	45 cm	0.0276	0.0047	0.0255	0.0328	0.0107	0.0072
Nitrate	15 cm	5.4152	6.2294	16.0273	22.4563	0.0921	0.0807
	25 cm	4.2269	6.9961	12.8282	15.5376	0.0932	0.3326
	35 cm	5.2402	14.0076	3.8663	4.4202	24.2231	31.5546
	45 cm	9.5118	11.4918	1.9142	6.0021	0.417	0.4754
Ammonium Nitrogen	15 cm	0.7797	1.0168	0.2074	0.2113	0.0422	0.0441
	25 cm	0.5452	0.7755	0.7561	0.9278	0.027	0.0248
	35 cm	0.6154	0.875	1.4036	1.8423	1.2271	1.3702
	45 cm	1.0365	1.4005	1.1279	0.8256	0.8747	0.8622

Table 6 shows that the increase in the fluctuation range of water level can effectively enlarge solute fluctuation.

Water level fluctuations have different effects on the fluctuation range of nitrogen concentration in the three media. When the fluctuation range of water level increases by 5 cm, the fluctuation range of nitrogen in the coarse sand medium increases by 37.52% on average, compared to the fluctuation experiment scenario; the nitrogen concentration in the medium sand medium is increased by 37.52%. The fluctuation range increased by 31.40% on average; the fluctuation range of the nitrogen concentration in the fine sand medium increased by 21.14% on average.

Table 6 shows that in the coarse sand, the fluctuation of the nitrogen concentration changes most significantly with the increase in the fluctuation range of the water level, followed by the medium sand, and the fine sand has the slightest change.

The impact of water level fluctuations on the three solutes is also different. The fluctuation range of the water level is expanded by 5 cm, and the fluctuation range of nitrite nitrogen, nitrate nitrogen and ammonium nitrogen in the coarse sand medium increases by 7.90%, 67.17%, and 37.49%, respectively. In the medium sand medium, the fluctuation range increased by 14.66%, 72.28%, and 7.26%, respectively. The fluctuation range of nitrite nitrogen in the fine sand medium decreased by 10.34%, while the fluctuation range of nitrate nitrogen and ammonium nitrogen increased by 72.17% and 1.62%, respectively.

Table 6 shows that among the three media, the fluctuation of the water level has the most significant effect on the fluctuation of nitrate nitrogen, and the influence on the fluctuation of nitrite nitrogen and ammonium nitrogen is relatively small.

3.2.2. Scenario of Reduced Water Level Fluctuation

Scenario 2 is set to: keep the initial pollutant concentration unchanged, and reduce the fluctuation of the water level. That is, the initial water level is set to 30 cm, and the height is raised by 5 cm each time, and then rises two times until the water level reaches 40 cm; it then starts to drop by 5 cm each time, and drops four times until the water level drops to 20 cm; finally, the water level rises again two times, by 5 cm each time, until it reaches 30 cm, which is the initial water level. The comparison chart of the water level changes is shown in Figure 11. It is recorded as situation B.

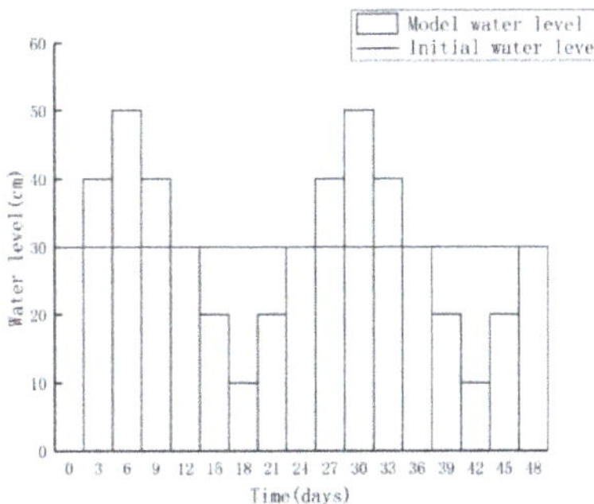

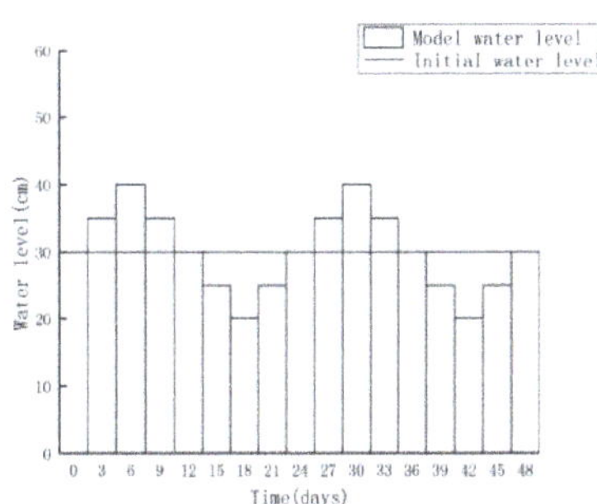

Figure 11. Comparison of Water Level Changes.

Similarly, the simulation period is set to 2. By increasing the fluctuation range of the water level, the variation range of the solute concentration in the water body is increased. The model is built and run according to the scenario, and the simulation result is shown in Figures 12–14.

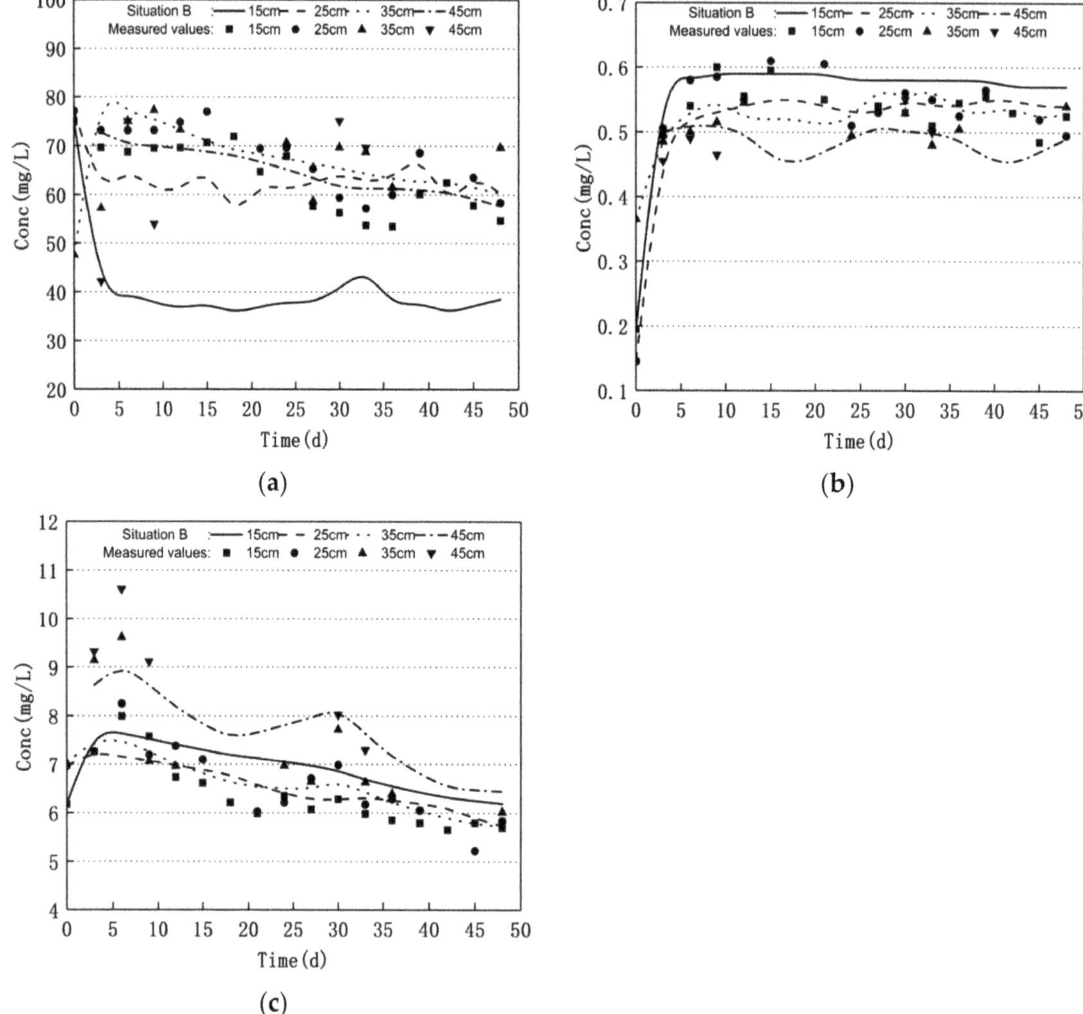

Figure 12. The measured results of coarse sand and the process of dynamic change of nitrogen in situation B ((**a**) nitrate nitrogen; (**b**) nitrite nitrogen; (**c**) ammonium nitrogen).

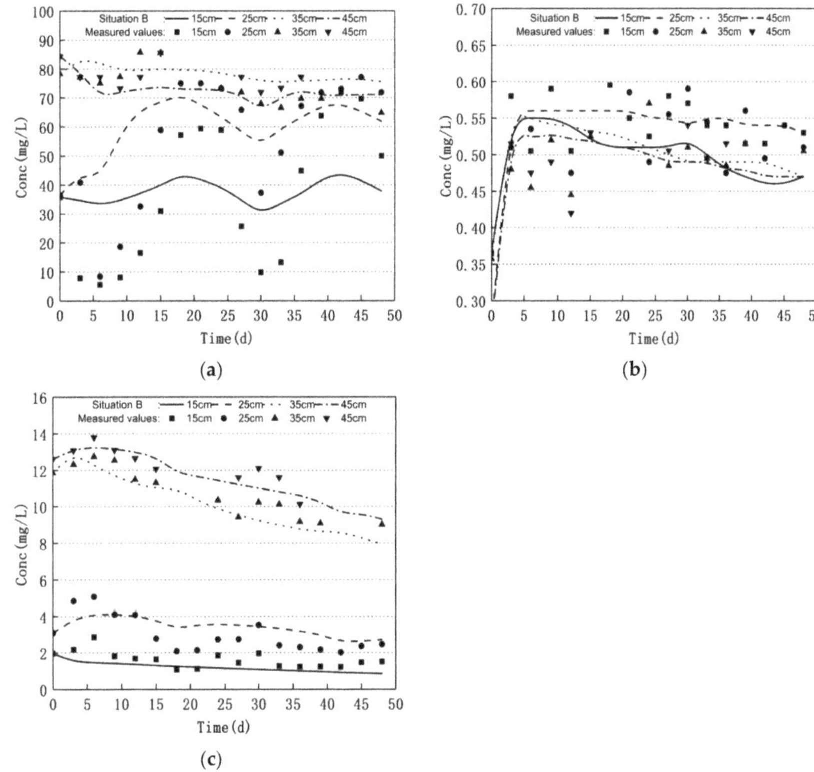

Figure 13. The measured results of medium sand and the dynamic process of nitrogen in situation B ((**a**) nitrate nitrogen; (**b**) nitrite nitrogen; (**c**) ammonium nitrogen).

The comparison of the standard deviation of the nitrogen concentration obtained by the three media simulations is shown in Table 7.

Table 7. Scenario B model standard deviation comparison.

	Observation Hole	Coarse Sand		Medium Sand		Fine Sand	
		Fluctuation Experiment	**Situation B**	**Fluctuation Experiment**	**Situation B**	**Fluctuation Experiment**	**Situation B**
Nitrite Nitrogen	15 cm	0.0249	0.0063	0.0348	0.0316	0.0838	0.0758
	25 cm	0.0197	0.0143	0.0141	0.0099	0.1037	0.1139
	35 cm	0.0159	0.0183	0.0261	0.0266	0.0166	0.0148
	45 cm	0.0276	0.0206	0.0255	0.0219	0.0107	0.0112
Nitrate	15 cm	5.4152	2.1756	16.0273	4.0429	0.0921	0.1069
	25 cm	4.2269	3.337	12.8282	8.1699	0.0932	0.0601
	35 cm	5.2402	6.0552	3.8663	2.4465	24.2231	15.1679
	45 cm	9.5118	4.4877	1.9142	2.658	0.417	0.3406
Ammonium Nitrogen	15 cm	0.7797	0.4898	0.2074	0.2042	0.0422	0.0391
	25 cm	0.5452	0.4511	0.7561	0.4871	0.027	0.029
	35 cm	0.6154	0.5448	1.4036	1.4785	1.2271	1.0044
	45 cm	1.0365	0.7817	1.1279	1.3127	0.8747	0.7587

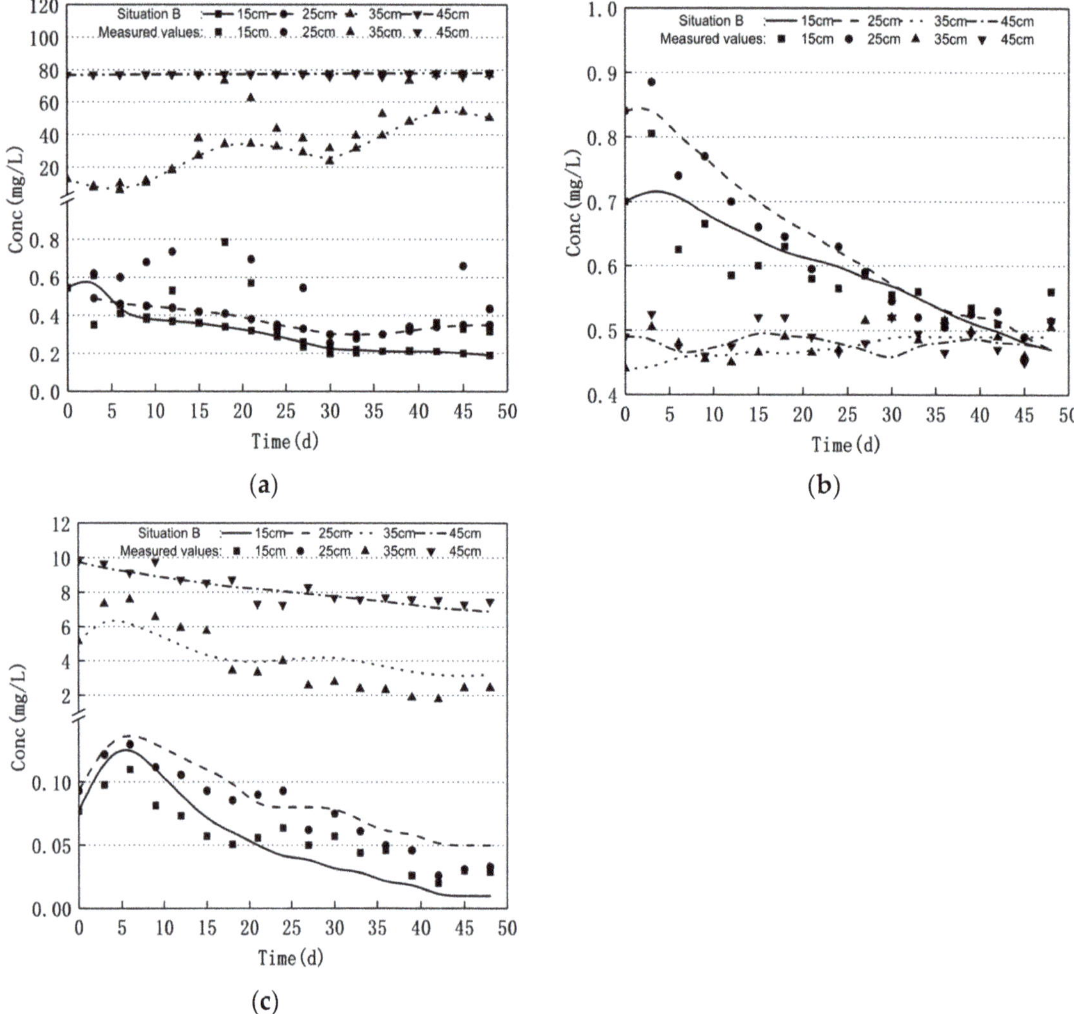

Figure 14. The measured results of fine sand and the dynamic process of nitrogen in situation B ((**a**) nitrate nitrogen; (**b**) nitrite nitrogen; (**c**) ammonium nitrogen).

Table 7 shows that the decrease in water level fluctuation can effectively reduce solute fluctuation.

Water level fluctuations have different effects on the fluctuation range of nitrogen concentration in the three media. When the fluctuation range of the water level decreases by 5 cm, the fluctuation range of nitrogen in the coarse sand medium is reduced by 36.74% on average, compared to the fluctuation experiment scenario. In the medium sand medium, the fluctuation range of the concentration of nitrogen decreased by 14.70% on average, while the fluctuation range of the nitrogen concentration in the fine sand medium decreased by 9.39% on average.

Table 7 shows that in the coarse sand, the fluctuation of the nitrogen concentration changes most significantly with the decrease in the fluctuation range of the water level, followed by the medium sand, and the fine sand demonstrates the most minor change.

The impact of water level fluctuations on the three solutes is also different. When the fluctuation range of the water level is reduced by 5 cm, the fluctuation ranges of nitrite nitrogen, nitrate nitrogen and ammonium nitrogen in the coarse sand medium are reduced by 28.07%, 29.55%, and 22.62%, respectively. In the medium sand medium, the average fluctuation range is reduced by 13.00%, 27.23%, and 3.85%, respectively. In the fine sand model, the fluctuation range is reduced by an average of 1.51%, 18.79%, and 7.86%, respectively.

Table 7 shows that among the three media, water level fluctuations have the most significant impact on the fluctuations of nitrate nitrogen, have less impact on the fluctuations of nitrite nitrogen and ammonium nitrogen, and the difference is the most obvious in the fine sand medium, followed by medium sand. This difference is not great in the coarse sand.

4. Conclusions

In this paper, through the experiment examining nitrogen migration and transformation in the groundwater fluctuating zone, we analyzed the nitrogen migration and transformation process. A numerical model of nitrogen migration and transformation in the groundwater level fluctuating zone was established with the help of the HYDRUS-1D model. The paper obtained the following main conclusions:

- Groundwater level fluctuations can significantly affect the nitrogen transport and transformation patterns in soil–groundwater. The nitrate nitrogen concentration increased and the ammonium nitrogen mass concentration decreased when the water level decreased. Moreover, the nitrate nitrogen mass concentration decreased and the ammonium nitrogen mass concentration increased when the water level increased; the nitrite nitrogen did not change significantly.
- In this study, indoor soil column experiments were combined with the Hydrus-1D software simulation prediction to simulate the theoretical values of the tri-nitrogen transformation process in indoor soil columns. Although simple, the software does not take into account the existence of non-homogeneous changes in the soil column, is not accurate enough, there are certain limitations, and the influence of biological processes on the transformation of pollutants has not been taken into account. Furthermore, there is still a certain gap with the actual results.
- As an important factor of the hydrological mechanism, the water level plays a vital role in the groundwater system, and the effect of the level and intensity of fluctuation of the groundwater level on the migration and transformation of nitrogen cannot be ignored. The change in water level will affect the water content in the soil, the state of water movement, the physical and chemical properties of soil and the state of the microorganisms, which in turn, will affect the migration and transformation of nitrogen in the soil.

Author Contributions: Conceptualization, Y.L. and J.Q.; methodology, Y.L., L.W. and X.Z.; software, L.W. and X.Z.; validation, Y.L., L.W. and J.Q.; writing—original draft preparation, Y.L. and L.W.; writing—review and editing, Y.L. and J.Q.; visualization, G.B. All authors have read and agreed to the published version of the manuscript.

Funding: This research was funded by the Doctoral Research Fund of North China University of Water Resources and Electric Power (40651) and the Key Project of Science and Technology Research of Henan Education Department (14A170006), and by the Key R&D and Promotion Projects in Henan Province (202102310012).

Institutional Review Board Statement: Not applicable.

Informed Consent Statement: Not applicable.

Data Availability Statement: Data are contained within the article.

Conflicts of Interest: The authors declare no conflict of interest.

References

1. Matiatos, I. Nitrate source identification in groundwater of multiple land-use areas by combining isotopes and multivariate statistical analysis: A case study of Asopos basin (Central Greece). *Sci. Total Environ.* **2016**, *541*, 802–814. [CrossRef] [PubMed]
2. Fu, J.; Shan, J.; Yuan, Y.; Gao, Y.; Wu, R.; Yang, J. Comprehensive traceability of nitrogen pollution in the Shenfu section of Hunhe River Basin. *Acta Sci. Circumstantiae* **2018**, *38*, 2560–2567.
3. van der Schans, M.L.; Harter, T.; Leijnse, A.; Mathews, M.C.; Meyer, R.D. Characterizing sources of nitrate leaching from an irrigated dairy farm in Merced County, California. *J. Contam. Hydrol.* **2009**, *110*, 9–21. [CrossRef] [PubMed]
4. Rosenstock, T.S.; Liptzin, D.; Dzurella, K.; Fryjoff-Hung, A.; Hollander, A.; Jensen, V.; King, A.; Kourakos, G.; McNally, A.; Pettygrove, G.S.; et al. Agriculture's Contribution to Nitrate Contamination of Californian Groundwater (1945–2005). *J. Environ. Qual.* **2014**, *43*, 895–907. [CrossRef] [PubMed]
5. Meghdadi, A.; Javar, N. Quantification of spatial and seasonal variations in the proportional contribution of nitrate sources using a multi-isotope approach and Bayesian isotope mixing model. *Environ. Pollut.* **2018**, *235*, 207–222. [CrossRef] [PubMed]
6. Cui, N.; Liu, S. Comprehensive evaluation of soybean meal replacement based on RIAM model. *Feed. Res.* **2020**, *43*, 126–130.
7. Xiaoliang, J.; Xie, R.; Yun, H.; Lu, J. Quantitative identification of nitrate pollution sources and uncertainty analysis based on dual isotope ap-proach in an agricultural watershed. *Environ. Pollut.* **2017**, *229*, 586–594.
8. Burow, K.R.; Nolan, B.T.; Rupert, M.G.; Dubrovsky, N.M. Nitrate in groundwater of the United States, 1991–2003. *Environ. Sci. Technol.* **2010**, *44*, 4988–4997. [CrossRef]
9. Gu, B.; Ge, Y.; Chang, S.X.; Luo, W.; Chang, J. Nitrate in groundwater of China: Sources and driving forces. *Glob. Environ. Chang.* **2013**, *23*, 1112–1121. [CrossRef]
10. Stuart, M.; Gooddy, D.; Bloomfield, J.; Williams, A. A review of the impact of climate change on future nitrate concentrations in groundwater of the UK. *Sci. Total Environ.* **2011**, *409*, 2859–2873. [CrossRef]
11. Cheong, J.-Y.; Hamm, S.-Y.; Lee, J.-H.; Lee, K.-S.; Woo, N. Groundwater nitrate contamination and risk assessment in an agricultural area, South Korea. *Environ. Earth Sci.* **2011**, *66*, 1127–1136. [CrossRef]
12. Zhao, T.; Zhang, C.; Du, L.; Liu, B.; An, Z. Investigation of nitrate content in groundwater of seven provinces (cities) around the Bohai Sea. *Acta Agri-Environ. Sci.* **2007**, *2*, 779–783.
13. Zhao, Y.; Shao, M. Experimental study on nitrate-nitrogen migration in farmland under different fertilization conditions. *J. Agric. Eng.* **2002**, *18*, 37–40.
14. Filippis, G.D.; Ercoli, L.; Rossetto, R. A Spatially Distributed, Physically-Based Modeling Approach for Estimating Agricultural Nitrate Leaching to Groundwater. *Hydrology* **2021**, *8*, 8. [CrossRef]
15. Wild, L.M.; Mayer, B.; Einsiedl, F. Decadal Delays in Groundwater Recovery from Nitrate Contamination Caused by Low O2Reduction Rates. *Water Resour. Res.* **2018**, *54*, 9996. [CrossRef]
16. European Commission. Establishing a Framework for the Community Action in the Field of Water Policy. In *Directive 2000/60/EC of the European Parliament and of the Council of 23 October 2000*; European Commission: Brussels, Belgium, 2000.
17. Liu, M. *Migration and Transformation of Petroleum Hydrocarbons in the Envelope Zone with Numerical Simulation*; Jilin University: Jilin, China, 2014.
18. Guo, H.M.; Wang, Y.X.; Li, Y.M. Analysis of arsenic enrichment factors in groundwater in the arsenic poisoning area of Shan Yin water. *Environ. Sci.* **2003**, *4*, 60–67.
19. Sorensen, J.; Butcher, A.S.; Stuart, M.E.; Townsend, B.R. Nitrate fluctuations at the water table: Implications for recharge processes and solute transport in the Chalk aquifer. *Hydrol. Process.* **2015**, *29*, 3355–3367.
20. Hefting, M.J.C.; Clément, D.D.; Cosandey, A.C.; Bernal, S.; Cimpian, C.; Tatur, A.; Burt, T.P.; Pinay, G. Water table elevation controls on soil nitrogen cycling in riparian wetlands along a European climatic gradient. *Biogeochemistry* **2004**, *67*, 113–134.
21. Welch, H.L.; Green, C.; Coupe, R.H. The fate and transport of nitrate in shallow groundwater in northwestern Mississippi, USA. *Appl. Hydrogeol.* **2011**, *19*, 1239–1252. [CrossRef]
22. Jurado, A.; Borges, A.V.; Brouyère, S. Dynamics and emissions of N2O in groundwater: A review. *Sci. Total Environ.* **2017**, *584–585*, 207–218.
23. Yang, L.-P.; Li, N.; Zhang, J. Effect of pH on the transport and transformation of "tri-nitrogen" in shallow groundwater. *Chin. Agron. Bull.* **2017**, *33*, 56–60.
24. Zhang, Z.; Furman, A. Redox dynamics at a dynamic capillary fringe for nitrogen cycling in a sandy column. *J. Hydrol.* **2021**, *603*, 26899. [CrossRef]
25. Ashworth, D.J.; Shaw, G. Effects of moisture content and redox potential on in situ Kd values for radioiodine in soil. *Sci. Total Environ.* **2006**, *359*, 244–254. [CrossRef]
26. Li, X.; Yang, T.; Bai, S.; Xi, B.; Zhu, X.; Yuan, Z.; Wei, Y.; Li, W. Research on the Influence of Groundwater Level Fluctuation on the Transport of Nitrogen in Vadose Zone. *Agric. Environ. Sci. J.* **2013**, *32*, 2443–2450.
27. Zhang, D.; Fan, M.; Liu, H.; Wang, R.; Zhao, J.; Yang, Y.; Cui, R.; Chen, A. Effects of shallow groundwater table fluctuations on nitrogen in the groundwater and soil profile in the nearshore vegetable fields of Erhai Lake, southwest China. *J. Soils Sediments* **2019**, *20*, 42–51. [CrossRef]
28. Xin, L.; Rui, Z.; Li, M.; Peng, L.; Zhiping, L.; Zhukun, H.; Qiao, L.; Jinsheng, W. Research on Nitrate Nitrogen (Nitrate) Pollution Changes in the Process of Groundwater Level Rising. *China Environ. Sci.* **2021**, *41*, 232–238.

29. Jia, W.; Yin, L.; Zhang, M.; Zhang, J.; Zhang, X.; Gu, X.; Dong, J. Modified method for the estimation of groundwater evapotranspiration under very shallow water table con-ditions based on diurnal water table fluctuations. *J. Hydrol.* **2021**, *597*, 126193. [CrossRef]
30. Kamon, M.; Endo, K.; Katsumi, T. Measuring the k–S–p relations on DNAPLs migration. *Eng. Geol.* **2003**, *70*, 64–65. [CrossRef]
31. Nedrich, S.M.; Burton, G.A. Indirect effects of climate change on zinc cycling in sediments: The role of changing water levels. *Environ. Toxicol. Chem.* **2017**, *36*, 2456–2464. [CrossRef]
32. Davis, G.B.; Rayner, J.L.; Trefry, M.G.; Fisher, S.J.; Patterson, B.M. Measurement and Modeling of Temporal Variations in Hydrocarbon Vapor Behavior in a Layered Soil Profile. *Vadose Zone J.* **2005**, *4*, 265–273. [CrossRef]
33. Kamon, M.; Li, Y.; Endo, K.; Inui, T.; Katsumi, T. Experimental Study on the Measurement of S-p Relations of LNAPL in a Porous Medium. *Soils Found.* **2007**, *47*, 33–45. [CrossRef]
34. Xie, X.; Li, Y.; Xia, B.; Su, Y.; Gu, Q. Study on the hysteresis of saturation-capillary pressure relationship in sand media under fluctuating water level. *Acta Pedol. Sin.* **2011**, *48*, 286–294.
35. Li, X.; Li, J.; Xi, B.; Yuan, Z.; Zhu, X.; Zhang, X. Effects of groundwater level variations on the nitrate content of groundwater: A case study in Luoyang area, China. *Environ. Earth Sci.* **2015**, *74*, 3969–3983. [CrossRef]
36. Chen, A.; Lei, B.; Hu, W.; Wang, H.; Dan, Z. Temporal-spatial variations and influencing factors of nitrogen in the shallow groundwater of the nearshore vegetable field of Erhai Lake, China. *Environ. Ence Pollut. Res.* **2017**, *25*, 4858–4870. [CrossRef] [PubMed]
37. Farnsworth, C.E.; Voegelin, A.; Hering, J.G. Manganese oxidation induced by water table fluctuations in a sand column. *Environ. Sci. Technol.* **2012**, *46*, 277–284. [CrossRef] [PubMed]
38. Yang, Y. Research on Soil Pollutant Transport Model Considering Groundwater Level Fluctuations. Master's Thesis, Hebei Agricultural University, Baoding, China, 3 June 2015.
39. Zhang, X.; Wang, P.; Wang, T.; Yu, J.; Liu, X. Hydrochemical characteristics of shallow groundwater in Ejina Oasis and the relationship between water level and depth under water transport conditions. *South-to-North Water Divers. Water Conserv. Sci. Technol.* **2019**, *17*, 86–94.
40. Cao, W.; Yang, H.; Gao, Y.; Nan, T.; Wang, Z.; Xu, S. Prediction of groundwater quality evolution in Baoding Plain in the wa-ter-receiving area of the Middle Route of the South-to-North Water Transfer Project. *J. Hydraul. Eng.* **2020**, *51*, 924–935.
41. Tafteh, A.; Sepaskhah, A.R. Application of HYDRUS-1D model for simulating water and nitrate leaching from continuous and alternate furrow irrigated rapeseed and maize fields. *Agric. Water Manag.* **2012**, *113*, 19–29. [CrossRef]
42. Mo, X.; Peng, H.; Xin, J.; Wang, S. Analysis of urea nitrogen leaching under high-intensity rainfall using HYDRUS-1D. *J. Environ. Manag.* **2022**, *312*, 114900. [CrossRef]
43. Chang, C.; Entz, T. Nitrate Leaching Losses under Repeated Cattle Feedlot Manure Applications in Southern Alberta. *J. Environ. Qual.* **1996**, *25*, 145–153. [CrossRef]
44. Steffy, D.A.; Johnston, C.D.; Barry, D.A. Numerical Simulations and Long-Column Tests of LNAPL Displacement and Trapping by a Fluctuating Water Table. *J. Soil Contam.* **1998**, *7*, 325–356. [CrossRef]
45. Marshall, H.W.H. Ecology of Water-Level Manipulations on a Northern Marsh. *Ecology* **1963**, *44*, 331–343.
46. Wang, Y. Ammonia Nitrogen Adsorption-Desorption Characteristics of Sediments of Different Sizes in the Xiliao River. Master's Thesis, Liaoning Technical University, Liaoning, China, 15 September 2012.
47. Sasikala, S.; Tanaka, N.; Wah, H.W.; Jinadasa, K. Effects of water level fluctuation on radial oxygen loss, root porosity, and nitrogen removal in subsurface vertical flow wetland mesocosms. *Ecol. Eng.* **2009**, *35*, 410–417. [CrossRef]

Article

The Effects of the Long-Term Freeze–Thaw Cycles on the Forms of Heavy Metals in Solidified/Stabilized Lead–Zinc–Cadmium Composite Heavy Metals Contaminated Soil

Zhongping Yang [1,2,3,*], Jiazhuo Chang [1,2,3], Xuyong Li [1,2,3], Keshan Zhang [1,2,3] and Yao Wang [1,4]

1 School of Civil Engineering, Chongqing University, Chongqing 400045, China; changjiazhuo@cqu.equ.cn (J.C.); lixuyong@cqu.edu.cn (X.L.); zhangkeshan1999@163.com (K.Z.); a_yao0218@163.com (Y.W.)
2 Key Laboratory of New Technology for Construction of Cities in Mountain Area, Chongqing University, Ministry of Education, Chongqing 400045, China
3 National Joint Engineering Research Center for Prevention and Control of Environmental Geological Hazards in the TGR Area, Chongqing University, Chongqing 400045, China
4 China Machinery China Union Engineering Co., Ltd., Chongqing 400045, China
* Correspondence: yang-zhp@163.com; Tel./Fax: +86-(023)-65120728

Abstract: Heavy metals (HMs) exist in nature in different forms, and the more unstable the form of an HM, the higher its toxicity and bioavailability. The content of HMs in stable fractions can increase significantly through the stabilization/solidification (S/S) technology. Still, external environments such as freeze–thaw (F–T) cycles will affect the stability of HMs directly. Therefore, a long-term F–T study of S/S Pb–Zn–Cd composite HM-contaminated soil was conducted under six conditions (0, 3, 7, 14, 30, and 90 cycles) with each F–T cycle process up to 24 h. The improved Tessier method was employed, and the results show that the S/S technology makes HMs transform to a more stable fraction. Still, the transformation efficiency is different for each HM. More than 98% of lead and zinc were converted to stable forms, while for cadmium, there are only 75.1%. Meanwhile, the S/S HMs were rapidly transformed into unstable forms at 0–14 cycles, but after 14 cycles, the transformation speed was significantly reduced. Among stable forms, it is mainly that the carbonate-bound fraction of HMs changes to unstable forms, and the characteristic peaks of carbonate stretching vibration were found at 874 cm^{-1}, and 1420 cm^{-1} by Fourier infrared spectroscopy proves the presence of carbonate-bound substances. As a result of this study, the change trend of contaminated soil with S/S HMs under the effect of long-term F–T cycle was revealed, and the crisis point of pollution prevention and control was found, which provides some theoretical basis for the safety of soil remediation project.

Keywords: long-term freeze–thaw cycles; composite heavy metal contamination; morphological analysis; solidification/stabilization

Citation: Yang, Z.; Chang, J.; Li, X.; Zhang, K.; Wang, Y. The Effects of the Long-Term Freeze–Thaw Cycles on the Forms of Heavy Metals in Solidified/Stabilized Lead–Zinc–Cadmium Composite Heavy Metals Contaminated Soil. *Appl. Sci.* **2022**, *12*, 2934. https://doi.org/10.3390/app12062934

Academic Editors: Bing Bai and Dibyendu Sarkar

Received: 6 February 2022
Accepted: 10 March 2022
Published: 13 March 2022

1. Introduction

Because of considerable migration depth, complex migration mechanism, sizeable spatial variability, strong concealment, and resistance to degradation of heavy metals (HMs), its high-concentration and high-risk pollutants in the soil are extremely difficult to repair and manage and are called "chemical time bombs" [1–3]. In recent years, the remediation of soil HM pollution has been an environmental topic worthy of attention. Studies have shown that HM pollutants may cause changes in soil structure, such as colloidal corrosion between soil particles, decrease in cementation strength, increase in clay content, resulting in increased soil compression, reduced liquid, and plastic limits, reduced shear strength (the reduction can reach about 35%), increased permeability, and the decreased bearing capacity [4–6]. In addition, the non-degradability of HMs leads to their accumulation in plants, animals, and humans along the biological chain [7–11]. Both pose a threat to human safety and health; therefore, it is imperative to control soil

HM pollution [12]. There are two kinds of solutions to manage HM pollution in the soil. The first one is to directly reduce the concentration of HMs in the soil, and the second one is to reduce the possibility of HM migration. That means fixing them in place and reducing their biotoxicity by special treatments [13,14]. In this context, a series of treatment methods for HM-contaminated soils have emerged, such as S/S technology, displacement technology, soil leaching, incineration, solvent extraction, glass curing, and physical separation, emerged. Among them, solidification/stabilization (S/S) technology is the most widely used because of its cost-effective and simple operation [14,15].

The S/S technology of soil usually provides the ability to handle HMs through two aspects. On the one hand, the S/S technology has a direct effect on HMs. Chemical precipitation, physical encapsulation, adsorption, and other means occur between HMs and hydrations products of binders; on the other hand, the S/S technology directly affects soil, and indirectly affects HMs by changing the soil structure. The reduced permeability leads to reduced migration of HMs, then the stability of HMs in the soil can be achieved [15,16]. However, while enjoying the restorative power of S/S technology, one must also consider its environmental impact.

Considering this problem, three inorganic binders, cement, lime, and fly ash, were used in the current study to improve S/S efficiency while reducing carbon emissions and easing the burden of industrial waste disposal [17–19]. However, it should be noted that the use of binders should involve reasonable control of the amount: too little S/S has poor effect, but if the amount is too much it will not only increase the cost, but also change the physical and chemical properties of the soil, and even cause secondary pollution.

Starting from the mechanism of the S/S technology, it can be realized that this technology cannot reduce the concentration of HMs in the soil, but restricts the migration of pollution, or limits the effectiveness of pollutants. Based on this, it is necessary to track the follow-up progress of HMs after the use of S/S technology. In the internal system of S/S HM-contaminated soil, the type of soil [20], the type and content of HMs [21], the type and content of binders [22,23], the curing conditions [23], and the moisture content [24] all have an impact on the long-term stability of HMs. In various complex environments, such as carbonation [25], acid rain leaching [25–27], high salt groundwater infiltration [28,29], wetting and drying cycles [30–32], and F–T cycles [33–36], the stable forms of the contaminants in the S/S-contaminated soil is very likely to change or even fail, which causes the activation of HM pollutants in the soil, causing secondary pollution and affecting its engineering mechanical properties. Therefore, the environmental durability of S/S-contaminated soil is extremely sensitive to environmental changes. The impact of F–T cycles on S/S HM-contaminated soil will continue to be focused on in-depth in this study.

Freeze–thaw cycles mainly occur in seasonal frozen regions. The moisture in the soil constantly transforms between liquid and solid with the periodic change of temperature: through changes in the volume of water in different fractions, as well as the free migration of water, the pore structure and density of the soil are changed, which is one of the direct factors affecting the migration effect of HMs. In China, the area of seasonal permafrost is about 5.14 million square kilometers, accounting for 53 percent of the total land area [37]. According to the second national soil erosion remote sensing survey, China's F–T erosion area has reached 13.4% of the country's total area [38]. Because of the vast F–T area, it is of great practical significance to explore the influence of the F–T cycle on the S/S of HM-contaminated soil. After 12 F–T cycles, the leaching concentrations of Pb/Zn-contaminated soil and composite Zn–Pb-contaminated soil increased greatly [34]. The elastic modulus of silty clay and modified clay can be significantly reduced after four F–T cycles [39]. After conducting seven F–T cycles on silty clay, Wang et al. concluded that the number of F–T cycles had a great influence on the mechanical properties of soil [40]. Studies have confirmed that under the long-term effect of simple F–T cycles (180 cycles, 180 days), the unconfined compressive strength loss rate of cemented contaminated soil can reach more than 30%, and the concentration of HM ions in HM leaching solution can increase by more than 20% [18,41,42].

This research has been improved on this basis, and "long-term" is the keyword of this. Studying the condition of HMs in the soil during the long F–T cycle period, observing the S/S effect of the HMs is a good way to ensure the long-term use of the S/S soil. Effective control of HMs in soil under adverse conditions is of great significance to prevent secondary pollution of HMs. A maximum of 90 F–T cycles (90 days) was set up under the same F–T mode to explore the stability of HMs under long-term F–T cycles in this research.

After making up for the problem of short-term F–T, the situation of HMs in the soil has also been improved, and the transformation from single HM pollution to compound HM pollution research has been realized. According to a large number of studies, one source of pollution often produces multiple types of HM pollution [41–43]. As, Cd, Cr, Cu, Mn, Ni, Pb, Zn, and other HMs exist in the farmland soil of the Qixia mining area in Nanjing, China [44]. There are many HMs in the weathered coal gangue, and the surrounding land is also harmed by HMs such as Zn, Cd, and Cr. Research on a single HM-contaminated soil can no longer meet the actual needs. Therefore, in this study, three pollutants, Pb, Zn, and Cd, were added to the soil at the same time.

The adsorption–desorption, precipitation reaction, complexation reaction, and oxidation–reduction reaction between HMs and soil allow HMs to exist in different forms in soil, and different forms of HMs exhibit different toxicity and environmental behavior [45]. When HMs are in a less bioavailable form or a more stable form, their risk in the environment is low. Therefore, the content of each form of HMs can be used as an observation indicator to assess the risk of HMs in soil. The improved Tessier method was used to extract seven fractions of HMs in the soil in this study.

Based on the above analysis, the improved Tessier method was adopted to detect the content of various forms of HMs (Pb, Zn, and Cd) in the soil that had undergone six F–T cycle conditions (0, 3, 7, 14, 30, and 90 cycles) up to 90 days, and the functional groups were analyzed by Fourier transform infrared spectroscopy analysis tests. Meanwhile, the study assumes that the HMs are uniformly distributed in the soil, by controlling the particles of the original soil less than 1 mm, the mixing time of the preparation of contaminated soil longer than 5 min, and the error of X-ray fluorescence analyses (XRF) of the same batch of three concentrations less than 3%.

In this study, the safety of the S/S technology is investigated by using a near-actual high concentration of complex hinged HM-contaminated soil, the morphological changes of HMs as a variable of interest, and the F–T cycle effects that may occur in most of the global land as the influencing factor. The conclusions and laws drawn from this study provide more rigorous theoretical support for predicting the safety of seasonally S/S HM-contaminated soil use and provide favorable conditions for post-remediation reuse of contaminated soil [18].

2. Materials and Methods

2.1. Material

2.1.1. Soil

The soil was collected from a site in Gaomiao Village, Chongqing, with dense soil, small porosity, brownish-red soil color, and large water content, but the soil was still in a plastic state. After bringing the soils back to the laboratory, they were placed into an electric constant temperature air blast drying box to dry and left at 100 °C for 24 h before all steps. Then, they were crushed and passed through a 1 mm nylon sieve after removing stones, tree roots, and other impurities.

Table 1 shows the basic physical properties of soil taken from the limit moisture content test, and the compaction test, which were carried out according to the Standards for Geotechnical Test Methods (GB/T 50123-2019). The soil was sieved to obtain a coarse-grained group containing more than 25% and not more than 50% and satisfying $w_L < 50\%$, $I_P \geq 7$, and $I_P \geq 0.73(w_L - 20)$. According to the engineering classification standard of soil (GB/T 50145-2007), the soil is low-liquid-limit clay. The sealed soil was assessed using the XRF to identify the chemical composition and relative content of each composition. The results are shown in Figure 1.

Table 1. Basic physical properties of undisturbed soil samples.

Physical Property Index	Soil Sample
Liquid Limit/%	27
Plastic Limit/%	12
Plasticity Index (I_P)	15
Optimum Moisture Content/%	11.8
Maximum dry density/g/cm^3	1.91

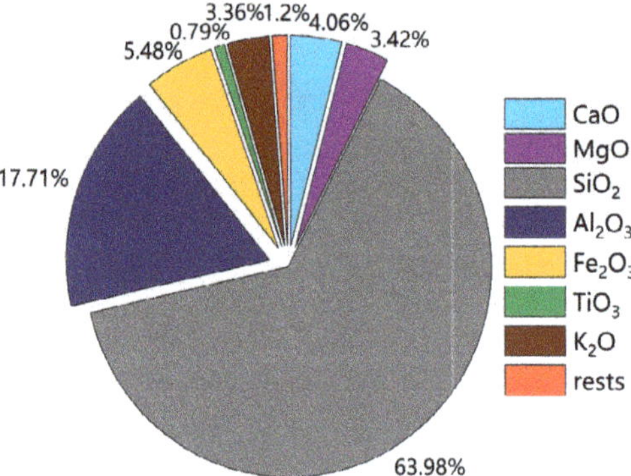

Figure 1. Main chemical components of undisturbed soil.

2.1.2. Binders

Cement, lime, and fly ash were used simultaneously in this study for the following three reasons. (1) Cement costs a lot of energy and releases a lot of carbon according to previous studies [45,46]. Therefore, the use of cement should be reduced; (2) the active substances in fly ash and lime react with the hydration products of cement to further improve properties of the soil and enhance the S/S effects of the HMs; (3) fly ash, as a kind of industrial byproduct, used in this way can reduce the pressure of waste treatment.

Figure 2 shows the results of the X-ray fluorescence analysis of cement, lime, and fly ash, containing the chemical composition and content of each component of three binders. Analytical-grade calcium oxide is the main component of quicklime, which was tested as high as 88.73 percent; the model of the ordinary Portland cement was 325; the fly ash was taken from the Chongqing power plant, and the grade of it was determined as two, according to GB/T 1594-2017. To ensure the uniform mixing of binders and soil, the binders needed to pass a 200-mesh sieve before use.

2.1.3. Heavy Metals

To minimize the influence of anions of the added HM compounds on the test results, analytical-grade $Zn(NO_3)_2 \cdot 6H_2O$, analytical-grade $Pb(NO_3)_2$, and high-purity $Cd(NO_3)_2 \cdot 4H_2O$ were used simultaneously as HM contaminants to provide the required HM ions. Reactions are difficult to take place between binders, hydration products, and nitrate, while the concentration of nitrate was not high in this research. The side effects of $Zn(NO_3)_2 \cdot 6H_2O$, $Pb(NO_3)_2$, and $Cd(NO_3)_2 \cdot 4H_2O$ were, thus, negligible [47,48].

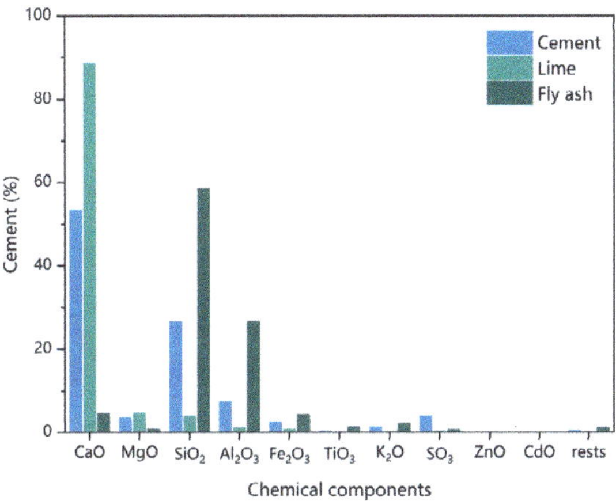

Figure 2. The chemical components of three binders.

2.2. Specimen Preparation

2.2.1. Preparation of Artificial Composite Contaminated Soil

In order to improve the observability of ion changes with F–T cycles in this study, as well as to make the test soils more closely resemble the form of contamination in the project, the tertiary standard limits of Pb, Zn, and Cd in the Soil Environmental Quality Standard (GB36000-2018) were enlarged by 16, 10, and 400 times, respectively, to obtain test yields of 8000 mg/kg, 5000 mg/kg, and 400 mg/kg for Pb, Zn and Cd, respectively.

At the same time, the soil moisture content was set at 120% of the optimal moisture content to ensure adequate F–T effects, and then, according to the soil water content and the HM mixture, a certain amount of deionized water, analytical-grade $Pb(NO_3)_2$, analytical-grade $Zn(NO_3)_2{\cdot}6H_2O$, and high-purity $Cd(NO_3)_2{\cdot}4H_2O$ was weighed and placed in a magnetic mixer for ten minutes to mix the solution well.

After that, the mixed solution was added to the sieved soil, stirred with a magnetic mixer for 10 min, then wrapped in a plastic bag and placed in a humidor with a temperature of 22 °C and a relative humidity of 95% for 30 days to make an artificial composite HMs-contaminated soil.

2.2.2. Stabilization/Solidification of Contaminated Soil

According to the previous results of our research group [41], the better designed mixing ratio of cement, lime, and fly ash on S/S of HMs-contaminated soil is 5%, 2.5%, and 2.5%, respectively, so this ratio was adopted in our current study. After weighing a certain amount of binder according to the binders' set ratio, the soil-added binders were stirred by a mechanical mixer for 10 min, then sealed by a plastic membrane and kept for 56 days in a standard curing chamber (22 °C, 95% relative humidity) for use.

2.3. Testing Methods

2.3.1. Freeze–Thaw Test

The alternately high- and low-temperature test chamber (Chongqing Tester Experimental Instrument Co., Ltd. Chongqing, China) is the instrument for F–T tests. Five kinds of F–T cycles were designed for the specimens, including 3, 7, 14, 30, and 90 F–T cycles (0 F–T cycle was set as the control group). The setting mode of one cycle is shown in Table 2 (24 h for each cycle), in which the freezing temperature of the soil was set at −10 °C, and the dissolution temperature was 20 °C. To ensure the complete freezing and dissolution of the soil, it was maintained for 11 h after reaching the specified temperature. The F–T cycled specimens were ground and pounded through a 200-mesh sieve to meet the experimental.

Table 2. The form of the F–T cycles.

Temperature Form	From 20° to − 10°	Remaining −10°	From − 10° to 20°	Remaining 20°
Time/h	0–1	2–12	12–13	14–24

Note: The first time the soil freezes, it starts to drop from room temperature.

2.3.2. The Improved Tessier Method

According to the Geological Survey Standards of China Geological Bureau (DD2005-03), the improved Tessier method was adopted to determine the distribution of seven fractions of three HMs of S/S Pb–Zn–Cd HM-contaminated soil under six conditions. This method is to further divide the organically bound fractions in the Tessier sequential extraction method proposed by Tessier into strongly organically bound fractions and humic-acid-bound fractions, and additionally add the water-soluble fraction [49,50]. The seven forms in the method are water-soluble fraction, ion exchange fraction, carbonate-bound fraction, iron and manganese oxidation fraction, weak organic fraction, strong organic fraction, and residue fraction.

The main reagents used are as follows, according to Technical Standards for Geological Survey of China Geological Survey (DD2005-03).

(a) Hydrochloric acid ($\rho(HCl) = 1.18$ g/mL);
(b) Hydrochloric acid in which concentrated hydrochloric acid and water are mixed in a volume ratio of 1:1;
(c) Nitric acid ($\rho(HNO_3) = 1.41$ g/mL);
(d) Nitric acid ($c(HNO_3) = 0.02$ mol/L);
(e) Perchlorate ($\rho(HClO_4) = 1.66$ g/mL);
(f) Hydrofluoric acid ($\rho(HF) = 1.15$ g/mL);
(g) Magnesium chloride ($c(MgCl_2{\cdot}6H_2O) = 1.0$ mol/L, pH $= 7.0 \pm 0.2$)

Five hundred and eight grams $MgCl_2 \cdot 6H_2O$ was weighed and dissolved in distilled water in a 2500 mL plastic bucket. The pH was adjusted by NaOH ($\omega(NaOH) = 10\%$).

(h) Sodium acetate ($c(CH_3COONa{\cdot}3H_2O) = 1.0$ mol/L, pH $= 5 \pm 0.2$)

Three hundred and forty grams $CH_3COONa \cdot 3H_2O$ was weighed and dissolved in distilled water in a 2500 mL plastic bucket. The pH was adjusted by CH_3COOH.

(i) Sodium pyrophosphate ($c(Na_4PO_7{\cdot}10H_2O) = 0.1$ mol/L, pH $= 10.0 \pm 0.2$)

One hundred and eleven and a half grams Na4$PO_7 \cdot 10H_2O$ was weighed and dissolved in distilled water in a 2500 mL plastic bucket, and the pH was adjusted to 10.0 ± 0.2 with HNO_3 where concentrated nitric acid and water are mixed in a volume ratio of 1:1.

(j) Ammonium hydroxide hydrochloride–hydrochloric acid mixture ($c(HONH_3Cl) = 0.25$ mol/L, $c(HCl) = 0.25$ mol/L)

Forty-three point four grams $HONH_3Cl$ was weighed and 104 mL hydrochloric acid was added, where concentrated hydrochloric acid and water are mixed in a volume ratio of 1:1 and dissolved in distilled water in a 2500 mL plastic bucket.

(k) Hydrogen peroxide ($\varphi(H_2O_2) = 30\%$, pH $= 2.0 \pm 0.2$)

The pH was adjusted by nitric acid where concentrated nitric acid and water are mixed in a volume ratio of 1:1.

(l) Ammonium acetate–nitric acid mixture ($c(CH_3COONH_4) = 3.2$ mol/L, $c(HNO_3) = 4.48$ mol/L)

Six hundred and sixteen point six grams CH_3COONH_4 was weighed and added 500 mL HNO_3 (c). Both were dissolved in a 2500 mL plastic bucket with distilled water.

Using the above reagent, the concentration of HMs in soil under seven forms could be accurately determined according to the following nine steps.

(1) A sample of 2.5000 g was accurately weighed and placed in a centrifuge tube.
(2) Extraction of water-soluble fraction.

Distilled water was used as the extraction agent for water-soluble HMs. First, 25 mL of distilled water was added to the beaker and shaken well, then the beaker was placed into the ultrasonic cleaner. Ultrasound was performed at the working mode of 5 min with 5 min intervals at a frequency of 40 kHz, and the total working time was 30 min, during which the water temperature in the ultrasonic cleaner was 25 ± 5 °C. Then, the material in the beaker was transferred into a centrifuge tube for centrifugation (20 min at a speed of 4000 r/min), the precipitated supernatant was filtered with a pore diameter of 0.45 μm, and then the liquid was poured into a 25 mL colorimetric tube. An Optima 8000 inductively coupled plasma emission spectrometer was used to detect the contents of lead, zinc, and cadmium in the extracted liquid.

(3) Residue cleaning

The remaining residue was added to about 100 mL distilled water to wash the precipitation, and centrifuged at 4000 r/min for 10 min. The aqueous phase was discarded, and the residue was left.

(4) Extraction of ion-exchange fraction

Twenty-five milliliters of magnesium chloride (g) solution were accurately added to the residue left in step (3), shaken well, and placed in an ultrasonic cleaner that had been placed into water. Ultrasound was performed at a frequency of 40 kHz for 30 min (ultrasonic was performed for 5 min every 5 min during this period, and the water temperature in the ultrasonic cleaner was controlled at 25 ± 5 °C). After the ultrasound, they were taken out, and centrifuged at 4000 r/min for 20 min. A total of 5 mL of liquid was taken into a 10 mL colorimetric tube and added to 0.5 mL of hydrochloric acid (a). After that, the distilled water was used for constant volume to scale. After shaking the colorimetric tube well, the ICP-OES theory was used to determine the content of HMs in liquid. The step of residue cleaning was repeated at last.

(5) Extraction of carbonate-bound fraction

Sodium acetate (h) was selected as the extraction agent of carbonate-bound fraction. A total of 25 mL of it was added to the residue which came from step (4), shaken well, and placed in an ultrasonic cleaner with a frequency of 40 kHz for 1 h (ultrasonic was performed for 5 min every 5 min during the period, and the water temperature in the ultrasonic cleaner was controlled at 25 ± 5 °C). Then they were taken out and centrifuged at 4000 r/min for 20 min. Separated 5 mL liquid was poured into a 10 mL colorimetric tube. After adding 0.5 mL HCl (a) to the colorimetric tube, the distilled water was brought to scale, and shaken well. An Optima 8000 inductively coupled plasma emission spectrometer was used to analyze the contents of lead, zinc, and cadmium, and step (3) was repeated finally.

(6) Extraction of humic-acid-bound fraction

Fifty milliliters sodium pyrophosphate solution (i) was accurately added to the residue obtained in step (5). After shaking well, it was taken to the ultrasonic cleaner with a frequency of 40 kHz for 40 min (ultrasonic 5 min every 5 min during this period, and the water temperature in an ultrasonic cleaner is controlled at 25 ± 5 °C) and placed for 2 h. Then, the solution was centrifuged at 4000 r/min for 20 min. A 50 mL beaker was filled with 10 mL clear liquid and 5 mL HNO_3 (c) and 1.5 mL $HClO_4$ were added (e). The surface dish was used to cover the beaker, and the beaker was heated and steamed on the electric heating plate until the occurrence of $HClO_4$ white smoke exhaust. Then, 1 mL HCl (b) was added, the surface dish was washed, and the dissolved salt was heated, removed, and cooled. Finally, a 10 mL colorimetric tube was scaled by distilled water. After shaking well, an Optima 8000 inductively coupled plasma emission spectrometer was used to analyze the contents of lead, zinc, and cadmium. The remaining residue was treated as described in step (3).

(7) Extraction of iron and manganese oxidized bond fraction

Fifty milliliters hydroxylamine hydrochloride solution (j) was added to residue obtained from step (6) accurately. After they were shaken well, the ultrasound was carried out in ultrasonic cleaner with a frequency of 40 kHz for 1 h (ultrasonic for 5 min every 5 min during this period, water temperature in an ultrasonic cleaner is controlled at 25 $\pm$ 5 °C). Then, the solution was removed and centrifuged at 4000 r/min for 20 min. Finally, 10 mL clear liquid was separated into a colorimetric tube. The contents of lead, zinc, and cadmium were analyzed by the ICP-OES method. The remaining precipitation was transferred to a 50 mL centrifuge tube with 50 mL distilled water and centrifuged for 10 min at a speed of 4000 r/min. The aqueous phase was discarded, and the residue was left after cleaning twice.

(8) Extraction of strong organic bonding fraction

Three milliliters HNO_3 (d) and 5 mL H_2O_2 (k) were accurately added to the residue obtained from step (7). The solution was shaken well and kept for 1.5 h in a constant temperature water bath at 83 $\pm$ 3 °C (stirred once every 10 min during the period). Then, the solution was continued to have 3 mL H_2O_2 added (k) and kept warm in the water bath for 1 h and 10 min (stirring once every 10 min during this period). After cooling to room temperature, 2.5 mL ammonium acetate–nitric acid solution (a) was added, and the samples were diluted to about 25 mL and stirred for 1 min, then placed at 25 $\pm$ 5 °C overnight, and centrifuged at 4000 r/min for 20 min. The clear liquid was poured into a 50 mL colorimetric tube, diluted with distilled water to scale, and shaken well. Then, a 50 mL beaker was used. First, 25 mL separated clear liquid, 10 mL HNO_3 (c), and 1 mL $HClO_4$ (e) was poured in, then, with a surface dish, it was heated on the electric heating plate at low temperature until it was nearly dry, and at high temperature until the concentrated white smoke was cleared. After this, 5 mL HCl (b) was added while hot, and the dish was washed. The beaker was heated at a low temperature until the salts dissolved. After the solution cooled, 25 mL of water was placed into a colorimetric tube and shaken well. A 5 mL solution was separated into 10 mL colorimetric tubes for ICP-OES analysis for Pb, Zn, and Cd. About 20 mL of distilled water was added to wash the sediment, and centrifuged for 10 min at a speed of 4000 r/min. The aqueous phase was discarded. The cleaning process was carried out twice, leaving the residue at the end.

(9) Extraction of residue fraction

The cleaned residue obtained from step (8) was air-dried, ground, weighed, and the correction factor was calculated, and then the sample of 0.2000 g was accurately taken out and placed in the PTFE crucible. The sample was wetted in the crucible by water and 5 mL of hydrochloric acid (f), nitric acid, and perchloric acid mixture in the ratio of 1:1:1 was added. The crucible was heated on an electric heating plate until the white smoke of perchloric acid was fully distributed. Then, 3 mL of hydrochloric acid (b) was added to the crucible. After the crucible wall was washed, the crucible continued to heat until the salt was dissolved, and then cooled to room temperature. Finally, the remaining product was transferred to a colorimetric tube, and the volume was kept at 25 mL and shaken well. The contents of lead, zinc, and cadmium were analyzed by the ICP-OES method.

2.3.3. Fourier Transform Infrared Spectrum (FTIR) Analysis Test

The attenuated total reflection Fourier transform infrared spectroscopy method was adopted in this study. Six kinds of soil were dried at 100 °C for 24 h and tested by a Fourier infrared spectrometer after passing through a 200-mesh sieve. The mid-infrared region was used for detection in this study. The following procedures were used to analyze the spectrum [51].

The measured mid-infrared was tentatively divided into a characteristic functional group area (4000–1333 cm^{-1}) and a fingerprint area (1333–400 cm^{-1}), then the characteristic functional group was further divided into three wavebands (4000–2400 cm^{-1}, 2400–2000 cm^{-1}, and 900–400 cm^{-1}), and the fingerprint area was finely divided into two wavebands (1333–900 cm^{-1} and 900–400 cm^{-1}). Infrared spectroscopies were taken as a preliminary analysis (compound type, functional group, the structural unit, etc.) based on the corresponding characteristic absorption bands to determine the possible functional groups and possible structural units.

Based on possible functional groups, possible structural units, the characteristic frequency table of the characteristic absorption bands of various compounds, and various factors affecting the movement of characteristic frequencies, the structural details were revealed. The standard spectra of related compounds determined according to the details were compared with the obtained spectra of soils under six environments to obtain specific functional group composition.

3. Results

3.1. Effect of Solidification/Stabilization on the Speciation of Different Heavy Metals

Among the seven different HM fractions extracted by the improved Tessier method, the order of bioavailability for different fractions of HMs in the soil is shown as follows: water-soluble state > ion-exchange state > carbonate-bound fraction > humic-acid-bound fraction > iron and manganese oxidized bond fraction > strong organic bonding fraction > residue fraction. Pb^{2+}, Zn^{2+}, and Cd^{2+} provided by $Pb(NO_3)_2$, $Zn(NO_3)_2{\cdot}6H_2O$, and $Cd(NO_3)_2{\cdot}4H_2O$ are in the ion exchange fraction. HMs whose bioavailability is greater than this form can be considered to have undergone a transition to the stable forms, and are referred to as the stable form, while the ion-exchange fraction and the water-soluble fraction are set as the unstable form.

Figure 3 shows the relative content of HMs in the S/S soil under a non-freeze–thaw condition in the unstable fraction, and the stable fraction. It is obvious that the content of Pb ion, Zn ion, and Cd ion supplied in the soil decreased sharped after the S/S by adding cement, lime, and fly ash. After S/S, the unstable forms of Pb, Zn, and Cd remain as only 1.19%, 1.09%, and 24.89%, respectively

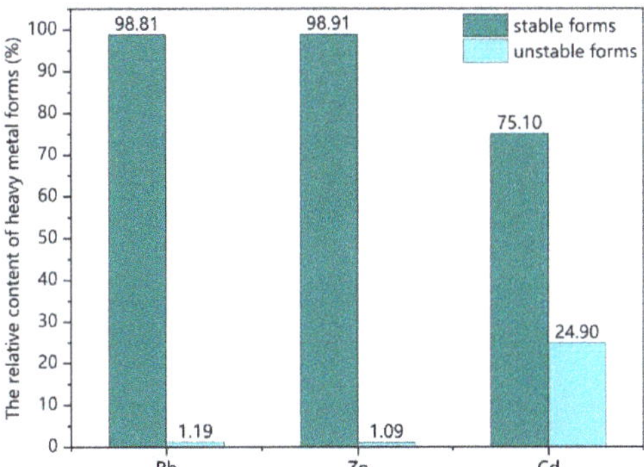

Figure 3. The relative content of heavy metals (HMs) in the solidified/stabilized (S/S) Pb–Zn–Cd composite HM-contaminated soil without F–T cycles.

However, Figure 3 also shows that the S/S efficiency varies for different HMs. The relative contents of the stable forms of Pb and Zn are much larger than that of the stable fractions of Cd. After the S/S process, 98.81% of lead and 98.91% of zinc were converted to a stable fraction, but only 75.10% of cadmium was converted to a stable fraction, with a relative content difference of more than 20%.

3.2. Effect of F–T Cycles on the Speciation of HMs

Figure 4 reveals the trend of the relative content of the two types of fractions of HMs in the S/S Pb–Zn–Cd composite HM-contaminated soil with the action of F–T cycles. The changing trend of the curve in it indicates that the F–T cycle causes HMs to transform from stable to unstable forms, but the reduction of stable HM becomes slow after reaching 14 cycles. After 90 F–T cycles, the relative contents of stable forms of Pb, Zn, and Cd in the soil decreased by 1.27%, 3.11%, and 8.16%, respectively, while the decrease reached more than 50% of the total decrease after 14 F–T cycles.

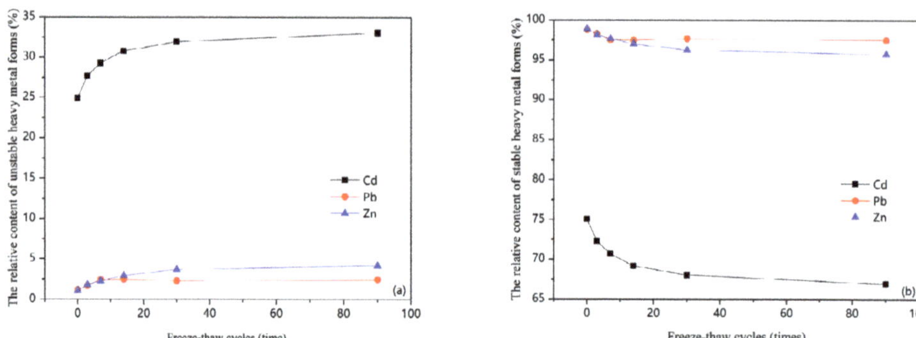

Figure 4. The variation of the relative content of HM forms with F–T cycles in the S/S Pb-Zn-Cd composite HM-contaminated soil. (**a**) unstable HM forms; (**b**) stable HM forms.

The variation of the relative content of stable fractions of three HMs in S/S soil with the number of F–T cycles is shown in Figure 5. It can be seen that the relative content of the carbonate-bound fraction of HMs decreases significantly with F–T cycles. The effect is significant within the 14 F–T cycles, the slope of the curve is large, and the rate of decline is fast. However, with the increase of the F–T cycles times, the rate of decline of carbonate-bound forms becomes relatively slow. After 14 F–T cycles, the relative contents of carbonate-bound fraction of Pb, Zn, and Cd decreased by 13.46%, 7.35%, and 8.68%, respectively, while the relative contents of corresponding types only decreased by 2.06% in the process from 14 F–T cycles to 90 F–T cycles, 4.58%, and 1.56%.

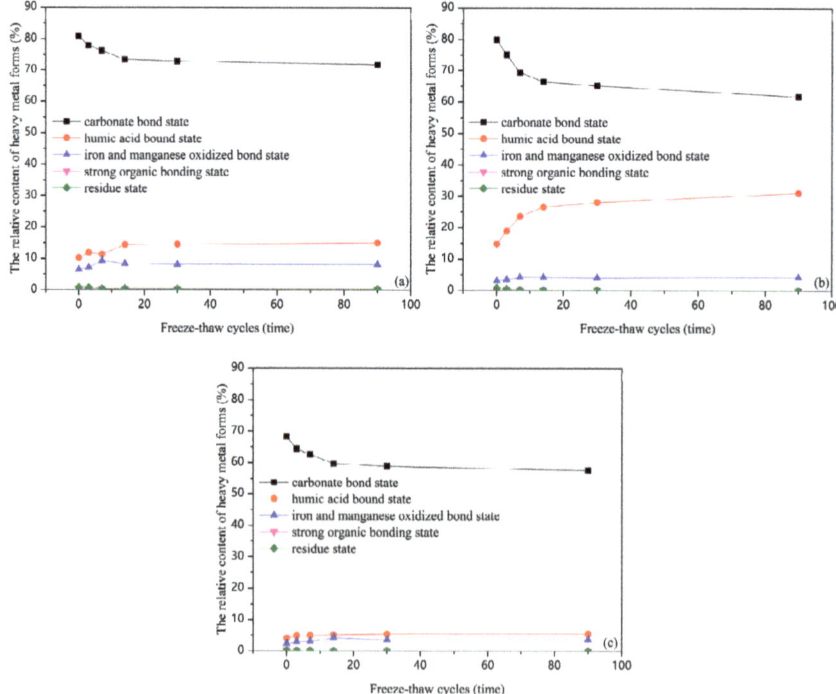

Figure 5. The variation of relative content of each speciation of HMs with F–T cycles in the S/S Pb–Zn–Cd composite HM-contaminated soil. (**a**) Zn; (**b**) Pb; (**c**) Cd.

3.3. Effect of F–T Cycles on the Functional Groups in S/S Soil

Figure 6 shows the FTIR spectra of S/S Pb–Zn–Cd composite HM-contaminated soil under six F–T cycle conditions (0, 3, 7, 14, 30, and 90 cycles). According to their own spectra, a large amount of literature data, and standard Fourier infrared spectrogram, the types of functional groups in each condition were distinguished.

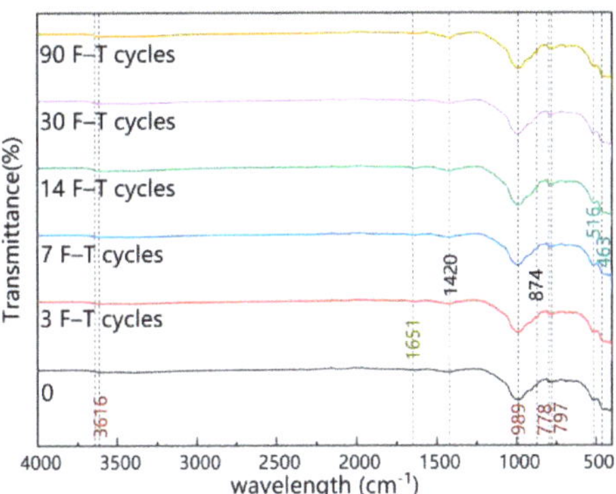

Figure 6. FTIR spectra of S/S Pb–Zn–Cd composite HMs-contaminated soil under 6 F–T cycle conditions (0, 3, 7, 14, 30, and 90 cycles).

The locations of the characteristic peaks of the FTIR spectra under the six F–T conditions are roughly the same, indicating that within 90 F–T cycles, the types of target substances (soil components, binder components, hydration products, etc.) detected in the S/S Pb–Zn–Cd composite HMs-contaminated soil did not change.

Seven hundred and ninety-seven per centimeter, and Seven hundred and eighty-seven per centimeter are caused by Si–O–Si stretching vibrations, which are typical double peaks of quartz minerals, which is an important part of the soil itself, and the added binders. The peaks at 516 cm^{-1} and 463 cm^{-1} are the Si–O vibration absorption peaks of the gel [52,53]. There are two main reasons for the existence of the characteristic peak of colloid. One is that the effective component of cement, tricalcium silicate, and dicalcium silicate, generates a large amount of hydrated calcium silicate colloid after hydration; the other is that the hydration product of cement, calcium hydroxide, continues to react with lime and fly ash to form colloids [54]; among them, there was residual calcium hydroxide, and the characteristic peak appears in 1651 cm^{-1}. The existence of colloids has made a great contribution to the S/S of HMs, such as the large amounts of adsorption of HM ions, and the filling of soil pores, which hinders the flow of HMs. The formation of calcium hydroxide also underwent a substitution reaction with HM ions, causing HM ions to precipitate.

Eight hundred and seventy-four per centimeter and one thousand four hundred and twenty per centimeter are caused by carbonate stretching vibration. These two obvious characteristic peaks appeared in the FTIR spectra under every F–T condition (0, 3, 7,14, 30, and 90 F–T cycles). This is mainly produced by the entry of carbon dioxide into the atmosphere and is affected by the pH of the soil. There is a huge relationship with the formation of HM carbonate-bound fraction.

4. Discussion

4.1. Effect of Solidification/Stabilization on the Speciation of Different Heavy Metals

The results in Section 3.1 show the transfer of HMs to stable forms after the addition of binders. After the hydration reaction of the binders, their hydration products will interact with HMs, increasing the inertness of HMs and reducing the migration ability of HMs.

The first mechanism for immobilization of HMs is a chemical reaction, in which lead, zinc, and cadmium react with $Ca(OH)_2$, calcium silicate hydrates, and so on, and are fixed to the binder components. Furthermore, the S/S technology also has a certain physical effect on the treatment process of HMs through the improvement of soil properties and the progress of the chemical reaction of the binders [55]. The decrease in soil permeability will hinder the migration of HMs, and the colloids produced by the curing agent will physically encapsulate the HMs [56]. Soil S/S technology, as a risk management technology, has a strong effect on reducing the risk of HMs. However, it can be clearly found that for different HMs in the soil, the conversion efficiency is different, with cadmium having the lowest conversion efficiency, with a difference of 20% compared to the other two HMs. The factors causing this phenomenon may have the following three aspects.

First, the properties of each HM are different [55]. Compared with cadmium ions, lead ions are more electronegative and have poor reaction activity [57]. Second, the S/S of HMs is closely related to the pH value of the soil, and different kinds of HM have different optimal pH values for S/S [58–60]. Most lead ions precipitate at a pH of about 9.8 and remain stable within a pH of 12 [61]. Zinc hydroxide begins to precipitate when pH is above 5 [62], and crystalline $Cd(OH)_2$ is precipitated in soil when pH is increased to above 6.2 [63]. Finally, the initial content of the three elements added to the undisturbed soil is different, the added cadmium is the least, and the progress of S/S will be less targeted.

4.2. Effect of F–T Cycles on the Speciation of HMs

The main reasons for this phenomenon are as follows. Through the continuous action of the F–T cycle, the porosity ratio in the soil keeps increasing, making it easier for carbon dioxide to enter the soil. Some studies have shown that the carbonization process of soil has adverse effects on the S/S of HMs. The pH value of the soil keeps decreasing with the carbonization, causing hydrolysis of the hydration products of many adhesives, and further activation of more HMs.

This is due to the fact that the carbonate-bound fraction of HMs is very sensitive to the change of pH. Under alkaline conditions, the carbonate-bound fraction is relatively stable. With the continuous decrease of soil pH under the action of the F–T cycle, the HM compounds in the carbonate-bound fraction decompose, while the rate of decomposition is associated with the change speed of pH because the degree of influence of carbon dioxide on soil Ph gradually decreases over time. During this time, the process of S/S of HM is still underway but does not play a dominant role.

As the F–T cycle progresses, the relative content of the humic-acid-bound fraction, iron and manganese oxidized bond fraction, strong organic bonding fraction, and residue fraction in the stable form increases continuously. However, due to their small content, they have little influence on the overall fraction of HMs, so the relative content of the stable form of HMs is still in a downward trend as the F–T cycle progresses. It may also be noted that these four HM forms are not sensitive to pH.

4.3. Effect of F–T Cycles on the Functional Groups in S/S Soil

According to Figure 6, a large number of Si–O–Si stretching vibrations, Si–O vibration absorption peaks, and carbonate stretching vibration can be found in the sample; this proves that substances that can enhance the stability of HMs, such as colloids and carbonate binders, are indeed present in the soil at this time [52,53]. This is why a large number of HMs have previously undergone a transition to the steady fractions.

Meanwhile, the types of functional groups contained in the soil are approximately the same in the six cases. The direct effect of the F–T cycle on the soil is mainly a physical reaction without the addition of new substances; therefore, the number of characteristic peaks did not increase. Meanwhile, the result of chemical speciation analysis shows that when the number of F–T cycles exceeded 14 times, the changing trend of the content of each HM forms was very slow, and gradually stabilized instead of a single decline. Therefore, there was no disappearance of the characteristic peak. Due to the above reasons, the F–T cycle did not change the types of functional groups in the system.

In addition, carbonate stretching vibration was detected in the soil. This is consistent with the fact that there are carbonate binders in the soil, and that when the pH environment in which the HMs are located changes, the HMs react with the carbonates to varying degrees. The F–T cycles, on the other hand, increase the amount of carbon dioxide entering the soil precisely by changing the porosity of the soil, causing changes in the carbonate-bound fraction of the soil.

Throughout the whole process of this study, it is clearly recognized that the change in the content of its stable fractions tends to slow down when the F–T cycles are performed several times. According to the experimental results, this node occurs at 30 cycles. When conducting an HM soil remediation project, the collected soil samples can be S/S in the laboratory and then subjected to a small trial test of 30 F–T cycles to test the morphological components, derive the instability limits, determine whether there is a safety hazard, and determine whether to keep the curing ratios or change the curing ratios.

5. Conclusions

The improved Tessier method and Fourier transform infrared spectrum (FTIR) analysis test were used to study the effects of long-term F–T on the S/S of Pb–Zn–Cd composite HM-contaminated soil in this study. The soil morphology and FTIR spectra were tested in six environments (without F–T cycles, with 3 cycles, 7 cycles, 14 cycles, 30 cycles, and 90 cycles) in HM-contaminated soil, and the following remarkable conclusions were obtained.

In the S/S Pb–Zn–Cd composite HM-contaminated soil, the addition of three binders, cement, lime, and fly ash, reduces the HM risk by making HMs exist in a more stable form. The characteristic peaks at 463 cm^{-1} and 516 cm^{-1} of the FTIR spectra indicate the presence of colloids that can adsorb and store HMs, and the characteristic peak of calcium hydroxide that can precipitate HMs appeared at 1651 cm^{-1}. However, for different types of HMs, the S/S efficiency is different. The relative content of HMs transferred to stable forms in lead and zinc is 20% greater than that in cadmium.

The content of HMs in a steady state decreased continuously with the increase of frequency of F–T cycles. There were drastic changes before 14 cycles, but with the continuous increase in the frequency of F–T cycles, the increasing trend gradually decreases, and the rate of decrease in the stable form of HMs became minimal after 30 F–T cycles. This is the main reason why F–T cycles would not change the types of functional groups.

Under the action of long-term F–T cycles, the decrease in the stable form content of HMs in the S/S composite HM-contaminated soil is mainly caused by the decrease in the carbonate-bound state content. Its peaks appeared at 874 cm^{-1} and 1420 cm^{-1}. Ninety F–T cycles caused the carbonate-bound content of Pb, Zn, and Cd to decrease by 18.04%, 8.92%, and 10.74%, respectively.

Through this experiment, it can be seen that after the number of F–T cycles reaches 30, the decrease rate of stable form HM content reaches the minimum; therefore, 30 F–T cycles can be used as the critical test point for indoor tests in actual projects to determine the danger zone. However, for different soils, exactly how the critical number changes is not yet known; in future research, it is recommended that the majority of scholars conduct further research on this part. At the same time, the huge change in the content of the carbonate-bound fraction of HMs during the F–T cycles must also recognize the tremendous influence of environmental pH on the S/S effect, and numerous studies have been conducted by domestic and foreign scholars on this content in the hope that a risk control method for the common assessment of various influencing factors can be established in the future.

Author Contributions: Conceptualization, Z.Y.; methodology, Z.Y. and Y.W.; investigation, Y.W., X.L., J.C. and K.Z.; writing—original draft preparation, J.C.; writing—review and editing, Z.Y.; supervision, Z.Y., X.L. and K.Z. All authors have read and agreed to the published version of the manuscript.

Funding: This research was funded by the National Natural Science Foundation of China, grant number 42177125 and the National Natural Science Foundation of China, grant number 41772306.

Institutional Review Board Statement: Not applicable.

Informed Consent Statement: Not applicable.

Data Availability Statement: Not applicable.

Acknowledgments: The authors acknowledge the financial support received from the National Natural Science Foundation of China (Grant No. 42177125, and No. 41772306). We also thank Ren Shupei and Wang Yao, graduate students, for their efforts in terms of conducting the laboratory tests.

Conflicts of Interest: The authors declare no conflict of interest. The funders had no role in the design of the study; in the collection, analyses, or interpretation of data; in the writing of the manuscript, or in the decision to publish the results.

References

1. Xue, Q.; Zhan, L.T.; Hu, L.M.; Du, Y.J. Environmental geotechnics: State-of-the-art of theory, testing and application to practice. *China Civil. Eng. J.* **2020**, *53*, 80–94.
2. Liu, S.Y. Geotechnical investigation and remediation for industrial contaminated sites. *Chin. J. Geotech. Eng.* **2018**, *40*, 1–37.
3. Kogbara, R.B. A review of the mechanical and leaching performance of stabilized/solidified contaminated soils. *Environ. Rev.* **2014**, *22*, 66–86. [CrossRef]
4. Cao, X.; Ma, R.; Zhang, Q.S.; Wang, W.B.; Liao, Q.X.; Sun, S.C.; Zhang, P.X.; Liu, X.L. The factors influencing sludge incineration residue (SIR)-based magnesium potassium phosphate cement and the solidification/stabilization characteristics and mechanisms of heavy metals. *Chemosphere* **2020**, *261*, 127789. [CrossRef] [PubMed]
5. Hao, H.Z.; Chen, T.B.; Jin, M.G.; Lei, M.; Liu, C.W.; Zu, W.P.; Huang, L.M. Recent advance in solidification /stabilization technology for the remediation of heavy metalscontaminated soil. *Chin. J. Appl. Ecol.* **2011**, *22*, 816–824.
6. Zhang, T.T.; Li, J.S.; Wang, P.; Huang, Q.; Xue, Q. Experimental study of mechanical and microstructure properties of magnesium phosphate cement treated lead contaminated soils. *Rock Soil Mech.* **2016**, *37*, 279–286.
7. Ngo, H.T.T.; Watchalayann, P.; Nguyen, D.B.; Doan, H.N.; Liang, L. Environmental health risk assessment of heavy metal exposure among children living in an informal e-waste processing village in Viet Nam. *Sci. Total Environ.* **2021**, *763*, 142982. [CrossRef]
8. Ustaoğlu, F.; Islam, M.S. Potential toxic elements in sediment of some rivers at Giresun, Northeast Turkey: A preliminary assessment for ecotoxicological status and health risk. *Ecol. Indic.* **2020**, *113*, 106237. [CrossRef]
9. Islam, M.S.; Ahmed, M.K.; Al-Mamun, M.H.; Eaton, D.W. Human and ecological risks of metals in soils under different land-use types in an urban environment of Bangladesh. *Pedosphere* **2020**, *30*, 201–213. [CrossRef]
10. He, L.Z.; Zhong, H.; Liu, G.X.; Dai, Z.M.; Brookes, P.C.; Xu, J.M. Remediation of heavy metal contaminated soils by biochar: Mechanisms, potential risks and applications in China. *Environ. Pollut.* **2019**, *252*, 846–855. [CrossRef]
11. Markus, J.; McBratney, A.B. A review of the contamination of soil with lead. *Environ. Int.* **2001**, *27*, 399–411. [CrossRef]
12. Buddhi, W.; Liu, A.; He, B.B.; Yang, B.; Zhao, X.; Godwin, A. Ashantha G Behaviour of metals in an urban river and the pollution of estuarine environment. *Water Res.* **2019**, *196*, 414911.
13. Shao, Y.Y.; Yan, T.; Wang, K.; Huang, S.M.; Yuan, W.Z.; Qin, F.G.F. Soil heavy metal lead pollution and its stabilization remediation technology. *Energy Rep.* **2020**, *6*, 122–127. [CrossRef]
14. Liu, S.J.; Miao, C.; Yao, S.S.; Ding, H.; Zhang, K. Soil stabilization/solidification (S/S) agent-water-soluble thiourea formaldehyde (WTF) resin: Mechanism and performance with cadmium. *Environ. Pollut.* **2021**, *272*, 116025. [CrossRef]
15. Gong, Y.Y.; Zhao, D.Y.; Wang, Q.L. An overview of field-scale studies on remediation of soil contaminated with heavy metals and metalloids: Technical progress over the last decade. *Water Res.* **2018**, *147*, 440–460. [CrossRef]
16. Qin, C.C.; Yuan, X.Z.; Xiong, T.; Tan, Y.Z.; Wang, H. Physicochemical properties, metal availability and bacterial community structure in heavy metal-polluted soil remediated by montmorillonite-based amendments. *Chemosphere* **2020**, *261*, 128010. [CrossRef]
17. Gao, P.; Yue, S.J.; Chen, H.T. Carbon emission efficiency of China's industry sectors: From the perspective of embodied carbon emissions. *J. Clean. Prod.* **2021**, *283*, 124655. [CrossRef]
18. Yang, Z.P.; Wang, Y.; Li, D.H.; Li, X.Y.; Liu, X.R. Influence of Freeze–Thaw Cycles and Binder Dosage on the Engineering Properties of Compound Solidified/Stabilized Lead-Contaminated Soils. *Int. J. Environ. Res. Public Health* **2020**, *17*, 1077. [CrossRef]
19. Wang, F.; Pan, H.; Xu, J. Evaluation of red mud based binder for the immobilization of copper, lead and zinc. *Environ. Pollut.* **2020**, *263*, 114416. [CrossRef]
20. Bai, B.; Rao, D.Y.; Chang, T.; Guo, Z.G. A nonlinear attachment-detachment model with adsorption hysteresis for suspension-colloidal transport in porous media. *J. Hydrol.* **2019**, *587*, 124080. [CrossRef]
21. Zha, F.S.; Zhu, F.H.; Xu, L.; Kang, B.; Yang, C.B.; Zhang, W.; Zhang, J.W.; Liu, Z.H. Laboratory study of strength, leaching, and electrical resistivity characteristics of heavy-metal contaminated soil. *Environ. Earth Sci.* **2021**, *80*, 184. [CrossRef]
22. Zha, F.S.; Qiao, B.R.; Kang, B.; Xu, L.; Chu, C.F.; Yang, C.B. Engineering properties of expansive soil stabilized by physically amended titanium gypsum. *Constr. Build. Mater.* **2021**, *303*, 124456. [CrossRef]
23. Wang, F.; Xu, J.; Yin, H.L.; Zhang, Y.H.; Pan, H.; Wang, L. Sustainable stabilization/solidification of the Pb, Zn, and Cd contaminated soil by red mud-derived binders. *Environ. Pollut.* **2021**, *284*, 117178. [CrossRef] [PubMed]

24. Bai, B.; Zhou, R.; Cai, G.Q.; Hu, W.; Yang, G.C. Coupled thermo-hydro-mechanical mechanism in view of the soil particle rearrangement of granular thermodynamics. *Comput. Geotech.* **2021**, *137*, 104272. [CrossRef]
25. Zha, F.S.; Liu, C.M.; Kang, B.; Xu, L.; Yang, C.B.; Chu, C.F.; Yu, C.; Zhang, W.; Zhang, J.W.; Liu, Z.H. Effect of Carbonation on the Leachability of Solidified/Stabilized Lead-Contaminated Expansive Soil. *Adv. Civil. Eng.* **2021**, *2021*, 8880818. [CrossRef]
26. Du, Y.J.; Jiang, N.J.; Shen, S.L.; Jin, F. Experimental investigation of influence of acid rain on leaching and hydraulic characteristics of cement-based solidified/stabilized lead contaminated clay. *J. Hazard. Mater.* **2012**, *225*, 195–201. [CrossRef]
27. Du, Y.J.; Wei, M.L.; Reddy, K.R.; Liu, Z.P.; Jin, F. Effect of acid rain pH on leaching behavior of cement stabilized lead-contaminated soil. *J. Hazard. Mater.* **2014**, *271*, 131–140. [CrossRef]
28. Stanforth, R.; Yap, C.F.; Nayar, R. Effects of weathering on treatment of lead contaminated soils. *J. Environ. Eng.* **2005**, *131*, 38–48. [CrossRef]
29. Wu, Z.L.; Deng, Y.F.; Chen, Y.G.; Gao, Y.F.; Zha, F.S. Long-term desalination leaching effect on compression/swelling behaviour of Lianyungang marine soft clays. *Bull. Eng. Geol. Environ.* **2021**, *80*, 8099–8107. [CrossRef]
30. Zha, F.S.; Liu, J.J.; Xu, L.; Cui, K.R. Effect of cyclic drying and wetting on engineering properties of heavy metal contaminated soils solidified/stabilized with fly ash. *J. Cent. South. Univ.* **2013**, *20*, 1947–1952. [CrossRef]
31. Kalkan, E. Impact of wetting–drying cycles on swelling behavior of clayey soils modified by silica fume. *Appl. Clay Sci.* **2011**, *52*, 345–352. [CrossRef]
32. Zha, F.S.; Liu, S.Y.; Du, Y.J.; Cui, K.R. Behavior of expansive soils stabilized with fly ash. *Nat. Hazards* **2008**, *47*, 509–523. [CrossRef]
33. Hafsteinsdottir EG, White DA, Gore DB, Stark SC Products and stability of phosphate reactions with lead under freeze-thaw cycling in simple systems. *Environ. Pollut.* **2011**, *159*, 3496–3503. [CrossRef] [PubMed]
34. Wei, M.L.; Du, Y.J.; Reddy, K.R.; Wu, H.L. Effects of freeze-thaw on characteristics of new KMP binder stabilized Zn- and Pb-contaminated soils. *Environ. Sci. Pollut. Res. Int.* **2015**, *22*, 19473–19484. [CrossRef]
35. Zhou, Z.W.; Ma, W.; Zhang, S.J.; Mu, Y.H.; Li, G.Y. Effect of freeze-thaw cycles in mechanical behaviors of frozen loess. *Cold Reg. Sci. Technol.* **2018**, *146*, 9–18. [CrossRef]
36. Liu, J.J.; Zha, F.S.; Xu, L.; Kang, B.; Yang, C.B.; Feng, Q.; Zhang, W.; Zhang, J.W. Strength and microstructure characteristics of cement-soda residue solidified/stabilized zinc contaminated soil subjected to freezing–thawing cycles. *Cold Reg. Sci. Technol.* **2020**, *172*, 102992.
37. Zhao, R.F.; Zhang, S.N.; Gao, W.; He, J.; Wang, J.; Jin, D.; Nan, B. Factors effecting the freeze thaw process in soils and reduction in damage due to frosting with reinforcement: A review. *Bull. Eng. Geol. Environ.* **2019**, *78*, 5001–5010. [CrossRef]
38. Su, Y.Y.; Li, P.; Ren, Z.P.; Xiao, L.; Zhang, H. Freeze-thaw effects on erosion process in loess slope under simulated rainfall. *J. Arid. Land* **2021**, *12*, 937–949. [CrossRef]
39. Wei, H.B.; Jiao, Y.B.; Liu, H.B. Effect of freeze–thaw cycles on mechanical property of silty clay modified by fly ash and crumb rubber. *Cold Reg. Sci. Technol.* **2015**, *116*, 70–77. [CrossRef]
40. Wang, T.L.; Liu, Y.J.; Yan, H.; Xu, L. An experimental study on the mechanical properties of silty soils under repeated freeze–thaw cycles. *Cold Reg. Sci. Technol.* **2015**, *112*, 51–65. [CrossRef]
41. Yang, Z.; Li, X.; Li, D.; Wang, Y.; Liu, X. Effects of Long-Term Repeated Freeze-Thaw Cycles on the Engineering Properties of Compound Solidified/Stabilized Pb-Contaminated Soil: Deterioration Characteristics and Mechanisms. *Int. J. Environ. Res. Public Health* **2020**, *17*, 1798. [CrossRef] [PubMed]
42. Yang, Z.P.; Chang, J.Z.; Wang, Y.; Li, X.Y.; Li, S. Effects of Long-Term Freeze-Thaw Cycles on the Properties of Stabilized/Solidified Lead-Zinc-Cadmium Composite-Contaminated Soil. *Int. J. Environ. Res. Public Health* **2021**, *18*, 6114. [CrossRef] [PubMed]
43. Chen, X.D.; Lu, X.W.; Yang, G. Sources identification of heavy metals in urban topsoil from inside the Xi'an Second Ringroad, NW China using multivariate statistical methods. *Catena* **2012**, *98*, 73–78. [CrossRef]
44. Yao, Y.R.; Li, J.; He, C.; Hu, X.; Yin, L.; Zhang, Y.; Zhang, J.; Huang, H.Y.; Yang, S.G.; He, H.; et al. Distribution Characteristics and Relevance of Heavy Metals in Soils and Colloids Around a Mining Area in Nanjing, China. *Bull. Environ. Contam. Toxicol.* **2021**, *107*, 996–1003. [CrossRef] [PubMed]
45. Liu, H.B.; Liu, S.J.; He, X.S.; Dang, F.; Tang, Y.Y.; Xi, B.D. Effects of landfill refuse on the reductive dechlorination of pentachlorophenol and speciation transformation of heavy metals. *Sci. Total Environ.* **2021**, *760*, 144122. [CrossRef]
46. McLellan, B.C.; Williams, R.P.; Lay, J.; van Riessen, A.; Corder, G.D. Costs and carbon emissions for geopolymer pastes in comparison to ordinary portland cement. *J. Clean. Prod.* **2011**, *19*, 1080–1090. [CrossRef]
47. Cuisinier, O.; Le Borgne, T.; Deneele, D.; Masrouri, F. Quantification of the effects of nitrates, phosphates and chlorides on soil stabilization with lime and cement. *Eng. Geol.* **2011**, *117*, 229–235. [CrossRef]
48. Yang, Z.P.; Li, X.Y.; Wang, Y.; Chang, J.Z.; Liu, X.R. Trace element contamination in urban topsoil in China during 2000-2009 and 2010-2019: Pollution assessment and spatiotemporal analysis. *Sci. Total Environ.* **2020**, *758*, 143647. [CrossRef]
49. PeÂrez-Cid, B.; Lavilla, I.; Bendicho, C. Application of microwave extraction for partitioning of heavy metals in sewage sludge. *Anal. Chim. Acta* **1999**, *378*, 201–210. [CrossRef]
50. Liu, D.D.; Miao, D.R.; Liu, F. Evaluation of the As, Cu and Pb Immobilizing Efficiency by Tessier, TCLP and SBET Method. *Adv. Mater. Res.* **2014**, *878*, 520–521. [CrossRef]
51. Linker, R.; Shmulevich, I.; Kenny, A.; Shaviv, A. Soil identification and chemometrics for direct determination of nitrate in soils using FTIR-ATR mid-infrared spectroscopy. *Chemosphere* **2005**, *61*, 652–658. [CrossRef] [PubMed]

52. Shi, T.; Yang, Z.P.; Zheng, L.W. FTIR spectra for early age hydration of cement-based composites incorporatted with CNTs. *Acta Mater. Compos. Sin.* **2017**, *34*, 653–660.
53. Yan, K.Z.; Guo, Y.X.; Li, N.J.; Cheng, F.Q. Effect and Mechanism Analysis of Calcium Oxide Additive on the Sodium Carbonate Activation of Fly Ash. *Bull. Chin. Ceram. Soc.* **2018**, *37*, 1003–1009.
54. Correia, A.A.S.; Matos, M.P.S.R.; Gomes, A.R.; Rasteiro, M.G. Immobilization of Heavy Metals in Contaminated Soils—Performance Assessment in Conditions Similar to a Real Scenario. *Appl. Sci.* **2020**, *10*, 7950. [CrossRef]
55. Guo, B.; Liu, B.; Yang, J.; Zhang, S.G. The mechanisms of heavy metal immobilization by cementitious material treatments and thermal treatments: A review. *J. Environ. Manag.* **2017**, *193*, 410–422. [CrossRef] [PubMed]
56. Zha, F.S.; Liu, C.M.; Kang, B.; Yang, X.H.; Zhou, Y.; Yang, C.B. Acid rain leaching behavior of Zn-contaminated soils solidified/stabilized using cement-soda residue. *Chemosphere* **2021**, *281*, 130916. [CrossRef]
57. Fei, Y.; Yan, X.L.; Zhong, L.R.; Li, F.S.; Du, Y.J.; Li, C.P.; Lv, H.Y.; Li, Y.H. On-Site Solidification/Stabilization of Cd, Zn, and Pb Co-Contaminated Soil Using Cement: Field Trial at Dongdagou Ditch, Northwest China. *Environ. Eng. Sci.* **2018**, *35*, 1329–1339. [CrossRef]
58. Zhang, W.L.; Zhao, L.Y.; Yuan, Z.J.; Li, D.Q.; Morrison, L. Assessment of the long-term leaching characteristics of cement-slag stabilized/solidified contaminated sediment. *Chemosphere* **2021**, *267*, 128926. [CrossRef]
59. Yoon, D.H.; Choi, W.S.; Hong, Y.K.; Lee, Y.B.; Kim, S.C. Effect of chemical amendments on reduction of bioavailable heavy metals and ecotoxicity in soil. *Appl. Biol. Chem.* **2019**, *62*, 1–7. [CrossRef]
60. Bai, B.; Nie, Q.K.; Zhang, Y.K.; Wang, X.L.; Hu, W. Cotransport of heavy metals and SiO2 particles at different temperatures by seepage. *J. Hydrol.* **2021**, *597*, 125771. [CrossRef]
61. Ouhadi, V.R.; Yong, R.N.; Deiranlou, M. Enhancement of cement-based solidification/stabilization of a lead-contaminated smectite clay. *J. Hazard. Mater.* **2021**, *403*, 123969. [CrossRef] [PubMed]
62. Velichenko, A.B.; Portillo, J.; Sarret, M.; Muller, C. Zinc dissolution in ammonium chloride electrolytes. *J. Appl. Electrochem.* **1999**, *29*, 1119–1123. [CrossRef]
63. Mitani, T.; Ogawa, M. Cadmium leaching by acid rain from cadmium-enriched activated sludge applied in soil. *J. Environ. Health* **1998**, *33*, 1569–1581. [CrossRef]

Article

Molecular Dynamics Simulation of Nanoscale Elastic Properties of Hydrated Na-, Cs-, and Ca-Montmorillonite

Lianfei Kuang [1], Qiyin Zhu [1,*], Xiangyu Shang [2] and Xiaodong Zhao [1]

1 State Key Laboratory for Geomechanics and Deep Underground Engineering, China University of Mining and Technology, Xuzhou 221116, China; kuanglf@cumt.edu.cn (L.K.); zxd@cumt.edu.cn (X.Z.)
2 School of Mechanics and Civil Engineering, China University of Mining and Technology, Xuzhou 221116, China; xyshang@cumt.edu.cn
* Correspondence: qiyin.zhu@cumt.edu.cn

Abstract: The knowledge of nanoscale mechanical properties of montmorillonite (MMT) with various compensation cations upon hydration is essential for many environmental engineering-related applications. This paper uses a Molecular Dynamics (MD) method to simulate nanoscale elastic properties of hydrated Na-, Cs-, and Ca-MMT with unconstrained system atoms. The variation of basal spacing of MMT shows step characteristics in the initial crystalline swelling stage followed by an approximately linear change in the subsequent osmotic swelling stage as the increasing of interlayer water content. The water content of MMT in the thermodynamic stable-state conditions during hydration is determined by comparing the immersion energy and hydration energy. Under this stable hydration state, the nanoscale elastic properties are further simulated by the constant strain method. Since the non-bonding strength between MMT lamellae is much lower than the boning strength within the mineral structure, the in-plane and out-of-plane strength of MMT has strong anisotropy. Simulated results including the stiffness tensor and linear elastic constants based on the assumption of orthotropic symmetry are all in good agreement with results from the literature. Furthermore, the out-of-plane stiffness tensor components of C_{33}, C_{44}, and C_{55} all fluctuate with the increase of interlayer water content, which is related to the formation of interlayer H-bonds and atom-free volume ratio. The in-plane stiffness tensor components C_{11}, C_{22}, and C_{12} decrease nonlinearly with the increase of water content, and these components are mainly controlled by the bonding strength of mineral atoms and the geometry of the hydrated MMT system. Young's modulus in all three directions exhibits a nonlinear decrease with increasing water content.

Keywords: molecular dynamics; mechanical properties; montmorillonite; basal spacing

Citation: Kuang, L.; Zhu, Q.; Shang, X.; Zhao, X. Molecular Dynamics Simulation of Nanoscale Elastic Properties of Hydrated Na-, Cs-, and Ca-Montmorillonite. *Appl. Sci.* **2022**, *12*, 678. https://doi.org/10.3390/app12020678

Academic Editor: Bing Bai

Received: 17 December 2021
Accepted: 10 January 2022
Published: 11 January 2022

1. Introduction

Montmorillonite (MMT) consists of well-defined layers separated by interlayer spaces and has great water absorption potential [1–4]. Thus, it is often used as clay barriers or adsorption material in environmental geotechnical engineering, for example, the backfill in the nuclear waste disposal and the cushion in the landfill [5–10]. It is very challenging to directly measure the fundamental mechanical properties of MMT [11,12]. Atomic force acoustic microscopy is used to determine the elastic properties of clay mineral aggregates by measuring adhesion forces [13–15]. Nanoindentation and ultrasonic pulse velocity technology (UPV) are also used to measure the elastic stiffness constants of clays and shale samples [16–18]. However, these techniques are powerless to measure the anisotropic mechanical properties of MMT, due to the complications in sample preparation and even the testing process.

In contrast to difficulty in experimental methods, molecular dynamics (MD) simulation is a supplemental way to model and understand the properties of hydrated clay minerals on an atomic scale [1,19–21]. With the development of polymer–clay nanocomposites in recent years, several studies on the nanomechanical properties of MMT minerals or hydrated MMT systems have emerged. The authors in [18] reported for the first time the stress–strain response of rock-forming minerals by MD simulation. The authors in [22] performed MD simulations to study the flexibility and the mechanical behavior of a single clay layer in a completely exfoliated state and found that the critical stress of a clay layer bending fracture is 0.8 MPa in both in-plane directions. In [23], the authors presented results for the elastic properties of a single lamella of MMT by MD simulation and further pointed out that the elastic constants characteristic of a thin plate is closely related to the thickness of the nanoplate. Based on the analogous minerals and modulus-density relations, ref. [24] deduced a convergence of opinion in the range 178–265 GPa for the elastic moduli of smectite clay platelets. The authors in [25] systematically simulated the structure and the thermomechanical properties of an isolated clay nanoplate and MMT crystals intercalated by water or polyethylene oxide. In [26], the authors clarified the linear elastic properties including tensile moduli, shear moduli, and potential failure mechanisms as a function of cation density and stress for the minerals pyrophyllite, MMT, and mica. In [27], the authors examined the mechanism of bending, the stored energy, and the failure of several clay minerals. They revealed that molecular contributions to the bending energy include bond stretching and bending of bond angles in the mineral as well as rearrangements of alkali ions on the surface of the layers. The authors in [28] further simulated the elastic and structural properties of muscovite as a function of temperature, pressure, and strain. The results demonstrate that the elastic properties of muscovite depend on both temperature and pressure. MD with the CLAYFF force field was used in [29] to simulate isothermal isobaric water adsorption of interlayer MMT, and nanoscale elastic properties of the clay-interlayer water system are calculated from the potential energy of the model system. Similar simulations have also been carried out on MMT by implementing the elastic bath method [30]. The mechanical properties of Na-MMT were investigated in [31] from the view of water content by MD, and the hysteresis phenomena of elastic constant during the swelling and shrinking was found. In addition, researchers found that the mechanical behavior of MMT crystal exhibits a clear dissymmetry between compression and tension and an important dependency on mean stress [32].

The above simulation studies are generally based on the strain method, that is, the stress is calculated by changing the size of the simulation box. The stress method and the large-scale fluctuation method are also usually adopted. For example, using steered molecular dynamics (SMD) technology, the mechanical response of the interlayer of hydrated MMT was evaluated by [33]. Using grid-computing technology, ref. [34] simulated MMT clay large-scale systems containing up to approximately ten million atoms; large-scale systems exhibit emergent behavior with increasing size, and the material mechanical properties were calculated based on thermal bending fluctuations. From the surveyed status of nanoscale mechanical properties of clay minerals, it can be found that the stress–strain response of minerals can be revealed by different methods (constant strain, SMD, or the large-scale fluctuation method) based on different force fields and different platforms. However, the moisture content under most simulated hydration conditions is usually arbitrarily selected and generally restricts some movement directions of clay plates or even the interlayer water molecules which affect the mechanical response characteristics of the hydrated mineral system.

The goal of this study is to simulate the nanoscale mechanical properties of hydrated MMT minerals with three cations (Na, Ca, and Cs) at variable water contents. In the simulation, all degrees of freedom of atoms are released, and the water content in different hydration states is determined according to stable-state thermodynamic conditions. To the best of our knowledge, no such treatments have been reported to study the mechanical properties of the MMT mineral up to now. The quantitative values of the basal spacing for different compensation cationic MMTs are firstly studied. The relative stabilities of different states are determined by comparing the immersion energy and hydration energy between MMT systems, and, then, the corresponding moisture content of thermodynamics stable states is obtained. Finally, the nanoscale elastic properties of MMT are further simulated by the constant strain method under the stable state water content.

2. Model and Methods

2.1. Molecular Models

Wyoming-type MMT, which has been widely simulated, was selected as the model clay in this study, as shown in Figure 1. The initial unit cell parameters are a = 5.23 Å and b = 9.06 Å, and the c-axis increases with increasing water content, the unit cell angles are $\alpha = \gamma = 90°$, $\beta = 99°$. The x and y axes were chosen parallel to the clay layer, while the z-axis was orthogonal to the clay layer. The unit cell of MMT was replicated (4a × 2b × 1c) in the x, y, and z directions so that the supercell contained 8-unit cells with approximate dimensions in the XY plane of 21 Å × 18 Å and 9.6 Å in the z-axis. One of every 8 octahedral Al^{3+} atoms was substituted with an Mg^{2+} atom and one of every 4 tetrahedral Si^{4+} atoms was substituted with an Al^{3+} atom. The isomorphic substitutions were distributed randomly obeying Lowenstein's rule. This extent of substitution led to a layer charge of −0.75 e per unit cell and the unbalanced charge will be compensated by Na^+, Ca^{2+}, and Cs^+ which are commonly adopted in engineering applications [12,35]. In addition, the interlayer space was randomly filled with a given number n of water molecules, the number n ranged from 0 to 160, and a total of 23 clay–water-ion systems will be evaluated. The structural formula of the simulated layer of Na- and Cs-MMT is $M_{0.75}\{Si_{7.75}Al_{0.25}\}[Al_{3.5}Mg_{0.5}]O_{20}(OH)_4 \cdot nH_2O$ (M = Na or Cs), whereas for the Ca-MMT, it can be expressed as $Ca_{0.375}\{Si_{7.75}Al_{0.25}\}[Al_{3.5}Mg_{0.5}]O_{20}(OH)_4 \cdot nH_2O$.

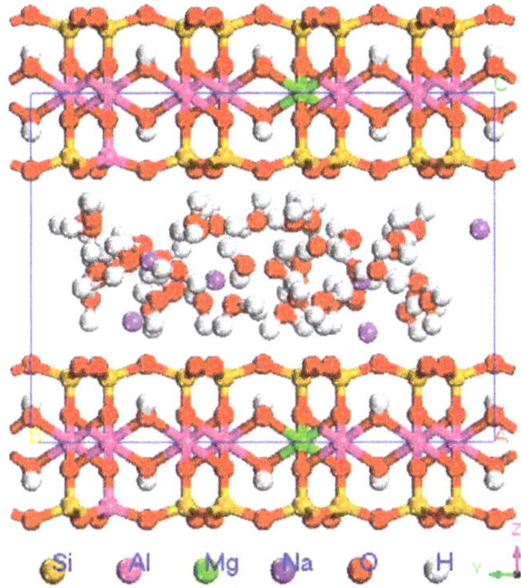

Figure 1. Super-cell model of Na-montmorillonite (Mg octahedral are green, Al octahedral and tetrahedral purple, Si tetrahedral brown, O atoms red, H atoms white, and Na ions cyan).

2.2. Simulation Protocol

MD simulations were conducted in Forcite module in the Material studio 8.0 simulation package and the CLAYFF force field, developed by [36]. The latter is a flexible model which describes hydrated mineral systems through nonbonded electrostatic and Lennard–Jones terms. It can accurately reproduce structural and spectroscopic properties of clays, as well as dynamical and energetic properties of clay interlayer and aqueous interfaces. The interlayer water molecules were described with the flexible version of the simple point charge (SPC) potential. The potential energy of the system was evaluated with an 8.5 Å cutoff, a 0.5 Å spline, and buffer width for short-range van der Waals interaction. The Ewald summation for the electrostatic interaction was calculated with a precision of

10^{-3} kcal/mol. Periodic boundary conditions were imposed on three dimensions to avoid interface effects.

The initial conformation of hydrated MMT needs optimization of the geometry first by searching for the minimum potential energy of the system. A smart algorithm was chosen for the optimization with a maximum number of 5000 cycles, and the convergence thresholds for the specified maximum energy and displacement changes were set as 1.0×10^{-4} kcal/mol and 5.0×10^{-5} Å, respectively. Following this, the final configuration from optimization was used as the initial configuration for the equilibration stage simulation in the NPT (isothermal–isobaric) ensemble at P = 1 atm and T = 298 K. Each system was equilibrated for 100 ps (105 steps) with a time step of 1 fs. The temperature and the pressure were controlled by a Nosé–Hoover–Langevin thermostat of 1 ps and a Berendsen barostat of 0.1 ps, respectively. Finally, following the thermal equilibration configurations, a microcanonical NVE ensemble was then performed for a 1 ns production simulation, and the time step was still set as 1 fs (NVE: Each system in the ensemble has the same energy, and the number of particles and volume of each system are usually the same). The data from all 1 ns trajectories of the NVE simulation runs were recorded every 500 fs and were used to analyze the microstructure and thermodynamic properties of hydrated MMT. By comparing the relative thermodynamic energy of different systems, the steady-state water content of hydrated MMT can be determined. Furthermore, the mechanical properties of the steady-state system will be further simulated by the constant strain method. The constant strain method obtains the elastic constants via minimizing the energy of the system and deforming the supercell parameters in 12 directions in order. To improve the statistical accuracy, the simulations of mechanical properties are carried out based on the production stage trajectory files obtained above, that is, 20 frames of instant conformation are extracted from the corresponding trajectory files. The strain step is set as 10, and the maximum strain in each strain step is set to 1% according to the linear elastic assumption. The final mechanical parameters are obtained by the geometric average of the 20 frames.

2.3. Methods for Calculating the Elastic Constants

Acting on external forces, the hydrated MMT system will be in a state of stress. If the system is in equilibrium, the external forces must be exactly balanced by internal stress. In general, stress is a second rank tensor with nine components as in Equation (1):

$$\vec{\sigma} = [\sigma_{ij}] = \begin{bmatrix} \sigma_{xx} & \sigma_{xy} & \sigma_{xz} \\ \sigma_{xy} & \sigma_{yy} & \sigma_{yz} \\ \sigma_{xz} & \sigma_{yz} & \sigma_{zz} \end{bmatrix} \tag{1}$$

In an atomistic calculation, the internal stress tensor can be obtained using the so-called virial expression as in Equation (2):

$$\vec{\sigma} = -\frac{1}{V_0}\left(\sum_{i=1}^{N} m_i(\vec{v}_i \vec{v}_i^{\,T}) + \sum_{i<j} \vec{r}_{ij} \vec{F}_{ij}^{\,T}\right) \tag{2}$$

where index i runs over all particles 1 through N; m_i, $\vec{v}_i$, and $\vec{F}_i$ denote the mass, velocity, and force acting on particle i, respectively; V_0 denotes the initial system volume.

The application of stress to the hydrated MMT system results in a change in the relative positions of particles within the system expressed quantitatively via the strain tensor as in Equation (3):

$$\vec{\varepsilon} = [\varepsilon_{ij}] = \begin{bmatrix} \varepsilon_{xx} & \varepsilon_{xy} & \varepsilon_{xz} \\ \varepsilon_{xy} & \varepsilon_{yy} & \varepsilon_{yz} \\ \varepsilon_{xz} & \varepsilon_{yz} & \varepsilon_{zz} \end{bmatrix} \tag{3}$$

For a parallelepiped (for example, a periodic simulation cell) characterized by the three column vectors $[\vec{a}_0\ \vec{b}_0\ \vec{c}_0]$ in some reference state, and by the vectors $[\vec{a}\ \vec{b}\ \vec{c}]$ in the deformed state, the strain tensor is given by Equation (4):

$$\vec{\varepsilon} = \frac{1}{2}\left[\left(\vec{h}_0^{T}\right)^{-1}\vec{G}\,\vec{h}_0^{-1} - 1\right] \tag{4}$$

where $\vec{h}_0$ denotes the matrix formed from the three column vectors $[\vec{a}_0\ \vec{b}_0\ \vec{c}_0]$, $\vec{h}$ denotes the corresponding matrix forms from $[\vec{a}\ \vec{b}\ \vec{c}]$, T denotes the matrix transpose, and $\vec{G}$ denotes the metric tensor $\vec{h}^{T}\vec{h}$.

Therefore, the elastic stiffness coefficients, relating the various components of stress and strain, are defined by Equation (5):

$$\vec{C} = \left[C_{ijkl}\right] = \left.\frac{\partial \sigma_{ij}}{\partial \varepsilon_{kl}}\right|_{T,\varepsilon_{kl}} = \left.\frac{1}{V_0}\frac{\partial^2 A}{\partial \varepsilon_{ij}\partial \varepsilon_{kl}}\right|_{T,\varepsilon_{ij},\varepsilon_{kl}} \tag{5}$$

where A denotes the Helmholtz free energy, and T denotes temperature.

For small deformations, the relationship between the stresses and strains may be expressed in terms of a generalized Hooke's law:

$$\sigma_{ij} = C_{ijkl}\varepsilon_{kl} \tag{6}$$

and, alternatively:

$$\varepsilon_{ij} = S_{ijkl}\sigma_{kl} \tag{7}$$

where C_{ijkl} and S_{ijkl} denote stiffness tensor and compliance tensor, respectively.

Since both the stress and strain tensors are symmetric, it is often convenient to simplify these expressions by making use of Voigt vector notation. The stress is represented as in Equation (8):

$$\begin{bmatrix} \sigma_{xx} & \sigma_{xy} & \sigma_{xz} \\ \sigma_{xy} & \sigma_{yy} & \sigma_{yz} \\ \sigma_{xz} & \sigma_{yz} & \sigma_{zz} \end{bmatrix} \Rightarrow \begin{bmatrix} \sigma_1 & \tau_6 & \tau_5 \\ & \sigma_2 & \tau_4 \\ & & \sigma_3 \end{bmatrix} \tag{8}$$

The strain is represented as in Equation (9):

$$\begin{bmatrix} \varepsilon_{xx} & \varepsilon_{xy} & \varepsilon_{xz} \\ \varepsilon_{xy} & \varepsilon_{yy} & \varepsilon_{yz} \\ \varepsilon_{xz} & \varepsilon_{yz} & \varepsilon_{zz} \end{bmatrix} \Rightarrow \begin{bmatrix} \varepsilon_1 & \gamma_6 & \gamma_5 \\ & \varepsilon_2 & \gamma_4 \\ & & \varepsilon_3 \end{bmatrix} \tag{9}$$

The generalized Hooke's law is thus often written as in Equation (10):

$$\begin{bmatrix} \sigma_1 \\ \sigma_2 \\ \sigma_3 \\ \tau_4 \\ \tau_5 \\ \tau_6 \end{bmatrix} = \begin{bmatrix} C_{11} & C_{12} & C_{13} & C_{14} & C_{15} & C_{16} \\ & C_{22} & C_{23} & C_{24} & C_{25} & C_{26} \\ & & C_{33} & C_{34} & C_{35} & C_{36} \\ & & & C_{44} & C_{45} & C_{46} \\ & & & & C_{55} & C_{56} \\ & & & & & C_{66} \end{bmatrix} \begin{bmatrix} \varepsilon_1 \\ \varepsilon_2 \\ \varepsilon_3 \\ \gamma_4 \\ \gamma_5 \\ \gamma_6 \end{bmatrix} \tag{10}$$

alternatively:

$$\sigma_i = C_{ij}\varepsilon_j \tag{11}$$

where C_{ij} denotes the stiffness tensor and is a symmetric matrix of 6×6. The number of independent elastic constants for any system is 21. The mineral of MMT crystals belongs to the monoclinic system ($a \neq b \neq c$, $\alpha = \beta = 90^\circ \neq \gamma$) and has C2/M symmetry. It, then, only needs 13 independent parameters to describe its stiffness tensor. However, according to the study in [23], the MMT crystal can be considered approximately as orthotropic symmetry, thus further reducing the stiffness tensor parameters to 9 to characterize the sheet. These 9 independent elastic constants include 3 Young's moduli (E_1, E_2, E_3), 3 shear moduli (G_{23}, G_{31} and G_{12}), and 3 Poisson's ratios (μ_{12}, μ_{13} and μ_{23}). In terms of these parameters, the stiffness tensor C_{ij} can be further expressed as in Equation (12):

$$\begin{bmatrix} C_{11} & C_{12} & C_{13} & 0 & 0 & 0 \\ C_{21} & C_{22} & C_{23} & 0 & 0 & 0 \\ C_{31} & C_{32} & C_{33} & 0 & 0 & 0 \\ 0 & 0 & 0 & C_{44} & 0 & 0 \\ 0 & 0 & 0 & 0 & C_{55} & 0 \\ 0 & 0 & 0 & 0 & 0 & C_{66} \end{bmatrix} = \begin{bmatrix} \frac{1-\mu_{23}\mu_{32}}{E_2E_3\Delta} & \frac{\mu_{21}+\mu_{31}\mu_{23}}{E_2E_3\Delta} & \frac{\mu_{31}+\mu_{21}\mu_{32}}{E_2E_3\Delta} & 0 & 0 & 0 \\ \frac{\mu_{12}+\mu_{13}\mu_{32}}{E_3E_1\Delta} & \frac{1-\mu_{31}\mu_{13}}{E_3E_1\Delta} & \frac{\mu_{32}+\mu_{31}\mu_{12}}{E_3E_1\Delta} & 0 & 0 & 0 \\ \frac{\mu_{13}+\mu_{12}\mu_{23}}{E_1E_2\Delta} & \frac{\mu_{23}+\mu_{13}\mu_{21}}{E_1E_2\Delta} & \frac{1-\mu_{12}\mu_{21}}{E_1E_2\Delta} & 0 & 0 & 0 \\ 0 & 0 & 0 & 2G_{23} & 0 & 0 \\ 0 & 0 & 0 & 0 & 2G_{31} & 0 \\ 0 & 0 & 0 & 0 & 0 & 2G_{12} \end{bmatrix} \quad (12)$$

where $\Delta = \frac{1-\mu_{12}\mu_{21}-\mu_{23}\mu_{32}-\mu_{31}\mu_{13}-2\mu_{12}\mu_{23}\mu_{31}}{E_1E_2E_3}$.

Therefore, the key to characterizing the elastic mechanical properties of the MMT–water-ion system at the nanoscale level is to obtain the parameters of the stiffness tensor or the compliance tensor.

3. Results and Discussion

3.1. Swelling and Thermodynamics Analysis

To obtain the water content corresponding to the thermodynamic steady-state in the hydration process of MMT, the strategy of simulating 23 different water content systems was adopted. The interlayer space of the MMT lamellae increases with the increase of water content, and this swelling process can be described by using the parameter of basal spacing. The basal spacing can be obtained from the view of water content by averaging the system volume during the production stage of simulation and defined as in Equation (13):

$$d_{001} = \frac{\langle V \rangle}{\langle a \rangle \langle b \rangle \sin\langle \alpha \rangle} \quad (13)$$

where $<V>$ denotes the statistically averaged volume, and $<a>$ $<b>$ $<\alpha>$ are the statistically averaged parameters of the simulation supercell.

Figure 2 illustrates the variation of basal spacing of different hydrated montmorillonite systems as a function of water content from this study and available results from the literature. In the early stage of crystalline swelling, the basal spacing increases from step to step, while the step characteristic (plateau) disappears, and shows approximately linear variation in the following stage of osmotic swelling. The swelling curves of MMT with different compensation cations are different. The basal spacing of Cs-MMT is greater than Na- and Ca-MMT under the same water content. However, there is almost no difference between the latter two. This is mainly due to the fact that Cs^+ ions are a heavy metal ion and their ion radius is about 1.7 times that of Na^+ and Ca^{2+} ions, while the latter two's ion radii are almost the same. The simulated results can well penetrate the distribution range of other available measurements.

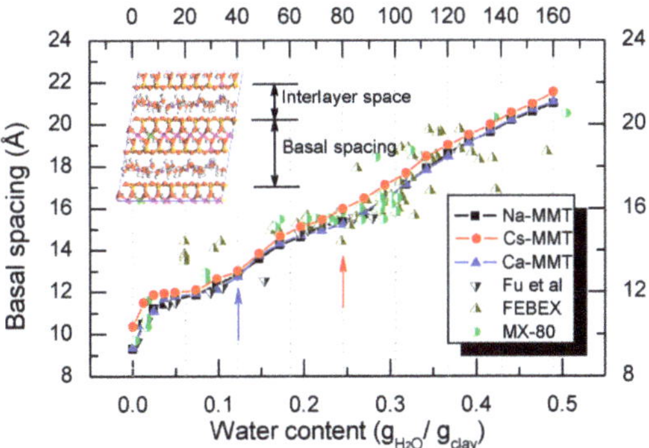

Figure 2. The simulated and experimental basal spacing of montmorillonite as a function of water content (*g* denotes molecular mass).

It has been shown that the energy contribution to free energy dominates clay swelling and the entropy term only played a minor role. The local minima of the swelling energy curve correspond to the stable hydration state. Hence, the relative stabilities of different states can be determined by immersion energy and hydration energy. The immersion energy is defined as in Equation (14):

$$Q = \langle E(N) \rangle - \left\langle E\left(N^0\right) \right\rangle - \left(N - N^0\right) E_{bulk} \tag{14}$$

Here $<E(N)>$ is the average potential energy of the hydrate with water content N, and $<E(N^0)>$ is the average energy of the referential hydration state (the state with the highest water content, 49%, was selected). E_{bulk} is the mean interaction potential of the bulk flexible SPC water; $E_{bulk} = -41.5392$ kJ/mol. The immersion energy Q is the energy released when MMT with water content N transforms into the referenced state with water content N^0 by adsorbing water from bulk water.

The hydration energy is defined as in Equation (15)

$$\Delta E = \frac{\langle E(N) \rangle - \langle E(0) \rangle}{N} \tag{15}$$

Here $<E(0)>$ is the average potential energy of the dry MMT. The hydration energy ΔE evaluates the energy change associated with water uptake by the dry clay.

The simulated hydration energy and immersion energy curves of MMT with different compensation cations are presented in Figure 3. For Na- and Cs-MMT the local minima occur when the interlayer water molecular number is 40 and 80 in the hydration energy curves, while Ca-MMT has only a global minimum at low water content. Correspondingly, in the immersion energy curves of Na- and Cs-MMT, local minima occur when the interlayer water molecular numbers are also 40 and 80. The global minima of Cs-MMT is found at $n = 80$, while Na-MMT global minima is located at $n = 160$, as shown by the enlarged illustration in Figure 3. Meanwhile, the variation of the Ca-MMT immersion energy curve decreases monotonically with the increase of water content, and the global minima occurs also at $n = 160$. Then, since the local minima of the swelling energy corresponds to the thermodynamic stable states, it can be deduced that the Na- and Cs-MMT will form the single- and double-layer stable hydrated states when the number of interlayer water molecules reach 40 and 80 respectively. The double-layer hydrate for Cs-MMT is in the most stable state. Thus, the Cs^+ ions can inhibit swelling. The global minima of Na-MMT is located at $n = 160$. After forming the first and second hydration layer, Na-MMT will continue expansion to the osmotic swelling stage, namely that the Na^+ ions promote swelling in contrast with Cs^+. For Ca-MMT, its role in promoting expansion is even more pronounced. The intermediate state of the first or second hydration layer will not form during the swelling process, and the osmotic swelling stage will rapidly and directly develop. The nanoscale elastic properties of these stable state water content systems will be further simulated in the following sections.

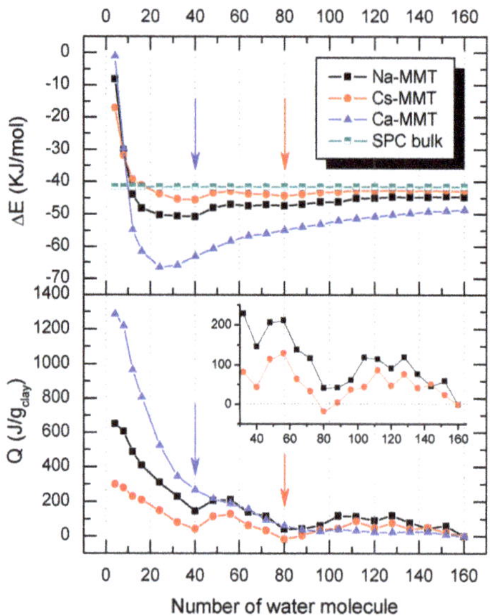

Figure 3. Hydration energy and immersion energy of montmorillonite as a function of increasing interlayer water molecule number.

3.2. Nanoscale Elastic Properties

To fully demonstrate the yield behavior of the MMT mineral, the maximum strain of each direction is expanded to 10% and the corresponding calculation step is increased to 100 steps. The stress–strain curves of the typical MMT system obtained from simulation are shown in Figure 4, where the interlayer compensation cation is Ca^{2+} ion, and the interlayer water number n is 80, corresponding to bilayer hydration.

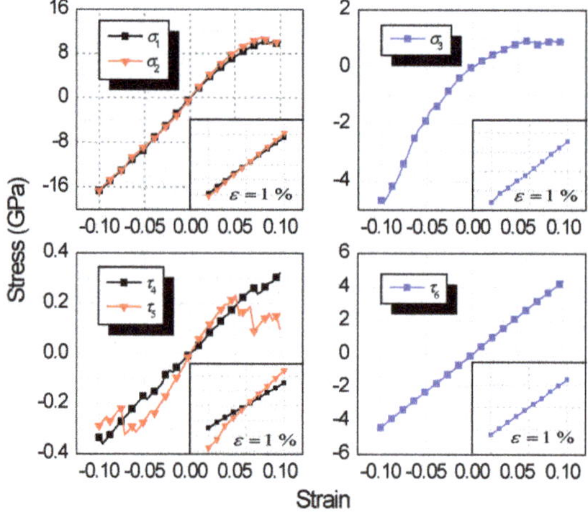

Figure 4. Typical stress–strain behavior for hydrated Ca-montmorillonite.

The micromechanical properties of hydrated MMT show obvious anisotropy. The in-plane uniaxial tensile and compressive strength of MMT is approximately equal (for σ_1 and σ_2), while the tensile strength is slightly less than the compressive strength. When the tensile strain reaches about 7%, the material yields. Meanwhile, the compressive stress–strain relationship shows a good linear relationship within the maximum compressive strain (10%). The tensile and compressive strength (σ_3) of MMT orthogonal to the mineral plane is much lower than that in-plane strength (σ_1 and σ_2). This is mainly due to the appearance of the interlayer water in MMT lamellae, and the bonding strength between the MMT crystal and water molecule through non-bonding is much lower than that of covalent or ionic bonding in the mineral plane. In addition, the compression of MMT in the Z direction shows obvious non-linear characteristics and hardening under high pressure, which is different from the linear characteristics in the crystal plane.

The hydration molecular layer that mixed within MMT lamellae has a certain shear strength ($\tau4$ and $\tau5$) and also shows anisotropy along with different directions of the MMT crystal plane. The shear stiffness of the interlayer water along the Y direction ($\tau5$) is slightly larger than that in the X direction ($\tau4$), while its shear strength is slightly lower. The shear capability of MMT lamellae is very high ($\tau6$) and shows good linear characteristics in the simulated strain ranges.

By comparing the quantitative values of in-plane strength, it was found that the in-plane compressive strength of hydrated MMT is the largest, followed by tensile strength and shear strength. Similarly, this variation also applies to the strengths that are perpendicular to the mineral plane.

All of the stress responses show good linearity in the range of 1% small strain, and this is illustrated in the lower right of each block in Figure 4. This also shows that the linear elastic constants calculated by the small strain of 1% are reliable. The stiffness tensor and linear elastic constants of typical MMT in the 1% strain range corresponding to Figure 4 are listed in Table 1.

Table 1. Typical stiffness coefficients and elastic constants of hydrated Ca-MM (GPa).

C_{ij}	1	2	3	4	5	6
1	189.6710 ± 1.0140	81.7433 ± 0.8172	10.1290 ± 0.4701	1.2430 ± 0.3777	−10.4597 ± 0.2771	0.4964 ± 0.4787
2		212.5464 ± 1.7814	8.4054 ± 0.4993	3.5123 ± 0.4520	−3.4307 ± 0.3618	1.1526 ± 1.0715
3			28.4335 ± 0.5057	0.3278 ± 0.2660	1.3468 ± 0.1654	0.1060 ± 0.2955
4				4.2630 ± 0.1881	−0.6221 ± 0.1371	−2.7480 ± 0.2758
5		Symmetric			5.4343 ± 0.1049	0.3864 ± 0.2236
6						45.4441 ± 0.6942
Young's modulus	E_1	139.1605	E_2	174.2115	E_3	27.0289
Shear modulus	G_{23}	2.1315	G_{31}	2.7172	G_{12}	22.7221
Poisson's ratios	μ_{12} 0.3463	μ_{21} 0.4335	μ_{13} 0.3436	μ_{31} 0.0667	μ_{23} 0.1199	μ_{32} 0.0186
Bulk modulus			70.1340			
Compressibility (1/TPa)			38.6319			

The simulated stiffness tensor components in Table 1 are in good agreement with simulations and measurements in the literature. The in-plane components C_{11} and C_{22} of the MMT with water bilayer at 300 K are 189 GPa and 191 GPa, respectively, from [25], which are almost equal to the results of this study that C_{11} = 189 GPa and C_{22} = 212 GPa. Furthermore, there are differences in the in-plane Young's moduli E_1 and E_2 which are derived from the stiffness tensor, and these differences are induced by the different topologies of trivalent cations in dioctahedral structure of MMT, as shown in Figure 5. However, the in-plane Young's moduli (E_1 and E_2) are all much higher than the Young's modulus perpendicular to the plane (E_3). On the contrary, the in-plane shear moduli (G_{23} and G_{31}) are lower than the shear modulus perpendicular to the plane (G_{12}). The larger difference of Poisson's ratio in different directions once again shows the anisotropy of the MMT structure. The boning strength in the mineral plane is much higher than that of non-bonding strength between mineral lamellae.

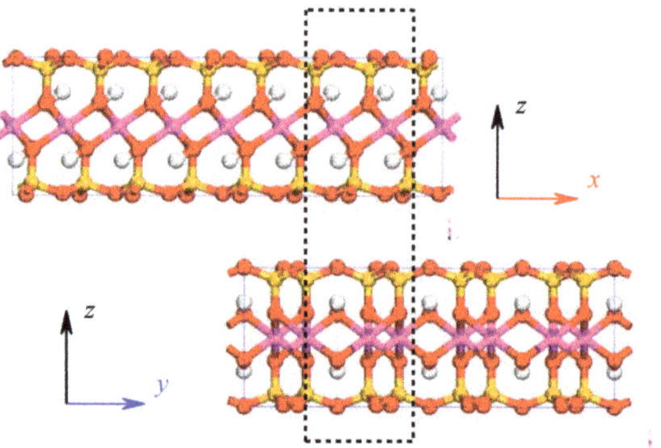

Figure 5. Different topologies of Al atom in the dioctahedral structure of montmorillonite.

The evolution of elastic properties of hydrated MMT under different stable state water content will be further illustrated by analyzing the changes of some components in the stiffness tensor. To validate the study, the simulation results from [29] and UPV test results from [17] are adopted for comparison due to the high correlation with the work in this study. The authors in [29] gave the variation of stiffness components during the hydration process of Na-Wyoming MMT with a water content of 0–40%. The authors in [17] obtained the range of elastic stiffness constants of smectite minerals on the microscale based on UPV test results of clay composition in shale.

For compression of clay under high pressure, the most concerning component of the stiffness tensor is C_{33}, that is, the ability to resist deformation in the direction of perpendicular to the mineral plane. Figure 6 shows the variation of C_{33} obtained upon hydration by [29], Na-, Cs-, Ca- MMT in this study under different stable state conditions, and the test range results of [17]. Since the stiffness tensor elements in this study are obtained by a statistical average of 20 frame trajectory files, the statistical standard deviation of data is also given in Figure 6.

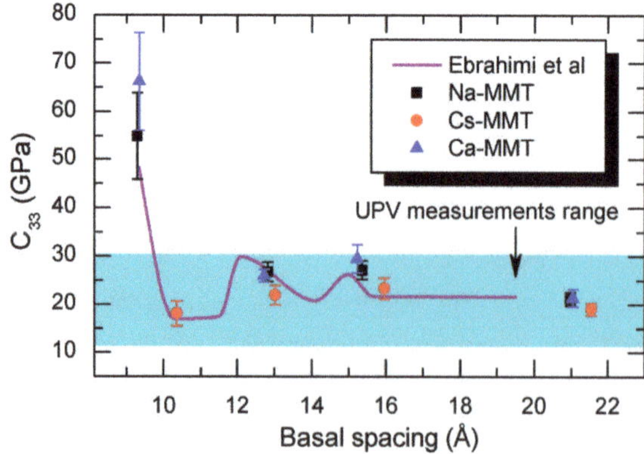

Figure 6. The C_{33} elastic constant as a function of basal layer spacing.

Ebrahimi et al. (2012) revealed that C_{33} decreased sharply from dry MMT to the formation of the first hydrated layer between lamellae, then increased to a local maximum with the formation of the stable hydrated layer, and finally maintained a constant until the basic spacing d > 15.6 Å [29]. The values of C_{33} obtained in this study are in good agreement with the results of [29], apart from the first hydrated layer. The C_{33} of Ca-MMT is slightly higher than that of Na-MMT, while the C_{33} of Cs-MMT is the lowest among all the same hydrated states. Apart from the dry MMT, all of the simulated C_{33} fall well into the given range of UPV measured by [17] (C_{33} = 11.5~30.4 GPa).

The in-plane stiffness tensor components C_{11}, C_{22}, and C_{12}, which reflect the Poisson effect in the plane, are shown in Figure 7 with basal spacing. C_{11}, C_{22}, and C_{12} all decrease nonlinearly with the increase of water content. These components are mainly controlled by the ionic or covalent bonding strength of mineral atoms and the geometric conformation of the hydrated MMT system. In addition, the simulation results of C_{11} and C_{12} are in good agreement with those of [29], but C_{12} is slightly lower. The reason may lie in the difference between our simulation structure and the octahedral isomorphic substitutions of [29]. However, the measured range of UPV given by [17] (C_{11} = 13.8~46.1 GPa and C_{12} = 6.9~17.8 GPa) is much lower than the simulated results. This is because the MMT mineral lamellae are very thin, and the relative slip between lamellae easily occurs in the UPV test. However, the simulated results in [25] (C_{11} = 189~231 GPa) and test results by Brillouin scattering on mica minerals in [26] (C_{11} = 181 GPa and C_{22} = 178 GPa) both show that the results of this study are credible.

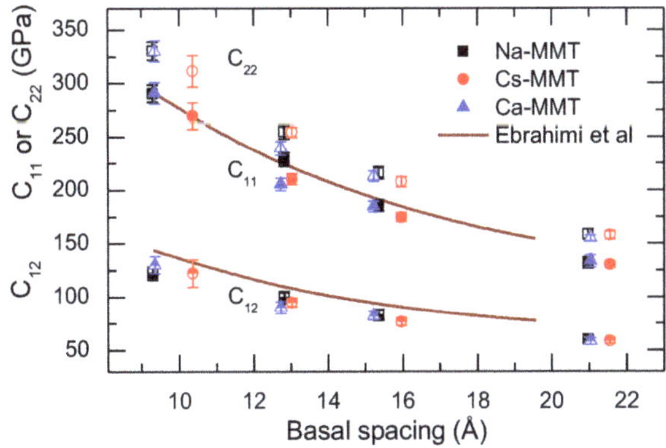

Figure 7. The C_{11}, C_{22}, and C_{12} elastic constants as a function of basal layer spacing.

For the shear of clay under high pressure, the most concerned is the shear resistance of limited interlayer water molecules in the hydrated MMT system, i.e., the in-plane shear stiffness tensor components C_{44} and C_{55}. Their variations with basal spacing are shown in Figure 8; C_{44} and C_{55} fluctuate with the increase of water content which is like the variation of C_{33}. Therefore, the mechanism of shear strength of interlayer water is closely related to the formation of H-bonds and atom-free volume. The shear strength between dry MMT lamellae is much larger than that of hydrated MMT lamellae, that is to say, the addition of water plays a lubricating role in interlayer friction, which makes C_{44} and C_{55} decrease sharply. However, these limited interlayer water molecules do not completely show the zero shear strength characteristics of free water. As water content increases, the in-plane shear strength increases to a local maximum when the system approaches the thermodynamic stable state and minimum atom free volume occurs. In the range of simulated water content, C_{44} and C_{55} eventually tend to be constant. Furthermore, the results in Figure 8 shows that the components C_{44} and C_{55} are in good agreement with those obtained by [29], and the in-plane shear strength between different MMT systems is close except for the dry MMT system. Meanwhile, the simulated results also fall well into the UPV measurement range of [17] (C_{44} = 2.9~8.9 GPa).

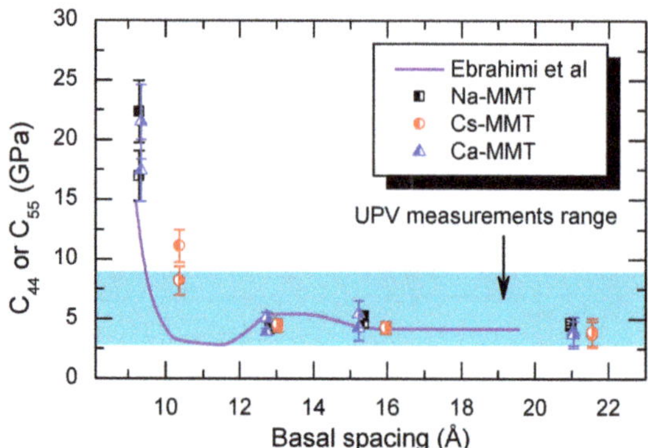

Figure 8. The C_{44} and C_{55} elastic constants as a function of basal layer spacing.

The shear of hydrated MMT along the direction perpendicular to the mineral plane is mainly controlled by the relative torsion of bonded atoms. The variations of C_{66} with basal spacing are shown in Figure 9. With the increase of interlayer water content, C_{66} decreases continuously, which is mainly due to the increase of interlayer water contribution in the weak shear zone. Considering that the ratio of width to height of a true MMT lamella is very large (about 100:1 to 1000:1), shear along the direction of orthogonal to the mineral plane is almost impossible, so the variations of C_{66} will not be discussed here.

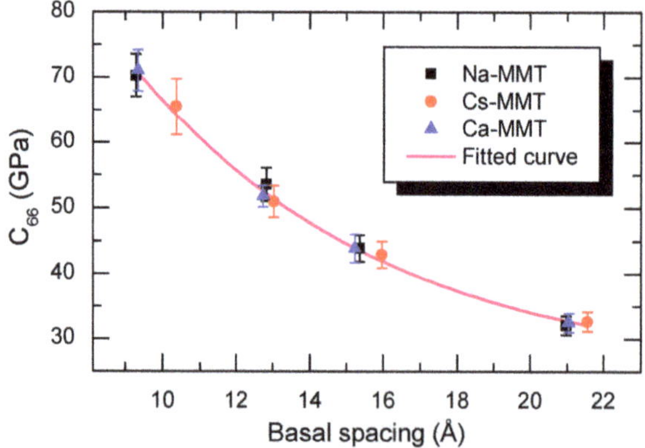

Figure 9. The C_{66} elastic constant as a function of basal layer spacing.

Young's moduli of the different MMT systems under various hydration can be obtained by further conversion of the stiffness tensor, and the variation of Young's moduli with basal spacing is shown in Figure 10. The in-plane Young's moduli E_1 and E_2 are consistent with C_{11} and C_{12} that decrease nonlinearly with the increase of water content. For the Young's modulus of E_3 which is perpendicular to the mineral plane, its value for the dry MMT system is 2–3 times that of the hydrated system for Na-MMT and Ca-MMT. E_3 of the hydrated MMT system also decreases nonlinearly with the increase of water content. Interestingly, for Cs-MMT, the Young's modulus, E_3, of different hydrated MMT systems, even dry MMT systems, is almost the same. Preliminary analysis shows that this may be related to the larger radius of Cs^+ ions. For the dry MMT system, the interlayer spacing is large, which induces a low compression strength. After hydration, the Cs^+ ions always form an inner-sphere complex, which has little effect on the strength, thus making all Cs-MMT systems of E_3 almost the same within the range of simulated water content. However, this interesting phenomenon has not yet been reported in the literature, so the above corollary needs further study in the future.

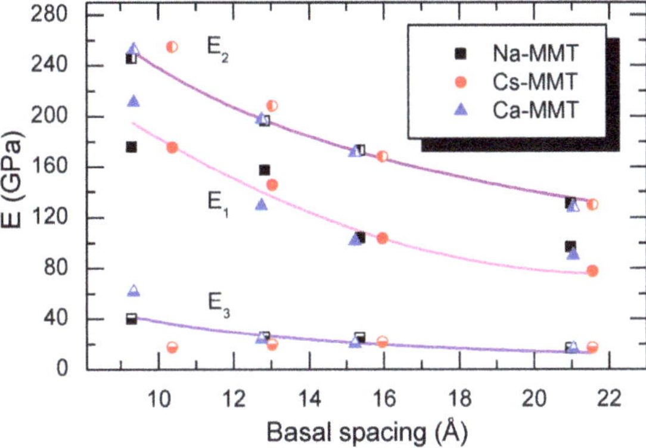

Figure 10. Young modulus as a function of basal layer spacing.

From the above comparisons, molecular microscopic testing techniques, such as UPV, can only qualitatively analyze the strength of clay minerals and the test results are not accurate enough. Molecular dynamics simulation provides a good way to understand the properties of hydrated clay minerals on an atomic scale. In particular, unconstrained system techniques and thermodynamic stable-state water content analysis techniques are used to make the calculation results of molecular dynamics more reliable.

4. Conclusions

Molecular dynamic simulations were conducted to investigate the mechanical properties of montmorillonite with different compensation cations upon hydration with unconstrained system atoms, expanding further studies of nanoscale elastic properties. The conclusions are as follows:

(1) The basal spacing of all MMT systems increases from step to step in the initial crystalline swelling stage. While this step characteristic (plateau) disappears, the basal spacing shows approximately linear variation in the subsequent osmotic swelling stage. However, the quantitative values of the basal spacing for different compensation cationic MMTs are different, which is related to the ion radius of cations.

(2) The nanoscale elastic properties of hydrated MMT show obvious anisotropy. The tensile and compressive strength (σ_3) of MMT orthogonal to the mineral plane is much lower than the in-plane strengths (σ_1 and σ_2), and the hydration molecular layer mixed within MMT lamellae has certain shear strengths (τ_4 and τ_5). The order of strength values from large to small is the compressive strength, the tensile strength, and the shear strength, respectively.

(3) The components of C_{33}, C_{44}, and C_{55} all fluctuate with the increase of interlayer water content, while the variation of the in-plane stiffness tensor components C_{11}, C_{22}, and C_{12}, all decrease nonlinearly with the increase of water content. These components are mainly controlled by the bonding strength of mineral atoms and the geometric conformation of the hydrated MMT system. The C_{66} component reflects the shear of hydrated MMT along the direction perpendicular to the mineral plane, which is mainly controlled by the relative torsion of bonded atoms. Due to the increase of interlayer water contribution in the weak shear zone, C_{66} decreases continuously with the increase of interlayer water content. Young's moduli all decrease nonlinearly with the increase of water content.

The simulation results reveal that the water molecules linked to the mineral surface will be arranged in an ordered structure (H-bond formation and thermodynamic dense stable state) and their shear strength are not zero. Preliminary discussion shows that this is related to the strength mechanism of clay at the microscale. Building a bottom-up procedure toward a constitutive law for soil mechanical still needs further research.

Author Contributions: Conceptualization, L.K. and Q.Z.; Methodology, X.S.; Validation, X.Z.; Formal analysis, L.K.; investigation, Q.Z.; writing—original draft preparation, L.K Writing-Original Draft Preparation, L.K.; Writing—review and editing, Q.Z. and X.Z.; Visualization, X.S.; Supervision, Q.Z. All authors have read and agreed to the published version of the manuscript.

Funding: This research was funded by the Fundamental Research Funds for the Central Universities, grant number 2020ZDPYMS18.

Institutional Review Board Statement: Not applicable.

Informed Consent Statement: Not applicable.

Data Availability Statement: The data presented in this study are available on request from the corresponding author.

Acknowledgments: The study presented in this article was supported by the Fundamental Research Funds for the Central Universities (2020ZDPYMS18). This support is gratefully acknowledged.

Conflicts of Interest: The authors declare no conflict of interest.

References

1. Sposito, G. *The Chemistry of Soils*, 2nd ed.; Oxford University Press, Inc.: Cary, NC, USA, 2008.
2. Bai, B.; Yang, G.C.; Li, T.; Yang, G.S. A thermodynamic constitutive model with temperature effect based on particle rearrangement for geomaterials. *Mech. Mater.* **2019**, *139*, 10318. [CrossRef]
3. Frenkel, D.; Smit, B. *Understanding Molecular Simulation*, 2nd ed.; Academin Press: San Diego, CA, USA, 2002.
4. Bai, B.; Rao, D.Y.; Chang, T.; Guo, Z.G. A nonlinear attachment-detachment model with adsorption hysteresis for suspension-colloidal transport in porous media. *J. Hydrol.* **2019**, *578*, 124080. [CrossRef]
5. Gilmour, K.A.; Davie, C.T.; Gray, N. An indigenous iron-reducing microbial community from MX80 bentonite—A study in the framework of nuclear waste disposal. *Appl. Clay Sci.* **2021**, *205*, 106039. [CrossRef]
6. Xu, H.; Zheng, L.G.; Rutqvist, J.; Birkholzer, J. Numerical study of the chemo-mechanical behavior of FEBEX bentonite in nuclear waste disposal based on the Barcelona expansive model. *Comput. Geotech.* **2021**, *132*, 103968. [CrossRef]
7. Wu, P.X.; Tang, Y.N.; Wang, W.M.; Zhu, E.W.; Li, P.; Wu, J.H.; Dang, Z.; Wang, X.D. Effect of dissolved organic matter from Guangzhou landfill leachate on sorption of phenanthrene by Montmorillonite. *J. Colloid Interf. Sci.* **2011**, *361*, 618–627. [CrossRef]
8. Ray, S.; Mishra, A.K.; Kalamdhad, A.S. Hydraulic performance, consolidation characteristics and shear strength analysis of bentonites in the presence of fly-ash, sewage sludge and paper-mill leachates for landfill application. *J. Environ. Manag.* **2022**, *302*, 113977. [CrossRef]
9. Bai, B.; Zhou, R.; Cai, G.Q.; Hu, W.; Yang, G.C. Coupled thermo-hydro-mechanical mechanism in view of the soil particle rearrangement of granular thermodynamics. *Comput. Geotech.* **2021**, *137*, 104272. [CrossRef]
10. Bai, B.; Nie, Q.K.; Zhang, Y.K.; Wang, X.L.; Hu, W. Cotransport of heavy metals and SiO_2 particles at different temperatures by seepage. *J. Hydrol.* **2021**, *597*, 125771. [CrossRef]
11. Zhu, B.; Wang, Y.M.; Liu, H.; Ying, J.; Liu, C.T.; Shen, C.Y. Effects of interface interaction and microphase dispersion on the mechanical properties of PCL/PLA/MMT nanocomposites visualized by nanomechanical mapping. *Compos. Sci. Technol.* **2020**, *190*, 108048. [CrossRef]
12. Shang, X.Y.; Zhou, G.Q.; Kuang, L.F.; Cai, W. Compressibility of deep clay in East China subjected to a wide range of consolidation stresses. *Can. Geotech. J.* **2015**, *52*, 244–250. [CrossRef]
13. Prasad, M.; Kopycinska, M.; Rabe, U.; Arnold, W. Measurement of Young's modulus of clay minerals using atomic force acoustic microscopy. *Geophys. Res. Lett.* **2002**, *29*, 1172. [CrossRef]
14. Vanorio, T.; Prasad, M.; Nur, A. Elastic properties of dry clay mineral aggregates, suspensions and sandstones. *Geophys. J. Int.* **2003**, *155*, 319–326. [CrossRef]
15. Kopycinska-Müller, M.; Prasad, M.; Rabe, U.; Arnold, W. Elastic properties of clay minerals determined by atomic force acoustic microscopy technique. *Acoust. Imaging* **2007**, *28*, 409–416.
16. Yang, C.; Xiong, Y.Q.; Wang, J.F.; Li, Y.; Jiang, W.M. Mechanical characterization of shale matrix minerals using phase-positioned nanoindentation and nano-dynamic mechanical analysis. *Int. J. Coal Geol.* **2020**, *229*, 103571. [CrossRef]
17. Ortega, J.A. Micropormechanical Modeling of Shale. Ph.D. Thesis, Massachusetts Institute of Technology, Cambridge, MA, USA, 2010.

18. Seo, Y.S.; Ichikawa, Y.; Kawamura, K. Stress-strain response of rock-forming minerals by molecular dynamics simulation. *Mater. Sci. Res. Int.* **1999**, *5*, 13–20. [CrossRef]
19. Mitchell, J.K.; Soga, K. *Fundamentals of Soil Behavior*, 3rd ed.; John Wiley & Sons, Inc.: Hoboken, NJ, USA, 2005.
20. Mao, H.; Huang, Y.; Luo, J.Z.; Zhang, M.S. Molecular simulation of polyether amines intercalation into Na-montmorillonite interlayer as clay-swelling inhibitors. *Appl. Clay Sci.* **2021**, *202*, 105991. [CrossRef]
21. Wei, P.C.; Zhang, L.L.; Zheng, Y.Y.; Diao, Q.F.; Zhuang, D.Y.; Yin, Z.Y. Nanoscale friction characteristics of hydrated montmorillonites using molecular dynamics. *Appl. Clay Sci.* **2021**, *210*, 106155. [CrossRef]
22. Sato, H.; Yamagishi, A.; Kawamura, K. Molecular simulation for flexibility of a single clay layer. *J. Phys. Chem. B* **2001**, *105*, 7990–7997. [CrossRef]
23. Manevitch, O.L.; Rutledge, G.C. Elastic properties of a single lamella of montmorillonite by molecular dynamics simulation. *J. Phys. Chem. B* **2004**, *108*, 1428–1435. [CrossRef]
24. Chen, B.Q.; Evans, J.R.G. Elastic moduli of clay platelets. *Scr. Mater.* **2006**, *54*, 1581–1585. [CrossRef]
25. Mazo, M.A.; Manevich, L.I.; Balabaev, N.K. Molecular dynamics simulation of thermo-mechanical properties of montmorillonite crystal. *Nanotechnol. Russ.* **2009**, *4*, 676–699. [CrossRef]
26. Zartman, G.D.; Liu, H.; Akdim, B.; Pachter, R.; Heinz, H. Nanoscale tensile, shear, and failure properties of layered silicates as a function of cation density and stress. *J. Phys Chem. C* **2010**, *114*, 1763–1772. [CrossRef]
27. Fu, Y.T.; Zartman, G.D.; Yoonessi, M.; Drummy, L.F.; Heinz, H. Bending of layered silicates on the nanometer scale: Mechanism, stored energy, and curvature limits. *J. Phys Chem. C* **2011**, *115*, 22292–22300. [CrossRef]
28. Teich-McGoldrick, S.L.; Greathouse, J.A.; Cygan, R.T. Molecular dynamics simulations of structural and mechanical properties of muscovite: Pressure and temperature effects. *J. Phys Chem. C* **2012**, *116*, 15099–15107. [CrossRef]
29. Ebrahimi, D.; Pellenq, R.J.-M.; Whittle, A.J. Nanoscale elastic properties of montmorillonite upon water adsorption. *Langmuir* **2012**, *28*, 16855–16863. [CrossRef] [PubMed]
30. Carrier, B.; Vandamme, M.; Pellenq, R.J.-M.; van Damme, H. Elastic properties of swelling clay particles at finite temperature upon hydration. *J. Phys Chem. C* **2014**, *118*, 8933–8943. [CrossRef]
31. Zheng, Y.; Zaoui, A. Mechanical behavior in hydrated Na-montmorillonite clay. *Phys. A* **2018**, *505*, 582–590. [CrossRef]
32. Zhu, L.P.; Shens, W.Q.; Shao, J.F.; He, M.C. Insight of molecular simulation to better assess deformation and failure of clay-rich rocks in compression and extension. *Int. J. Rock Mech. Min.* **2021**, *138*, 104589. [CrossRef]
33. Schmidt, S.R.; Katti, D.R.; Ghosh, P.; Katti, K.S. Evolution of mechanical response of sodium montmorillonite interlayer with increasing hydration by molecular dynamics. *Langmuir* **2005**, *21*, 8069–8076. [CrossRef]
34. Suter, J.L.; Coveney, P.V.; Greenwell, H.C.; Thyveetil, M.A. Large-scale molecular dynamics study of montmorillonite clay: Emergence of undulatory fluctuations and determination of material properties. *J. Phys. Chem. C* **2007**, *111*, 8248–8259. [CrossRef]
35. Liu, X.D.; Lu, X.C.; Wang, R.C.; Zhou, H.Q. Effects of layer-charge distribution on the thermodynamic and microscopic properties of Cs-smectite. *Geochim. Et Cosmochim. Acta* **2008**, *72*, 1837–1847. [CrossRef]
36. Cygan, R.T.; Liang, J.J.; Kalinichev, A.G. Molecular models of hydroxide, oxyhydroxide, and clay phases and the development of a general force field. *J. Phys. Chem. B* **2004**, *108*, 1255–1266. [CrossRef]

Article

Experimental Study on Microstructure of Unsaturated Expansive Soil Improved by MICP Method

Xinpei Yu [1], Hongbin Xiao [1,*], Zhenyu Li [1], Junfeng Qian [2], Shenping Luo [1] and Huanyu Su [1]

[1] School of Civil Engineering, Central South University of Forestry and Technology, Changsha 410018, China; 20191100320@csuft.edu.cn (X.Y.); 13787024735hdlizhenyu@163.com (Z.L.); 20191100323@csuft.edu.cn (S.L.); 20201100370@csuft.edu.cn (H.S.)

[2] Department of Civil Engineering, Monash University, Clayton, VIC 3800, Australia; junfeng.qian@monash.edu

* Correspondence: t20090169@csuft.edu.cn; Tel.: +86-135-7482-6532

Abstract: The soil water characteristic curve and microstructure evolution of unsaturated expansive soil improved by microorganisms in Nanning, Guangxi were studied by means of filter paper method and scanning electron microscope imaging (SEM). Based on Fredlung & Xing model, the influence law of different cement content on the soil water characteristic curve of improved expansive soil is proved. According to the analysis of SEM test results, the influence mechanism of MICP method on the engineering characteristics of improved expansive soil is revealed. The results show that with the increase of cement content, the saturated water content and residual water content of the improved expansive soil gradually increased. At the same time, the water stability gradually increased while the air inlet value gradually decreased. The improved expansive soil changes from the superposition of flat particles and flake particles to the contact between spherical particles and flake particles, which indicates that the aggregate increases significantly. With the increase of the content of cement solution, the contact between particles tends to be smooth and the soil pores gradually tend to be evenly distributed. The particle size and microstructure of soil particles was changed and the connection between particles was enhanced in the improved expansive soil. Eventually the strength and water stability of expansive soil were improved. The conclusions above not only provide a theoretical basis for the in-depth study of engineering characteristics of unsaturated expansive soil improved by MICP method, but also offer theoretical evidence for perfecting engineering technology of expansive soil improved by MICP method.

Keywords: microbially induced carbonate precipitation (MICP); unsaturated soil; soil-water characteristic curves; matrix suction; microstructure

Citation: Yu, X.; Xiao, H.; Li, Z.; Qian, J.; Luo, S.; Su, H. Experimental Study on Microstructure of Unsaturated Expansive Soil Improved by MICP Method. *Appl. Sci.* **2022**, *12*, 342. https://doi.org/10.3390/app12010342

Academic Editor: Bing Bai

Received: 7 December 2021
Accepted: 23 December 2021
Published: 30 December 2021

1. Introduction

Expansive soil is special catastrophic clay composed of strong hydrophilic minerals such as montmorillonite, illite, and kaolinite, which usually exhibit significant over-consolidation, multiple fissures, and swell-shrink characteristics. Expansive soils are distributed in a wide range in China and exist to varying degrees in more than 20 provinces such as Guangxi, Yunnan, and Sichuan. After expansive soil is exposed to water, it is highly susceptible to swelling deformation, softening and strength decay, and dry shrinkage cracks when water is lost. This repeated expansion and contraction deformation will lead to crack, subgrade uplift and slope sliding which will cause major engineering disasters [1–3]. Therefore, any construction work in expansive soil area must be treated and improved to eliminate potential engineering hazards. Traditional methods widely used in engineering for improving expansive soils are usually divided into physical and chemical improvement methods [4,5]. These methods of improving expansive soil could reduce expansive soil's swell-shrink characteristic effectively and also improve its strength. However, the traditional methods of improvement are not only labor-intensive and time-consuming, but also pollute the surrounding environment, and may even have negative impacts on the long-term ecosystem. Therefore, the academic and engineering circles urgently need a new, sustainable, and environmentally friendly alternative method to improve the poor engineering properties of expansive soil.

With the interpenetration of modern biotechnology and engineering science, bio-geotechnical engineering has become a new and hot research field. In recent decades, the Microbially Induced Carbonate Precipitation (MICP) method was used by scholars at home and abroad to repair fractures, seepage control, sand consolidation and soft soil improvement with good results [6–9]. In terms of applying the MICP method to improve the engineering characteristics of soft soils, scholars from home and abroad have mainly researched in improving sandy soils and made great progress [10–14]. In recent years, a few scholars have used the MICP method to improve expansive soil and other soft clays, which has made gratifying research progress. Bai studied the coupled migration of ions in soil [15]. Liu demonstrated that the MICP method can effectively improve the strength of loess through unconfined compressive strength test, calcium carbonate content test and scanning electron microscope test, and found the optimal reaction conditions and cementation liquid content for the improvement [16]. Safdar and Gowthaman studied the MICP method for improving peat soils. It was found that *Bacillus licheniformis* was more effective in improving peat soils, and that scallop powder was effective in increasing the activity of microorganisms and promoting the cementation of peat soil particles [17,18]. Some scholars investigated the swelling properties, strength properties, and microstructure of MICP modified expansive soils through a series of tests such as one-dimensional swelling tests, tensile tests, consolidation tests and compression tests. The MICP method of improved expansive soils can control the swelling and shrinking characteristics of the soil and improve the strength of the soil, and the calcite produced by the reaction can also provide a linkage to the soil particles [19–21]. Expansive soil in nature is usually unsaturated. Based on the theory of unsaturated soil mechanics, scholars carried out research on the engineering properties of unsaturated soil [22–26]. Soil-water Characteristic Curves (SWCC) obtained from unsaturated soil tests can be used to study the hydrophilicity, swell-shrinkage and strength characteristics of expansive soil improved by the MICP method.

In this paper, a series of experimental studies were carried out on expansive soil improved by the MICP method using the filter paper method. The SWCC of the expansive soil was measured after the improvement of different solutions with different cementation liquid content. The relationship between the moisture content of the expansive soil and the matric suction was obtained. Following comparison of four empirical models, Fredlund & Xing model was chosen to analyze the SWCC of the expansive soil after improvement by different schemes. The air entry value and water stability of the expansive soil were obtained.

The particle shape and pore space changes in the improved expansive soil were gotten by Scanning Electron Microscope (SEM) tests. From the perspective of microstructural changes in soil samples, the mechanism of the effect of different cementation liquid content on the water stability and water sensitivity of improved expansive soil during the MICP process was revealed.

2. SWCC Test Materials and Methods

2.1. Test Materials

The expansive soil used in the study was taken from Nanning, Guangxi, and was off-white. The particle size distribution of the unimproved expansive soil was obtained by taking the expansive soil for particle analysis tests. The particle size distribution of the expansive soils, as shown in Table 1.

Table 1. Particle size distribution of expansive soil.

≤0.005	(0.005,0.075]	(0.075,0.025]	(0.025,0.5]	(0.5,2]	(2,20)
15.4	62.9	7.1	6.9	5.2	2.6

The obtained expansive soil was dried, crushed, adopted 2 mm sieve and a portion of the particles less than 2 mm in diameter was taken for testing. The basic physical properties of the expansive soil were obtained from the tests and are shown in Table 2.

Table 2. Indicators of physical property of expansive soil.

Liquid Limit (%)	Plastic Limit (%)	Plastic Index (%)	Natural Moisture Content (%)	Optimal Moisture Content (%)	Maximum Dry Density (%)
58.1	22.3	35.8	21.8	23.3	1.75

2.2. *Sample Preparation of MICP Improved Expansive Soil*

2.2.1. Bacteria and Culture Media

The microorganism used to improve the expansive soil was *Bacillus pasteurii*, purchased from the Chinese General Microbial Strain Collection (CGMCC) under the number ATCC11859. This strain is a nonpathogenic, Gram-positive bacteria, which has no adverse effects on humans or the biological environment, a good ability to secrete urease. and is present in natural soil [27]. The *Bacillus pasteurii* used for the test is shown in Figure 1. The culture medium consisted of 20 g/L urea, 20 g/L agar, 15 g/L casein peptone, 5 g/L soy peptone and 5 g/L sodium chloride. *Bacillus pasteurii* was inoculated in a slant medium and incubated at 30 °C for 24 h and stored at 4 °C. The culture was inserted into an agar-free medium and incubated for 36–48 h in a constant temperature shaking incubator at a temperature of 30 °C and a speed of 150 r/min. The absorbance OD_{600} value of the bacterial liquid at a wavelength of 600 nm was measured using a spectrophotometer. When the $OD_{600} \approx 1.6$, the bacterial solution is used to improve the expansive soil.

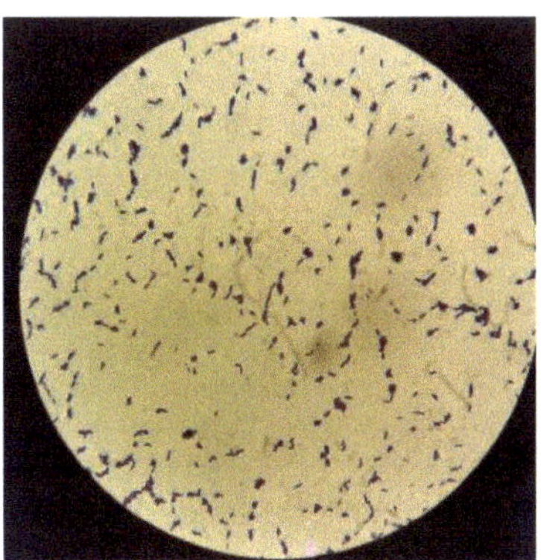

Figure 1. *Bacillus pasteurii* used in testing.

2.2.2. Preparation of Cementation Liquid

The cementation liquid is a source of calcium for the MICP process and provides the raw material and hydrolytic environment for the metabolic processes of the microorganisms. Based on the chemical reaction equation for the MICP process, it can be calculated that for every molecule of calcium carbonate produced, 1 molecule of calcium chloride and 1 molecule of urea are required [28]. Therefore, the ratio of calcium chloride to urea in the cementation liquid was determined to be 1:1 and the concentration of the cementation liquid was 1 mol/L. Each 100 mL of cementation liquid contained 21.91 g of $CaCl_2{\cdot}6H_2O$ and contained 6.01 g of urea.

2.2.3. Preparation of Soil Samples

The soil samples used in the test were divided into four groups, one of which was an unimproved expansive soil as a control group and the other three groups were samples that had been improved using the MICP method. Due to the high liquid limit and low permeability of expansive soil, conventional grouting and soaking methods cannot be used to improve expansive soil by the MICP method. In the tests, the expansive soil was amended by spraying and then mixing. The unimproved expansive soil was mixed with the cementation liquid and the bacterial fluid, respectively, and then kept in a constant temperature and humidity biochemistry cultivation cabinet at 30 °C for 14 days. During the curing process, holes need to be poked in the plastic film to provide the aerobic environment for bacterial metabolism to survive.

2.3. *Test Scheme and Procedure*

Four sets of tests were set up to investigate the effect of cementation liquid content on the SWCC of expansive soil during MICP. In addition to one group of unimproved expansive soil samples, three other groups of expansive soil samples were successively improved with cementation liquid and bacterial liquid. The improved soil samples were dehumidified and absorbed after 14 days of curing and weighed regularly to make sure they reached the moisture content required for the experimental design. The maximum dry density ρ_{dm} of the expansive soil is taken as the dry density of the specimen and the degree of compaction is 90%. Calculated the mass of loose soil required for the sample and made it into a ring-knife sample of size 61.8 × 20 mm. Based on relevant studies by Jiang [19] and Li [20] on the microbiological improvement of expansive soil, the admixture of cementation liquid and bacterial liquid in the experimental scheme was designed [19]. In the three groups of improved expansive soil samples, the content of cementation liquid was 100, 125, and 150 mL respectively, and the content of bacterial liquid was 50 mL. The test scheme for the MICP method of improving expansive soil is shown in Table 3.

Table 3. Test scheme of improved expansive soil by MICP method.

Total Sample	Sample Number	Bacterial Liquid Content (mL)	Cementation Liquid Content (mL)	Degree of Compaction (%)
112	A1, A2, A3	50	100, 125, 150	90

Dried filter paper ("Double Circle" brand, No. 203) is placed between two soil samples and three layers of filter paper are placed in each set of samples. The top and bottom layers are qualitative filter paper with a diameter of 60 mm and the middle layer is quantitative filter paper with a diameter of 50 mm. Use electrical insulation tape to seal the contact surface of the two ring knives, then wrap the test sample in plastic film and then wrap the plastic film tightly with electrical insulation tape. The wrapped samples are placed in the basin, sealed twice using plastic film and placed in a 20 °C thermostat. After 7 days, moisture equilibrium is reached between the filter paper and the expansive soil, at which point the matric suction of the filter paper and the expansive soil can be considered equal. The middle layer of the filter paper is removed with forceps and the mass of the wet filter paper is quickly weighed and recorded.

The test needs to be conducted under three assumptions. Firstly, it was assumed that *Bacillus pasteurii* with consistent absorbance would have the same activity and produce the same urease activity. Secondly, it is assumed that the expansive soils are well mixed during the treatment of the MICP method. Thirdly, it is assumed that the resting specimen is completely sealed during the filter paper method of testing. To ensure the reliability of the test results and to minimize errors arising during the test, a control group was designed in the experimental study. Measure the matric suction, as shown in Figure 2.

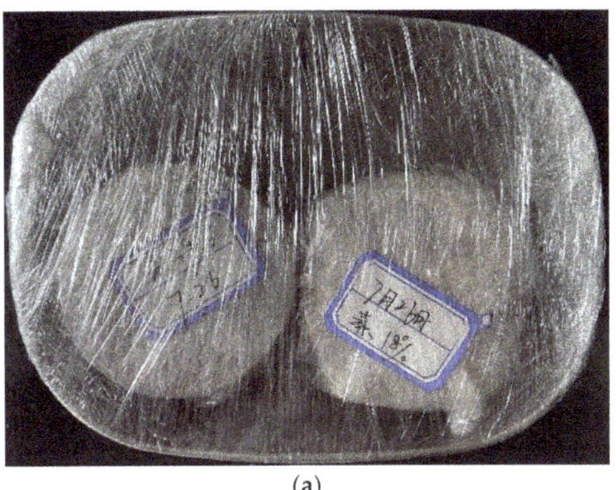

(a)

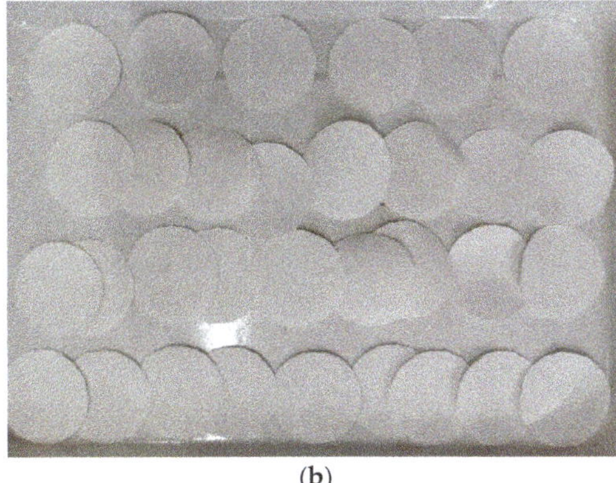

(b)

Figure 2. Measurement for matrix suction: (**a**) sealed storage; (**b**) dried filter paper.

According to the study of the calibration curve of the "double circle" brand filter paper by Bai Fuqing [29] and Zhang Hua [30], the matric suction of the soil sample is determined by using the calibration formula. The calibration result of matrix suction, as shown in Equation (1).

$$\begin{cases} \lg(S_m) = -0.0767\theta_e + 5.493 & \theta_e \leq 47\% \\ \lg(S_m) = -0.0120\theta_e + 2.470 & \theta_e > 47\% \end{cases} \quad (1)$$

where S_m represents the matrix suction, and θ_e represents the moisture content of the filter paper after equilibrium.

3. Results and Analysis of SWCC Test

3.1. Selection of the SWCC Fitting Model

To facilitate the analysis of the test data, empirical models of SWCC were used to fit the test results. Commonly used SWCC models include the Van Genuchten model (VG model) [31,32], the Gardner model [33] and the Fredlund & Xing model [34]. Among them, the VG model is in two- and three-parameter forms, and the commonly used empirical model equations are as follows.

$$\theta = \theta_r + \frac{\theta_s - \theta_r}{\left[1 + (\alpha\varphi)^n\right]^{(1-\frac{1}{n})}} \quad (2)$$

$$\theta = \theta_r + \frac{\theta_s - \theta_r}{\left[1 + (\alpha\varphi)^n\right]^{(m)}} \quad (3)$$

$$\theta = \theta_r + \frac{\theta_s - \theta_r}{1 + \left(\frac{\varphi}{\alpha}\right)^n} \quad (4)$$

$$\theta = \theta_r + \frac{\theta_s - \theta_r}{\left\{ln\left[e + \left(\frac{\varphi}{\alpha}\right)^n\right]\right\}^m} \quad (5)$$

where θ represents the soil volumetric moisture content, θ_s represents the saturated volumetric moisture content, θ_r represents the residual volumetric moisture content, e represents the base of the natural logarithm and $e \approx 2.71$, φ represents the matric suction, α, n, m all represents the fitted parameters.

The four models mentioned above were used to fit the test results of the expansive soils after improvement with different cementation liquid contents. The SWCC was obtained for the improved expansive soils with different fitting models, as shown in Figure 3.

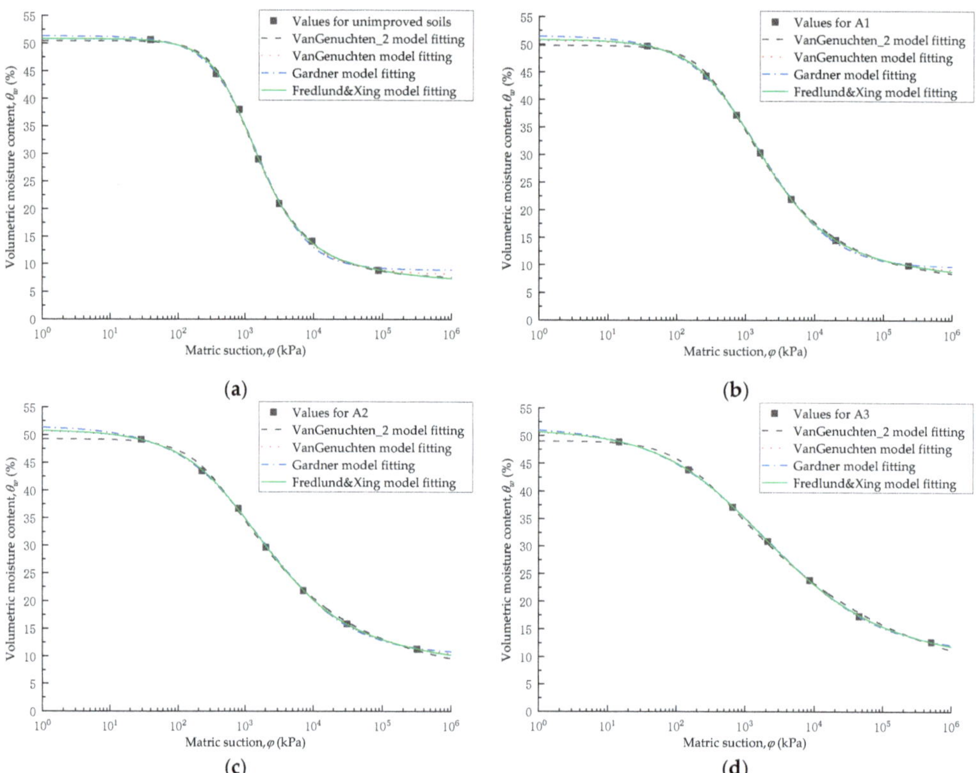

Figure 3. Soil-water characteristic curves obtained by model fitting: (**a**) fitted curve for unimproved expansive soil; (**b**) A1 fitted curve; (**c**) A2 fitted curve; (**d**) A3 fitted curve.

Figure 3 shows that the SWCC of the expansive soils is anti-"S" shaped after the different solutions. The transition section of SWCC in unimproved expansive soils is relatively steep, the rate of desorption in expansive soils is greater, matrix suction has a greater effect on the volumetric moisture content of the soil and the capacity to hold water is weaker. The transition section of the SWCC of the improved expansive soil tends to flatten out as the admixture of cementation liquid increases, the rate of desorption gradually decreases and the capacity to hold water gradually increases.

The saturated volumetric moisture content and residual volumetric moisture content of the improved expansive soils, and the fitting parameters of the different empirical models, as shown in Table 4.

Table 4. θ_s, θ_r, and correlation coefficient of fitting.

Sample Number	θ_s/%	θ_r/%	Correlation Coefficient R^2			
			VG_2 Model	VG Model	Gardner Model	F&X Model
Unimproved expansive soil	50.873	5.534	0.99743	0.99708	0.99710	0.99777
A1	50.904	5.901	0.99887	0.99999	0.99939	0.99999
A2	50.919	6.225	0.99758	0.99944	0.99932	0.99957
A3	51.179	6.467	0.99773	0.99980	0.99972	0.99972

From Table 4, the correlation coefficients of the four empirical models fitted above are all more than 0.99 when fitted by the models to the SWCCs of unsaturated expansive soils, indicating that the fitted results are reliable. The fitting of the various models, in descending order, is the Fredlund & Xing model, the three-parameter Van Genuchten model, the Gardner model and the two-parameter Van Genuchten model. The saturated volumetric moisture content of the expansive soil increased from 50.87% to 51.18% with an increase in the admixture of the cementation liquid at the same admixture of the bacterial liquid. Similarly, its residual volumetric moisture content increased from 5.53% to 6.47%. Furthermore, the increasing trend of both saturated and residual volumetric moisture content is non-linear. The Fredlund & Xing model with the optimum fit was selected to investigate the variation of the SWCC parameters of the improved expansive soils, to analyze the SWCC of the expansive soils under different cementation liquid content. It shows that the water holding capacity of the improved expansive soil is gradually increasing, and the water stability of the soil, which is positively correlated with the water-holding capacity also tends to increase.

3.2. Effect of Cementation Liquid Content on SWCC of Soil Samples

The Fredlund & Xing model was chosen to analyze the test results and the parameters of the model obtained after fitting, as shown in Table 5. As a result, the SWCC of the microbially improved expansive soil can be obtained for different cementation liquid content, as shown in Figure 4.

Table 5. Fitting parameters of Fredlund & Xing model.

Sample Number	α	n	m	Air Entry Value
Unimproved expansive soil	896.997	1.356	1.418	226.948
A1	818.070	0.959	1.424	155.666
A2	781.332	0.762	1.420	102.804
A3	760.739	0.585	1.462	64.240

The Fredlund & Xing model assumes that the parameters α, n and m are independent of each other. Where α is the soil parameter associated with the air entry value and can be used to characterize the value of matric suction at the inflection point of the soil-water characteristic curve [35]. As can be seen from Table 5, the air entry values for the expansive soils are 226.948, 155.666, 102.804 and 64.240 kPa from unimproved, A1 and A2 to A3 respectively. The air entry value of the improved expansive soil is significantly lower than that of the unimproved soil, and the air entry value of the improved soil tends to decrease slowly and non-linearly as the content of the cementation liquid increases. The reason for this is that there is a large amount of freedom Ca^{2+} in the cementation liquid, which displaces with the Na^+ in the expansive soil, reducing the thickness of the electric double layer, increasing the gravitational force between the particles and effectively promoting the coalescence of the particles. At the same time, the coagulation of the soil particles results in a significant reduction in the dispersion of the soil, an increase in the agglomerates and an increase in the porous between the soil particles, which causes a reduction in the air entry value. On the other hand, the larger porous produced by the microbially induced carbonate precipitation, adsorbed on the contact surface between soil particles and the particle surface, effectively enhance the interparticle linkage and reduce the air entry value of the expansive soil [36,37]. n is positively correlated with the slope at the inflection point of the SWCC, which can be used to characterize the water-holding capacity of the soil. The lower the slope at the SWCC inflection point, the better the water holding capacity of the soil [38].

As can be seen from Table 5, the values of the soil parameter n for the improved expansive soils have decreased compared to before the improvement. According to the value of n , the unimproved expansive soil has the weakest water-holding capacity. The value of n in the improved expansive soil decreases gradually from A1, A2 to A3 as the content of cementation liquid in the expansive soil increases under the condition of controlling the content of bacterial liquid of 50 mL.

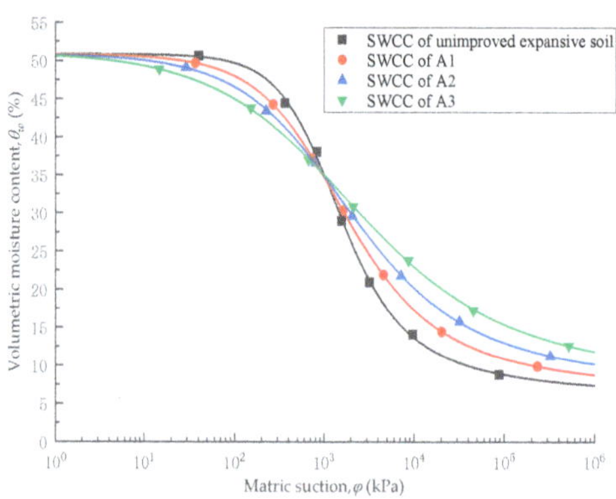

Figure 4. SWCCs of expansive soils under different cementing liquid content.

Figure 4 shows that the SWCC intersection of the expansive soils is located near the optimum moisture content under the conditions of the different improvement schemes. When low volumetric moisture content, the greater the content of cementation liquid, the greater the matric suction at the same volumetric moisture content. When the volumetric moisture content is higher than the moisture content at the SWCC intersection, the matric suction of the soil sample decreases with increasing the content of cementation liquid. From unimproved expansive soils to A1, A2, and A3, the SWCC of the expansive soils gradually levelled off, which means that the dehydration rate of the microbiological improved expansive soils gradually decreased, indicating that their water stability was gradually enhanced.

4. Microbial Improvement of the Microstructure of Expansive Soils

4.1. Scanning Electron Microscope Imaging

To investigate the microstructural evolution mechanism of the effect of different cementation liquid content on the soil-water characteristics of expansive soils under the same bacterial liquid content conditions. The TESCAN MIRA4 field emission scanning electron microscope was used to image and analyze the expansive soils before and after modification by the MICP method to investigate the changes in particle morphology and porous structure. The sample is an irregular sphere less than 1 cm in diameter and the sample was dried before the test. The surface of the sample is identified as the test surface, which is sprayed with gold to enhance conductivity. Secondary electron imaging was performed using a scanning electron microscope with an electron energy of 10 kev and a magnification of 5 kx.

4.2. Analysis of Test Results

Scanning electron microscope imaging tests were carried out to obtain SEM images of the expanded soil, as shown in Figure 5.

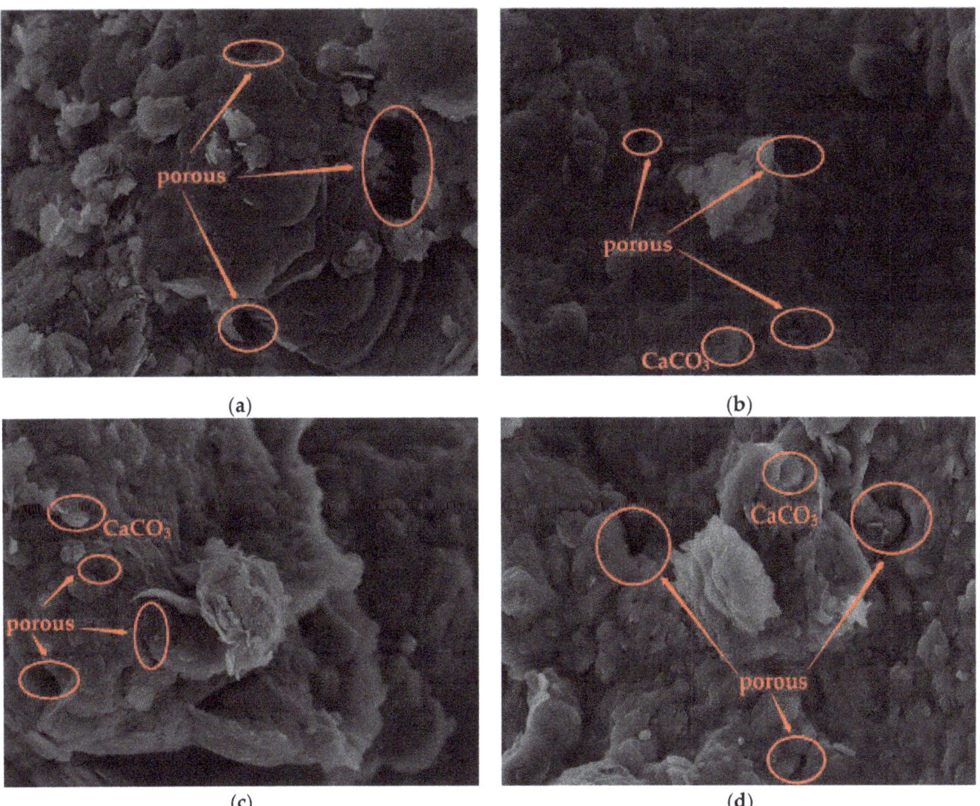

Figure 5. SEM images of expansive soil: (**a**) fitted curve for unimproved expansive soil; (**b**) A1 fitted curve; (**c**) A2 fitted curve; (**d**) A3 fitted curve.

Scanning electron microscope imaging tests were carried out to obtain SEM images of the expansive soil, as shown in Figure 5. The shape of the unimproved expansive soil particles is mainly flat and flake, the particles are mostly in edge-to-face or face-to-face contact, with large variability in pore size. The $CaCO_3$ crystals appeared at the contact points of the expansive soil particles after the MICP improved, and the morphology was mostly spherical particles. The contact between the particles tends to be smooth and the pore size distribution is relatively uniform. In the expansive soil improved by higher cementation liquid content, there are more $CaCO_3$ crystals and larger porous spaces, and the porous spaces are evenly distributed. At low levels of cementation liquid, the expansive soils contain mostly small or medium porous spaces that are uniformly distributed. Li carried out XRD analysis on the improved expansive soils, the study found that the calcite characteristic peaks increased significantly, and the main peak values were enhanced in the expansive soils treated by the MICP method [39]. The calcium carbonate precipitate produced by the test was of a different shape to the typical rhombohedral calcite. This phenomenon might be caused by the shape of the particles being obscured by the soil, or by some bacteria and fine clay particles remaining on the surface of the crystals. Based on the variation of the particle porous structure, a microscopic perspective reveals the internal mechanism by which the air entry value of expansive soils gradually decreases with increasing contents of the cementation liquid. Compared to that of the cohesion of unimproved expansive soils, which are strongly influenced by matric suction and are more water sensitive, matric suction no longer has a restraining effect on the structure after saturation [40]. After improvement by the MICP method, the $CaCO_3$ crystals produced better cementation between the soil particles. The greater the content of the cementation liquid, the more calcium carbonate precipitation is induced and the greater the contact surface of the particles formed, which promotes the agglomeration of the soil particles, and the strength of the expansive soil is improved. Moreover, due to the low solubility of calcium carbonate, it is less affected by water, thus significantly improving the water stability of the expansive soil.

The swelling-shrinkage characteristics of expansive soils in the unsaturated state are mainly related to the water film thickness on the surface of the soil particles when they react with water. The process of water swelling in expansive soils is in essence a process whereby hydrophilic soil particles are wrapped in water molecules under the influence of electric field forces to form a water film. The thickness of the water film gradually increases as the expansive soil absorbs water, resulting in the expansion of the soil particle lattice and the swelling of the soil [41–43]. The unimproved expansive soil particles are mostly in the form of tightly packed or stacked flakes, the lattice can only expand longitudinally outwards when exposed to water. With the increase in the content of the cementation liquid, the flaky particles in the improved expansive soil gradually decrease and the pore space gradually increases. When reacting with water, the lattice of the improved soil particles expands in every direction, balancing the expansion potential in the internal space of the soil before expanding outwards, effectively reducing the expansion deformation of the expansive soil. At the same time, the microbially induced formation of calcium carbonate adheres to the surface of the expansive soil particles, reducing the interaction area between hydrophilic soil particles and water, reducing the water film thickness and interparticle spacing, tightening the soil structure and improving the strength of the soil effectively.

5. Conclusions

The SWCC of the expansive soils improved by the MICP method was obtained using the filter paper method under different conditions of cementation liquid content. Based on the theory of unsaturated soil mechanics, the experimental study and theoretical analysis were carried out on the water stability of improved expansive soils. At the same time, scanning electron microscope imaging technology was used to test on expansive soil samples. Based on the analysis of SWCC, the microscopic mechanisms affecting the water sensitivity and strength properties of the improved expansive soil were revealed. The following conclusions were obtained.

(1) The content of the cementation liquid has a significant influence on the SWCC of the expansive soils improved by the MICP method. The saturated and residual water content of the improved expansive soils gradually increases with the content of cementation liquid increased. The air entry value decreases, and the water stability increases gradually.

(2) Scanning electron microscope imaging tests were carried out on the expansive soils before and after improvement. The change of particle morphology and porous structure of the expansive soils under different conditions of cementation liquid content was investigated. The study shows that, compared to that of the unimproved expansive soil, the particle composition of the improved expansive soil evolves from the mutual superposition of flattened and flaky particles to spherical particles in contact with flaky particles, which increased agglomerates significantly. At the same time, with the increase in cementation liquid content, the contact between the particles tends to be smoother and the soil porous tends to be uniformly distributed gradually.

(3) Based on the analysis of macroscopic and microscopic test results, it was found that the mineralization process of microorganisms changed the particle size of soil

particles and the porous structure of the soil. The microscopic mechanism affecting the water stability, swelling and shrinkage characteristics and strength properties of the improved expansive soil were revealed. After the improvement, the calcium carbonate formed by microbial induction precipitates on the surface of the soil particles and in the soil pores, enhancing the interparticle linkage, reducing the hydrophilicity and swelling-shrinkage of the soil, and improving the strength and water stability of the soil. The study also shows that the antierosion ability of the soil is improved significantly due to the increase of aggregates in the soil, the coarsening of the soil particles, and the decreasing water sensitivity of the soil.

The MICP method of improving expansive soils is effective in increasing the strength and water-holding capacity of the soil, but the ammonium ions produced during the reaction can also contaminate the soil [44,45]. The management of ammonium ions in the improvement process is an issue worthy of further study.

Author Contributions: Conceptualization, X.Y. and H.X.; methodology, X.Y. and H.X.; software, X.Y. and H.X.; validation, X.Y. and H.X.; formal analysis, X.Y., H.X., Z.L., J.Q., S.L. and H.S.; investigation, X.Y., H.X., Z.L., J.Q., S.L. and H.S.; resources, X.Y. and H.X.; data curation, X.Y. and H.X.; writing—original draft preparation, X.Y. and H.X.; writing—review and editing, X.Y., H.X. and Z.L.; visualization, X.Y., H.X., Z.L., J.Q., S.L. and H.S.; supervision, X.Y., H.X. and Z.L.; project administration, H.X.; funding acquisition, H.X. All authors have read and agreed to the published version of the manuscript.

Funding: This research was funded by the National Natural Science Foundation of China, grant number 50978097.

Institutional Review Board Statement: Not applicable.

Informed Consent Statement: Not applicable.

Data Availability Statement: The data presented in this study are available on request from the corresponding author.

Acknowledgments: The work described in this paper was supported by a grant from the National Natural Science Foundation of China (Project No. 50978097).

Conflicts of Interest: The authors declare no conflict of interest.

References

1. Xu, Y. Hydraulic Mechanism and Swelling Deformation Theory of Expansive Soils. *Chin. J. Geotech. Eng.* **2020**, *42*, 1979–1987.
2. Ye, Y.; Zou, W.; Han, Z.; Xie, P.; Zhang, J.; Xu, Y. Study on One-dimensional Swelling Characteristics of Expansive Soils Considering the Influence of Initial States. *Chin. J. Geotech. Eng.* **2021**, *43*, 1518–1525.
3. Ning, X.; Xiao, H.; Zhang, C.; He, B.; Xie, J. Study on the Nonlinear Creep Model of Expansive Soil. *J. Nat. Disasters* **2017**, *26*, 149–155.
4. Li, G.; Gong, Q.; Li, T.; Wu, J.; Chen, W.; Cao, X. Experimental Study on Properties of Weak Expansive Soil Improved by Disintegrated Sandstone. *J. Eng. Geol.* **2021**, *29*, 34–43.
5. Chuang, X.; Zhou, M.; Tao, G.; Zhou, R.; Peng, C.; Lin, W. Experimental Study of Dynamic Elastic Modulus and Damping Ratio of Improved Expansive Soil under Cyclic Loading by Expanded Polystyrene. *Rock Soil Mech.* **2021**, *42*, 2427–2436.
6. Bai, B.; Long, F.; Rao, D.; Xu, T. The effect of temperature on the seepage transport of suspended particles in a porous medium. *Hydrol. Processes* **2017**, *31*, 382–393. [CrossRef]
7. He, X.; Liu, H.; Han, F.; Ma, G.; Zhao, C.; Chu, Q.; Xiao, Y. Spatiotemporal Evolution of Microbial-induced Calcium Carbonate Precipitation Based on Microfluidics. *Chin. J. Geotech. Eng.* **2021**, *43*, 1861–1869.
8. Seung, H.C.; Hyeonyong, C.; Kyoungphile, N. Evaluation of Microbially Induced Calcite Precipitation (MICP) Methods on Different Soil Types for Wind Erosion Control. *Environ. Eng. Res.* **2021**, *26*, 123–128.
9. Bai, B.; Nie, Q.Z.; Wang, X.; Hu, W. Cotransport of heavy metals and SiO_2 particles at different temperatures by seepage. *J. Hydrol.* **2021**, *597*, 125771. [CrossRef]
10. Zhang, Y.; Wan, X.; Li, N.; Zhao, W.; Chen, J. Analysis of Influencing Factors and Mechanism of Microbial Clay Reinforcement. *J. China Inst. Water Resour. Hydropower Res.* **2021**, *19*, 246–254+261.
11. Li, X.; Wang, S.; He, B.; Shen, T. Permeability Condition of Soil Suitable for MICP Method. *Rock Soil Mech.* **2019**, *40*, 2956–2964.
12. Zhao, C.; Zhang, J.; Zhang, Y.; He, X.; Ma, G.; Liu, H.; Xiao, Y. Research Progress on Multi-scale Biocemented Soil. *J. Beijing Univ. Technol.* **2021**, *47*, 792–801.

13. Xiao, Y.; Wang, Y.; Wang, S.; Evans, T.M.; Stuedlein, A.W.; Chu, J.; Zhao, C.; Wu, H.; Liu, H. Homogeneity and Mechanical Behaviors of Sands Improved by A Temperature-controlled One-phase MICP Method. *Acta Geotech.* **2021**, *2021*, 1417–1427. [CrossRef]
14. Konstantinou, C.; Biscontin, G.; Jiang, N.; Soga, K. Application of Microbially Induced Carbonate Precipitation to Form Bio-cemented Artificial Sandstone. *J. Rock Mech. Geotech. Eng.* **2021**, *13*, 579–592. [CrossRef]
15. Bai, B.; Jiang, S.; Liu, L.; Li, X.; Wu, H. The transport of silica powders and lead ions under unsteady flow and variable injection concentrations. *Powder Technol.* **2021**, *387*, 22–30. [CrossRef]
16. Liu, X.; Fan, J.; Yu, J.; Gao, X. Solidification of Loess Using Microbial Induced Carbonate Precipitation. *J. Mt. Sci.* **2021**, *18*, 265–274. [CrossRef]
17. Safdar, M.U.; Mavroulidou, M.; Gunn, M.J.; Garelick, J.; Payne, I.; Purchase, D. Innovative Methods of Ground Improvement for Railway Embankment Peat Fens Foundation Soil. *Géotechnique* **2021**, *71*, 985–998. [CrossRef]
18. Gowthaman, S.; Chen, M.; Nakashima, K.; Kawasaki, S.; Frontiers, M. Effect of Scallop Powder Addition on MICP Treatment of Amorphous Peat. *Front. Environ. Sci.* **2021**, *9*, 690376. [CrossRef]
19. Jiang, W.; Zhang, C.; Xiao, H.; Luo, S.; Li, Z.; Li, X. Preliminary Study on Microbially Modified Expansive Soil of Embankment. *Geomech. Eng.* **2021**, *26*, 301–310.
20. Li, X.; Zhang, C.; Xiao, H.; Jiang, W.; Qian, J.; Li, Z. Reducing Compressibility of the Expansive Soil by Microbiological-Induced Calcium Carbonate Precipitation. *Adv. Civ. Eng.* **2021**, *2021*, 1–12. [CrossRef]
21. Tiwari, N.; Satyam, N.; Sharma, M. Micro-mechanical Performance Evaluation of Expansive Soil Biotreated with Indigenous Bacteria Using MICP Method. *Sci. Rep.* **2021**, *29*, 1–12. [CrossRef]
22. Abdelbaki, A.M. Assessing the Best Performing Pedotransfer Functions for Predicting the Soil-water Characteristic Curve According to Soil Texture Classes and Matric Potentials. *Eur. J. Soil Sci.* **2021**, *72*, 154–173. [CrossRef]
23. Bai, B.; Xu, T.; Nie, Q.; Li, P. Temperature-driven migration of heavy metal Pb^{2+} along with moisture movement in unsaturated soils. *Int. J. Heat Mass Transf.* **2020**, *153*, 119573. [CrossRef]
24. Bai, B.; Rao, D.; Chang, T.; Guo, Z. A nonlinear attachment-detachment model with adsorption hysteresis for suspension-colloidal transport in porous media. *J. Hydrol.* **2019**, *578*, 124080. [CrossRef]
25. Zhang, G.; Wang, L.; Cao, Z.; Li, Q. Comparative Study of Different Methods for Determining the Soil Water Characteristic Curve Model Considering Uncertainty. *J. Nat. Disasters* **2018**, *27*, 151–158.
26. Fredlund, D.G. State of Practice for Use of the Soil-Water Characteristic Curve (SWCC) in Geotechnical Engineering. *Can. Geotech. J.* **2019**, *56*, 1059–1069. [CrossRef]
27. Sarda, D.; Choonia, H.S.; Sarode, D.D.; Lele, S.S. Biocalcification by *Bsacillus pasteurii* Urease: A Novel Application. *J. Ind. Microbiol. Biotechnol.* **2009**, *36*, 1111–1115. [CrossRef]
28. Pei, D.; Liu, Z.; Hu, B.; Wu, W. Progress on Mineralization Mechanism and Application Research of *Sporosarcina pasteurii*. *Prog. Biochem. Biophys.* **2020**, *47*, 467–482.
29. Bai, F.; Liu, S.; Yuan, J. Measurement of SWCC of Nanyang Expansive Soil Using the Filter Paper Method. *Chin. J. Geotech. Eng.* **2011**, *33*, 928–933.
30. Zhang, H.; Zhang, F.; Yuan, M. Study of Temperature Effect on Domestic Quantitative Filter Paper's Calibration Curves. *Rock Soil Mech.* **2016**, *37*, 274–280.
31. van Genuchten, M.T. A Closed-form Equation for Predicting the Hydraulic Conductivity of Unsaturated Soils. *Soil Sci. Soc. Am. J.* **1980**, *44*, 892–898. [CrossRef]
32. Pan, D.; Ni, W.; Yuan, K.; Zhang, Z.; Wang, X. Determination of Soil-water Characteristic Curve Variables Based on VG Model. *J. Eng. Geol.* **2020**, *28*, 69–76.
33. Gardner, W.R. Some Steady-state Solutions of the Unsaturated Moisture Flow Equation with Application to Evaporation from A Water Table. *Soil Sci.* **1958**, *85*, 228–232. [CrossRef]
34. Fredlund, D.; Xing, A. Equations for the Soil-Water Characteristic Curve. *Can. Geotech. J.* **1994**, *31*, 521–532. [CrossRef]
35. Zhou, B.; Kong, L.; Chen, W.; Bai, H.; Li, X. Analysis of Characteristic Parameters of Soil-water Characteristic Curve (SWCC) and Unsaturated Shear Strength Prediction of Jingmen Expansive Soil. *Chin. J. Rock Mech. Eng.* **2010**, *29*, 1052–1059.
36. Li, C.; Shi, G.; Wu, H.; Wang, C.; Gao, Y. Experimental Study on Bio-mineralization for Dispersed Soil Improvement Based on Enzyme Induced Calcite Precipitate Technology. *Rock Soil Mech.* **2021**, *42*, 333–342.
37. Wang, Z.; Zhang, N.; Cai, G.; Jin, Y.; Ding, N.; Shen, D. Review of Ground Improvement Using Microbial Induced Carbonate Precipitation (MICP). *Mar. Georesour. Geotechnol.* **2017**, *35*, 1135–1146. [CrossRef]
38. Wang, X.; Xu, C.; Wang, S.; Li, X. Study on Overburden Pressure Effect of Expansive Soil-water Characteristic Curve. *Chin. J. Rock Mech. Eng.* **2020**, *39*, 3067–3075.
39. Li, X. Study on Microstrcture and Mechanical Propeities of Expansive Soil Improved by MICP Technology. Ph.D. Thesis, Central South University of Forestry and Technology, Changsha, China.
40. Yao, H.; Liu, S.; Cheng, C. Scopic Essence of Cohesion of a Natural Cemented Soil. *Chin. J. Rock Mech. Eng.* **2001**, *20*, 871–874.
41. Leng, T.; Tang, C.; Xu, D.; Li, Y.; Zhang, Y.; Wang, K.; Shi, B. Advance on the Engineering Geological Characteristics of Expansive Soil. *J. Eng. Geol.* **2018**, *26*, 112–128.
42. Bai, B.; Nie, Q.; Wu, H.; Hou, J. The attachment-detachment mechanism of ionic/nanoscale/microscale substances on quartz sand in water. *Powder Technol.* **2021**, *394*, 1158–1168. [CrossRef]

43. Tan, L. The Effect of Lattice Expansion and Contraction of Montmorillonite on its Volume Change. *Hydrogeol. Eng. Geol.* **1981**, *06*, 5–7+38.
44. Keykha, H.A.; Mohamadzadeh, H.; Asadi, A.; Kawasaki, S. Ammonium-Free Carbonate-Producing Bacteria as an Ecofriendly Soil Biostabilizer. *Geotech. Test. J.* **2019**, *42*, 19–29. [CrossRef]
45. Mohsenzadeh, A.; Aflaki, E.; Gowthaman, S.; Kawasaki, S.; Ebadi, T. A two-stage treatment process for the management of produced ammonium by-products in ureolytic bio-cementation process. *Int. J. Environ. Sci. Technol.* **2021**, *9*, 18313. [CrossRef]

Article

Experimental Investigation on the Effects of Ethanol-Enhanced Steam Injection Remediation in Nitrobenzene-Contaminated Heterogeneous Aquifers

Ruxue Liu [1,2,3], Xinru Yang [1,2,3], Jiayin Xie [1,2,3], Xiaoyu Li [1,2,3] and Yongsheng Zhao [1,2,3,*]

1 Key Laboratory of Groundwater Resources and Environment of Ministry of Education, College of New Energy and Environment, Jilin University, Changchun 130021, China; liurx18@mails.jlu.edu.cn (R.L.); xryang19@mails.jlu.edu.cn (X.Y.); xiejy20@mails.jlu.edu.cn (J.X.); xyl21@mails.jlu.edu.cn (X.L.)
2 Jilin Provincial Key Laboratory of Water Resources and Environment, Jilin University, Changchun 130021, China
3 National and Local Joint Engineering Laboratory for Petrochemical Contaminated Site Control and Remediation Technology, Jilin University, Changchun 130021, China
* Correspondence: zhaoyongsheng@jlu.edu.cn

Abstract: Steam injection is an effective technique for the remediation of aquifers polluted with volatile organic compounds. However, the application of steam injection technology requires a judicious selection of stratum media because the remediation effect of hot steam in heterogeneous layers with low permeability is not suitable. In this study, the removal effect of nitrobenzene in an aquifer was investigated through a series of two-dimensional sandbox experiments with different stratigraphic structures. Four types of alcohols were used during steam injection remediation to enhance the removal effect of nitrobenzene (NB)-contaminated heterogeneous aquifers. The principle of the removal mechanism of alcohol-enhanced organic compounds is that alcohols can reduce the surface tension of the contaminated water, resulting in Marangoni convection, thereby enhancing mass and heat transfer. The addition of alcohol may also reduce the azeotropic temperature of the system and enhance the volatility of organic compounds. The study revealed that all four alcohol types could reduce the surface tension from 72 mN/m to <30 mN/m. However, among these, only ethanol reduced the azeotropic temperature of NB by 15 °C, thereby reducing energy consumption and remediation costs. Therefore, ethanol was selected as an enhancing agent to reduce both surface tension and azeotropic temperature during steam injection. In the 2-D simulation tank, the interface between the low-and high-permeability strata in the layered heterogeneous aquifer had a blocking effect on steam transportation, which in turn caused a poor remediation effect in the upper low-permeability stratum. In the lens heterogeneous aquifer, steam flows around the lens, thereby weakening the remediation effect. After adding ethanol to the low-permeability zone, Marangoni convection was enhanced, which further enhanced the mass and heat transfer. In the layered and lens heterogeneous aquifers, the area affected by steam increased by 13% and 14%, respectively. Moreover, the average concentration of NB was reduced by 51% in layered heterogeneous aquifers and by 58% in low-permeability lenses by ethanol addition. These findings enhance the remediation effect of steam injection in heterogeneous porous media and contribute to improve the remediation efficiency of heterogeneous aquifers by steam injection.

Keywords: steam injection; ethanol; azeotropic temperature; heterogeneous aquifers; nitrobenzene

Citation: Liu, R.; Yang, X.; Xie, J.; Li, X.; Zhao, Y. Experimental Investigation on the Effects of Ethanol-Enhanced Steam Injection Remediation in Nitrobenzene-Contaminated Heterogeneous Aquifers. *Appl. Sci.* **2021**, *11*, 12029. https://doi.org/10.3390/app112412029

Academic Editor: Bing Bai

Received: 1 November 2021
Accepted: 24 November 2021
Published: 17 December 2021

1. Introduction

Steam injection is an important remediation technology that can effectively remove organic pollutants during site remediation [1–3]. In particular, steam injection technology can remediate contaminated sites containing multiple volatile or semi-volatile organic pollutants [4]. The remediation efficiency of steam injection technology primarily depends on the permeability of the stratum medium [2,5,6]. The volatility of organic compounds in water is closely related to temperature [7–10]. However, the transmission of steam may be limited by the soil particle size, resulting in a reduction in remediation efficiency [11,12]. In general, the remediation efficiency is the highest in homogeneous sand because of the large particle size and non-layered nature of the soil. Many researchers have used homogeneous sand to analyze the characteristics of steam injection [4,13,14]. However, a heterogeneous stratum structure is quite common at actual remediation sites [15,16]; thus, the impact of heterogeneous strata on the remediation effect should be considered.

Some studies have shown that the heterogeneity of the stratum can lead to unsatisfactory remediation effects [2]. Even in sandy soils, the heterogeneity of soil also affects the remediation effect. At present, systematic research on the migration, distribution, and remediation effects of steam in heterogeneous aquifers is lacking. In addition, only a few studies exist on the enhanced steam remediation of heterogeneous aquifers.

In the present study, a combination of thermal technologies such as steam injection with thermal conduction heating (TCH) [17] or steam injection with electrical resistance heating (ERH) was adopted to improve the remediation efficiency of steam injection in heterogeneous layers [18], with the aim of using steam injection in high-permeability zones and TCH or ERH in low-permeability zones. However, these methods are more expensive and complicated. Therefore, a method for strengthening steam injection to repair heterogeneous aquifers is needed.

Chao et al. proposed the addition of alcohol in the repair process of air sparging (AS) to enhance remediation efficiency [19]. N-pentanol and n-heptanol were selected as strengthening reagents as both could enhance the removal efficiency of organic compounds; however, the effect of n-heptanol was stronger than that of n-pentanol. The key mechanism is that alcohol molecules can interact with organic compounds at the interface, resulting in increased organic compound volatilization. Based on the theory of enhanced volatilization, it is assumed that alcohols can enhance volatilization more than surfactants [19,20]. Volatility is also a key factor in the process of steam injection and remediation of organic pollutants, which also determines the increased volatilization of organic compounds [21]. According to some studies, alcohol can be used as a surfactant to reduce the surface tension of the solution [22,23] and thus can be used to enhance the volatilization of organic compounds. Previous studies have shown that the addition of ethanol, n-propanol, and n-butanol can reduce surface tension and enhance mass transfer [24]. This study aimed to explore the effects of these alcohols on organic compounds as strengthening reagents.

During steam injection remediation, the removal rate of organic compounds increases with an increase in temperature, and the remediation efficiency increases significantly at boiling point [8,25,26]. Some organics can form azeotropes with water at temperatures lower than their boiling point [27,28]. Therefore, a lower azeotropic temperature can quickly remove organic matter. As alcohol and water can form azeotropes, this study investigated which alcohol could better reduce the azeotropic temperature of organic compounds using the common semi-volatile organic compound nitrobenzene (NB) in contaminated sites as the target pollutant.

Alcohol-enhanced steam injection has rarely been studied, therefore, this study investigated the effects of alcohol on the remediation of heterogeneous aquifers. The effects of the four alcohols (ethanol, n-butanol, n-pentanol, and n-heptanol) on the surface tension and azeotropic temperature of NB were studied. Heterogeneous layers with different permeability ratios (R) were selected to compare the temperature distribution and removal efficiency of organic pollutants under different permeability ratios. The migration and distribution mechanisms of steam in a lens heterogeneous aquifer were also studied. In addition, the effect of alcohol-enhanced remediation was studied for a stratum with a weak remediation effect. This study aims to select an alcohol to enhance the effect of steam injection to remediate contaminated heterogeneous aquifers, thus extending the application of steam injection technology.

2. Theoretical Background

2.1. Marangoni Effect

In a gas–liquid mass transfer process, the interface condition is important. Interfacial tension, molecular structure of the interface, intermolecular forces of combining gas and liquid molecules, and interface turbulence all have an important impact on the gas–liquid mass transfer process [29]. During gas–liquid mass transfer, the interfacial convection phenomenon is caused by the change in some physical properties of the liquid such as surface tension and density due to the mass transfer at the interface, which may lead to convective movement of the interfacial fluid [29–31]. Researchers attribute this to the Marangoni effect or Rayleigh effect caused by transmission, which in turn strengthens the transmission process [30,32]. During interphase mass transfer, when the temperature or concentration in different areas on the surface are different, a surface tension gradient is generated, resulting in the movement of liquid on the surface layer and liquid under the interface. This interfacial convection phenomenon is known as the Marangoni effect. The Marangoni effect not only has various geometric patterns on the flow structure, but also enhances the heat and mass transfer processes. Marangoni convection can be characterized by the dimensionless number Ma, which is expressed as follows:

$$\mathrm{Ma} = \frac{\mathrm{d}\Delta\sigma}{\mathrm{D}\mu} \quad (1)$$

where d is the characteristic length (m); $\Delta\sigma$ is the difference in surface tension between the surface fluid and the main fluid (N m^{-1}); D is the diffusion coefficient of the solute in the fluid (m^2 s^{-1}); and μ is the viscosity of the fluid (Pa s). In Equation (1), when Ma > 0 and exceeds a certain critical value, the surface tension gradient generated by the interface will cause Marangoni convection. In gas–liquid and liquid–liquid liquefaction processes, due to the changes in the physical and chemical properties of the system [33,34], the heat and mass transfer processes cause interfacial instability, and the surface tension is a function of concentration and temperature, which causes the temperature and concentration gradient of the interface to lead to the surface tension gradient of the interface, resulting in Marangoni convection.

2.2. Alcohol-and Heat-Enhanced Marangoni Convection

As the temperature change along the interphase surface can promote the surface tension gradient, it will also cause the Marangoni convection [35,36]. After heating, the surface tension decreases and the surface tension gradient increases, thereby enhancing the Marangoni convection and strengthening the mass transfer process [37,38]. The addition of alcohol reduces the surface tension, and with the evaporation of alcohol, the concentration of alcohol on the liquid film decreases, promoting the diffusion of alcohol molecules to the liquid film and enhancing the gradient of surface tension. This enhances the Marangoni convection [39,40]. The existence of interfacial convection can greatly enhance the interphase mass transfer process and reduce energy consumption.

3. Materials and Methods

3.1. Materials

Analytically pure (AR)-grade NB was obtained from Sinopharm Chemical Reagent Co. Ltd., and the alcohols used in this study are shown in Table 1. All the materials were used without further purification, and all aqueous solutions were prepared with deionized water.

3.2. Experimental Methods

3.2.1. Surface Tension

Because ethanol and water are miscible, the surface tension at different volume fractions of ethanol was measured using a surface tension meter. Different volumes of ethanol and water were added to 40 mL headspace bottles and mixed well. The surface tension of the mixtures was measured at room temperature (20 °C). Reagents were added as shown in Table 2.

Table 1. List of alcohols used in this study.

Alcohols	Purity	Manufacturer
Ethanol	AR	Tianjin Huihang Chemical Technology Co. Ltd.
N-butanol		Guangdong Guanghua Technology Co. Ltd.
N-pentanol		Beijing Chemical Plant Co. Ltd.
N-heptanol		Beijing Chemical Plant Co. Ltd.

Table 2. Specific plan of reagent addition.

Ethanol: Water	Ethanol Volume Fraction	Ethanol (mL)	Water (mL)
1	1	40	—
9:1	0.9	36	4
8:2	0.8	32	8
7:3	0.7	28	12
6:4	0.6	24	16
5:5	0.5	20	20
4:6	0.4	16	24
3:7	0.3	12	28
2:8	0.2	8	32
1:9	0.1	4	36
0	0	—	40

A total of 3 × 32 mL of ultrapure water was added into three 40 mL headspace bottles separately, followed by 8 mL of n-butanol, n-amyl alcohol, and n-heptanol. The samples were placed in a shaker for 24 h, and the speed of the shaker was set at 180 rpm/min. The headspace bottles were left to stand for 30 min, and then samples were extracted with 5 mL syringes to measure the surface tension by using a surface tension meter. Each experiment was performed in triplicate to avoid experimental errors.

3.2.2. Effect of Alcohol on Azeotropy

An azeotropic experimental device is shown in Figure 1. The azeotropic experiment primarily included the binary mixed system and ternary mixed system. The specific experimental scheme is presented in Table 3. The ratio of organic compound/water/alcohol was determined based on the optimal ratio of organic compound/water and the optimal ratio of water/alcohol obtained from previous studies on azeotropes of binary systems. The temperature measurement probes were placed in the upper, middle, and lower positions of the solution; subsequently, a condensate pipe was connected, and the solution was heated to 140 °C in an oil bath. The temperature reading was recorded every minute. Each experimental group was repeated thrice to avoid experimental errors.

3.2.3. Steam Remediation of Nitrobenzene-Contaminated Heterogeneous Aquifer

This part of the experiment was conducted in a 2-D simulation tank with the dimensions of 60 cm (length) × 50 cm (height) × 4 cm (width) (Figure 2). The steam injection well installed at the bottom of the tank for stem injection comprised a cylinder with a diameter of 4 cm and height of 7.5 cm, having many tiny pores at the top. The tiny pores were covered with gauze to prevent sand from entering the steam injection well. There were 24 sampling ports in front of the tank to monitor the temperature using a multi-channel temperature detector (Shenzhen Hua Xin Measuring Instrument Company, Shenzhen, China). Figure 2a,b illustrates the experimental schematic design and the layered heterogeneous aquifer device, and Figure 2c shows the lens heterogeneous aquifer. The physical properties of the quartz sand samples used in the tests are presented in Table 4. The experimental filling medium of the layered heterogeneous aquifer is presented in Table 5. The purpose of the experimental design was to simulate the influence of the interface on steam migration when the upper layer was a low-permeability medium and the lower layer was a high-permeability medium in the actual site. The R values of the two groups differed by an order of magnitude, which depicts the scenario of stratified heterogeneous aquifers.

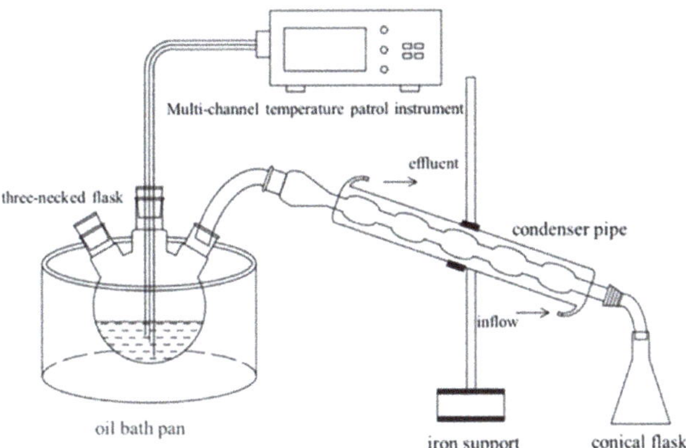

Figure 1. Schematic of the azeotropic test device.

Table 3. Specific scheme of the experiment.

Test No.	Reagent Composition	Reagent Addition
1	water + NB	100 mL + 30 mL
2	water + NB + ethyl alcohol	25 mL + 30 mL + 100 mL
3	water + NB + n-butyl alcohol	40 mL + 30 mL + 80 mL
4	water + NB + n-amyl alcohol	60 mL + 30 mL + 60 mL
5	water + NB + n-heptanol	90 mL + 30 mL + 30 mL

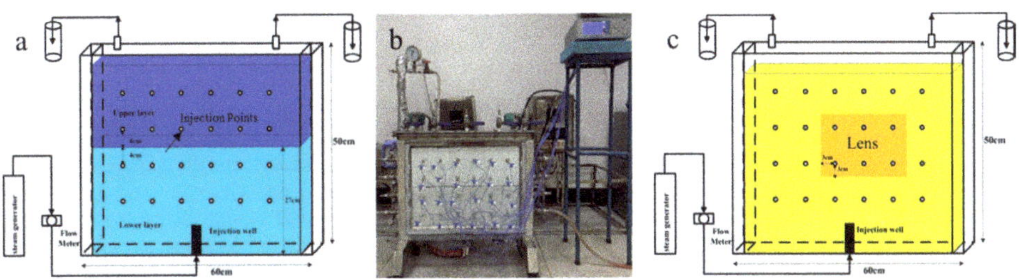

Figure 2. Schematic (**a**) and photo (**b**) of the experimental setup for the layered heterogeneous aquifer, and schematic of the experimental device for the lens heterogeneous aquifer (**c**).

Table 4. Properties of quartz sand samples in the tests.

Soil Type	Particle Size (mm)	Hydraulic Conductivity K (cm/s)
Gravel	3–5	6.1×10^{-1}
Coarse sand	0.4–0.85	6.5×10^{-2}
Medium sand	0.2–0.4	3.5×10^{-2}
Fine sand	0.1–0.2	1.6×10^{-3}

Table 5. Specific experimental plan for the layered heterogeneous aquifer experiment.

Test No.	Packing Configuration		R
1	Upper layer	Medium sand	17.4
	Lower layer	Gravel	
2	Upper layer	Fine sand	381
	Lower layer	Gravel	

Note: R = hydraulic conductivity of lower layer medium K_l/hydraulic conductivity of upper layer medium K_u.

The background medium of the lens heterogeneous aquifer was filled with coarse sand and the lens was filled with fine sand. Before the experiment, tap water was continuously injected at a rate of 0.5 m d^{-1} by a peristaltic pump to ensure uniformity of the porous media and to remove entrapped air. Subsequently, 300 mg/L NB was injected into the lower left water inlet until the concentration detected at the upper right water outlet was the same as the water inlet, which simulated the formation and stabilization of the NB contaminant plume. Steam was generated at a constant rate by a 3 kW steam generator (Norbest Machinery Manufacturing Co. Ltd., Wuhan, China), which was supplied by a constant flux of water producing saturated steam at a temperature of 120 °C and pressure of 200 kPa. The steam flow was set to 1 kg h^{-1}. The temperatures were continuously measured in the sandbox. Water samples were collected at selected intervals (0–270 min) using a disposable syringe. The concentration of NB in water was analyzed using liquid chromatography (HPLC, Agilent 1260, Santa Clara, CA, USA). The hydraulic conductivity ratio R is defined in the layered heterogeneous aquifer as follows:

$$R = K_l/K_u \quad (2)$$

where K_l is the hydraulic conductivity of the lower layer medium and K_u is the hydraulic conductivity of the upper layer medium.

3.2.4. Ethanol-Enhanced Steam Remediation of Nitrobenzene-Contaminated Layered Heterogeneous Aquifers

The experimental device shown in Figure 2 was the same as the previous experimental device. An experimental setting with a hydraulic conductivity ratio R of 381 in the layered heterogeneous aquifer remediation experiment was selected to accelerate the remediation rate experiment. On the completion of the experiment, samples were analyzed for the distribution of pollutants in the aquifer. Furthermore, 500 mL of ethanol at a flow rate of 5 mL/min with a peristaltic pump was injected above the interface to enhance the repair efficiency of the interface in the upper low-permeability zone. The injection positions are shown in Figure 2. In the lens heterogeneous aquifer, 300 mL ethanol was injected into the lens. The ethanol injection volume was calculated using Equation (3). Ethanol distribution stabilization for 2 h was conducted after injecting the ethanol. Later, the concentration distribution of the pollutants before remediation was analyzed. The concentration of NB was determined using liquid chromatography. After injecting hot steam, the sandbox temperature was continuously measured. Water samples were collected at 0–270 min-intervals using a disposable syringe. Ethanol injection volume was calculated using the equation:

$$V = A \times D \times \sigma \times 0.3 \quad (3)$$

where A is the affected area of the low-permeability (cm^2); D is the thickness of the simulated tank (cm); σ is the porosity of the medium; and 0.3 is the volume fraction of ethanol.

4. Results and Discussion

4.1. *Effect of Alcohol on Surface Tension*

After adding different volumes of ethanol, the surface tension of the mixture system gradually decreased, with surface tension values of 72.6 mN/m, 22.3 mN/m, and 30 mN/m at 0, 1, and 0.3 volume fractions of ethanol, respectively (Figure 3). Thus, when the volume of ethanol was >0.3 mL, the surface tension of the system significantly reduced, indicating that ethanol can reduce the surface tension of the solution and thus can be used as a surfactant. The optimal ratio of ethanol was 0.3.

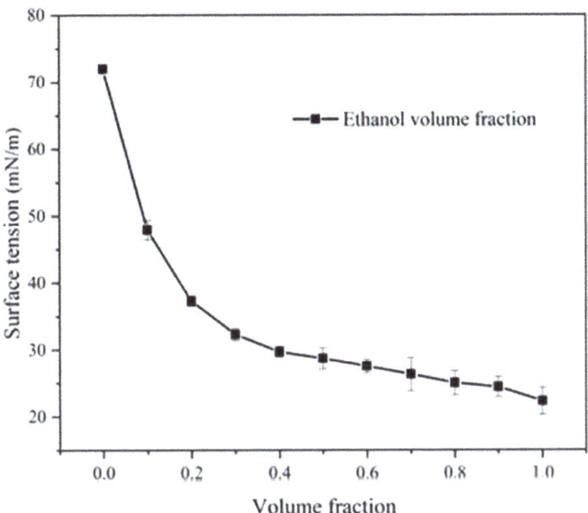

Figure 3. Change curve of surface tension under different ethanol volume fractions.

Figure 4 shows the surface tension of different alcohol–water mixtures. N-butanol, n-pentanol, and n-heptanol can greatly reduce the surface tension of the solution, reducing the surface tension of the mixture to below 30 mN/m. The surface tension of the mixed system was almost equal to that of pure alcohol, thereby proving that alcohol is a good agent for reducing surface tension.

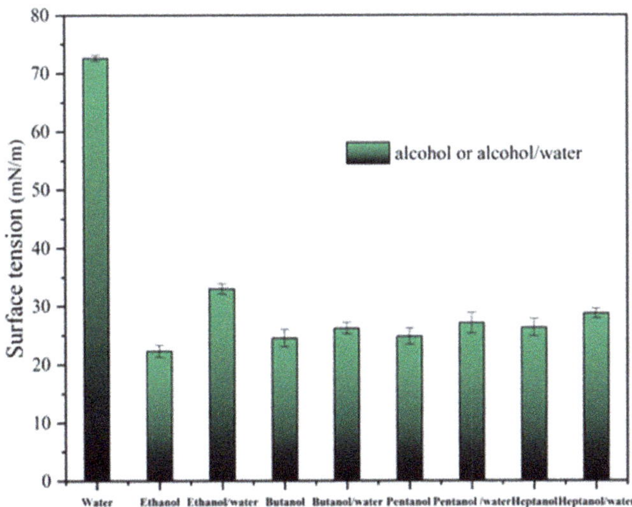

Figure 4. Surface tension of different alcohol–water systems, and the volume fraction of alcohol was 0.2.

4.2. Effect of Alcohol on Azeotrope

4.2.1. Azeotropic Temperature of Binary Mixture

Figure 5a shows the azeotropic temperature between NB and water. Initially, an azeotropic plateau occurred, during which the temperature growth rate was approximately 0, azeotropic time was approximately 8 min, and the azeotropic temperature was approximately 97.4 °C. As the temperature increased, the balance of the azeotropic ratio between the liquid phase and the gas phase was disturbed, azeotropic plateau period ended, and the temperature gradually increased to the boiling point of water at 100 °C. The short azeotropic time was primarily due to the low NB content, which did not allow the same proportion of gas and liquid mixture for a long time, thereby ending the azeotrope process. Furthermore, previous studies have shown that the organic compound content is only related to the azeotropic time.

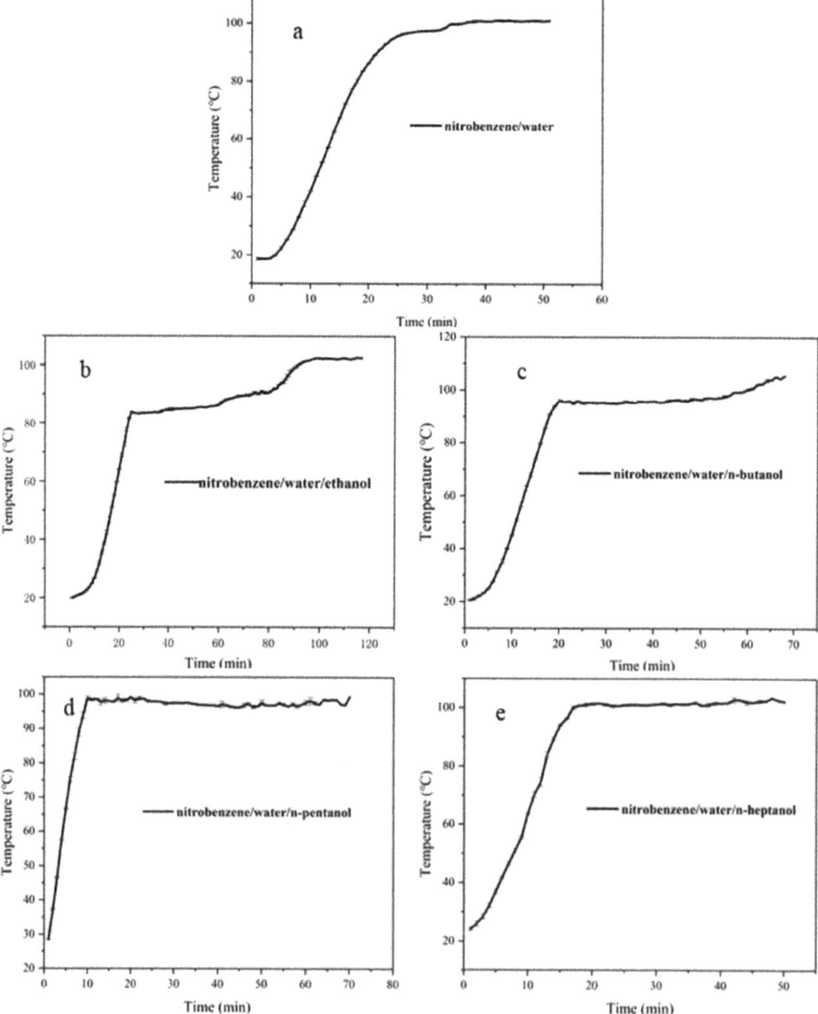

Figure 5. Temperature curves of nitrobenzene (NB)–water (**a**), NB/water/ethanol (**b**), NB/water/ n-butanol (**c**), NB/water/ n-pentanol (**d**), and NB/water/ n-heptanol (**e**).

4.2.2. Azeotropic Temperature of Ternary Mixture

The ternary mixture of NB/water/ethanol had an azeotropic plateau (Figure 5b), the azeotropic time lasted for 30 min, azeotropic temperature was 83.2 °C, and temperature increased at 56 min. Compared with the azeotrope of NB and water, adding ethanol could have reduced the azeotrope point by approximately 15 °C, which helped to advance the azeotrope stage by removing NB with hot steam and improving the removal effect. This indicated that ethanol could be used as an azeotropic agent to reduce the azeotropic temperature of NB. The reduction in azeotropic temperature can not only improve the removal efficiency of organic compounds, but also reduce the remediation costs.

The ternary azeotropic mixture of NB, water, and n-butanol entered the azeotropic platform period after 25 min of recording (Figure 5c). The azeotropic temperature was 95.8 °C, and the azeotropic time was 30 min. The azeotropic temperature of the ternary mixture after adding n-butanol was close to that of the binary mixture, which indicated that n-butanol was not effective in reducing the azeotropic temperature.

The azeotropic temperature of the NB/water/n-pentanol ternary azeotropic mixture was 97.7 °C (Figure 5d). N-heptanol was added to the binary mixture of NB/water to form the ternary mixture. The azeotropic temperature was 101.4 °C (Figure 5e). Compared to the boiling point of water, the addition of n-heptanol increased the boiling point of the binary mixture. This indicates that n-pentanol and n-heptanol cannot be used as azeotropic agents to reduce the azeotropic temperature of NB. Only ethanol among the four alcohols could reduce the azeotropic temperature of organic compounds, therefore, ethanol was selected as the enhanced repair reagent in subsequent experiments.

4.3. Steam Remediation of Heterogeneous Aquifers Contaminated by Nitrobenzene

4.3.1. Temperature Distribution of Layered Heterogeneous Aquifer

The temperature distribution of the layered heterogeneous aquifer is shown in Figure 6. Due to the high hydraulic conductivity of the lower medium and the low hydraulic conductivity of the upper medium, the upward transmission of steam was inhibited, resulting in heat accumulation below the interface. Therefore, when $R > 1$, a blocking interface is formed, resulting in the formation of steam override below the interface and high temperature in the area below the interface. The higher the R value, the higher the temperature below the interface, and the greater the lateral migration distance of the steam. The steam in the lower layer primarily relies on heat convection to transfer heat, whereas it primarily relies on heat conduction in the upper low-permeability zone.

4.3.2. Nitrobenzene Removal in 2-D Layered Heterogeneous Porous Media

When $R = 17.4$ and steam injection time was 3 h, the NB concentration in the center of the 2-D simulation tank (the position closest to the steam injection point) was the lowest, and the concentration on both sides was high (Figure 7a). This was primarily due to the effect of buoyancy. The vertical migration of steam was faster, resulting in the preferential removal of vertical pollutants. After 6 h of steam injection, the concentration distribution of NB was affected by the interface, and the concentration was the lowest below the interface, with a concentration of 60 mg/L. After 6 h, the NB removal rate in the simulated tank was 68%, of which the removal rates in gravel and medium sand were 70.3% and 65.3%, respectively, and the average NB concentration in medium sand was 1.13 times that of gravel.

When $R = 381$, the blocking effect of the interface was enhanced (Figure 7b). The NB concentration in the upper layer was evidently higher than that in the lower layer after 3 h of steam injection, and the difference was more obvious with the increase in steam injection time. After 6 h of steam injection, the lowest concentration of NB in gravel reached 28 mg/L, and the average concentration of NB in fine sand was 3.13 times that of gravel. The total removal rate of NB was 60.6%, among which the removal rates in gravel and fine sand were 79.9% and 37%, respectively. Compared with $R = 17.4$, the total removal rate of NB decreased, while the removal rate of NB in gravel increased.

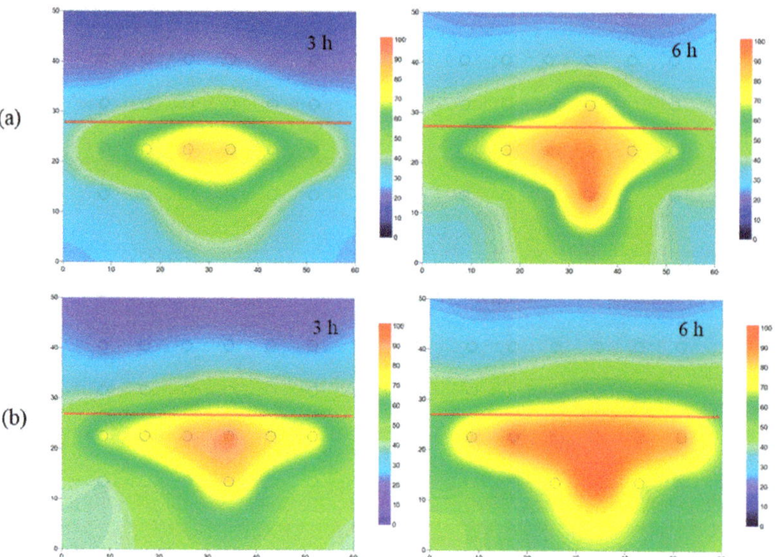

Figure 6. Temperature distribution of the layered heterogeneous aquifers (a) R = 17.4 (b) R = 381.

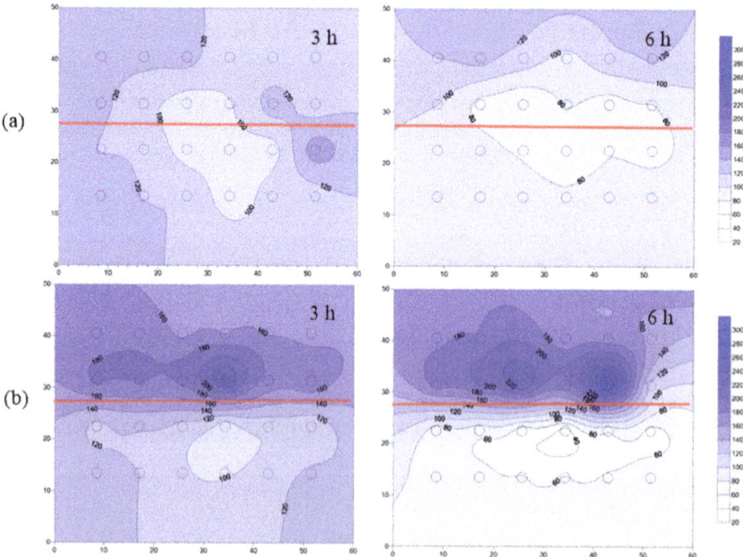

Figure 7. Isograms of NB concentrations with different R values: (a) R = 17.4; (b) R = 381.

4.3.3. Alcohol-Enhanced Steam Remediation of Layered Heterogeneous Aquifer Contaminated by Nitrobenzene

After adding ethanol, some heat was accumulated at the interface, but the temperature difference between the upper and lower layers decreased significantly, the average temperature increased by 15 °C, and the area affected by steam in the low-permeability zone increased by 35% (Figure 8). Alexeev et al. (2005) found that Marangoni convection could enhance liquid phase heat transfer through experimental and numerical simulation [41].

Therefore, the addition of ethanol produces Marangoni convection and enhances mass and heat transfer.

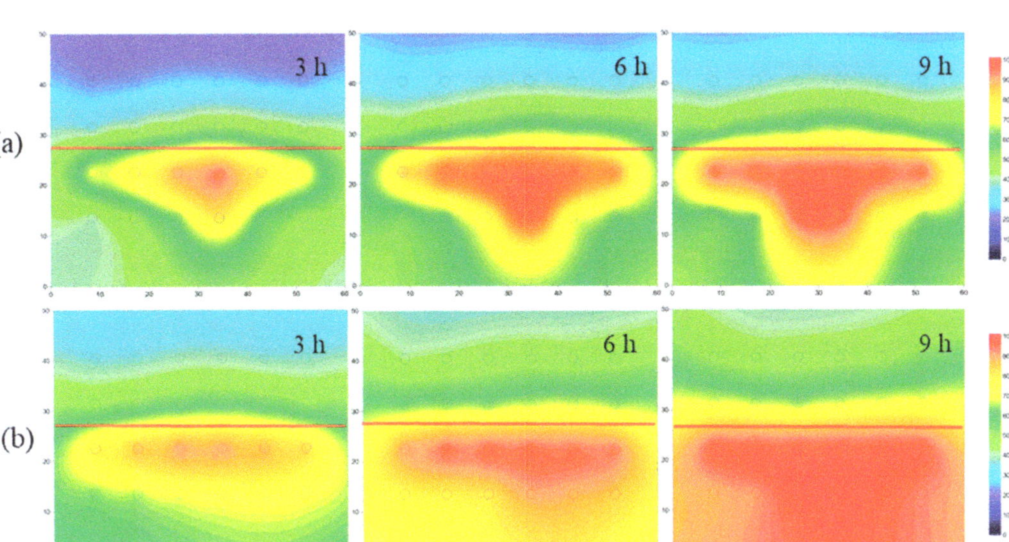

Figure 8. Temperature distribution without ethanol (**a**) and with ethanol (**b**) at R = 381.

After adding ethanol, the remediation effect of NB was significantly increased, particularly in the fine sand medium (Figure 9) and the average NB concentration in the simulated tank was reduced by 51%. The average concentration of NB in the lower gravel and upper fine sand decreased by 43.7% and 43.5%, respectively, indicating that ethanol addition enhanced the gas–liquid mass transfer. During the repair process, ethanol volatilized at the gas–liquid interface, forming a surface tension gradient at the interface and increasing the convective vortices in the flow field, resulting in the Marangoni convection [20,40,42]. The Marangoni convection can also enhance heat transfer, and an increase in temperature also enhances NB removal. In addition, ethanol reduces the azeotropic temperature of NB, promoting the boiling of NB at 83.2 °C, thus, improving the removal rate and efficiency of NB.

4.3.4. Alcohol-Enhanced Steam Remediation of Lens Heterogeneous Aquifer Contaminated with NB

Figure 10 shows the temperature distribution of the aquifers containing lenses with and without ethanol solution. Due to the existence of a low-permeability lens, a blocking interface was formed below the lens, and steam flowed around the lens. Only a small amount of steam could enter the fine-sand lens, causing the temperature inside the lens to increase. However, most steam will flow around the lens from the high-permeability area in the form of preferential flow, resulting in an uneven distribution of steam, which affects the removal effect of NB. After adding ethanol, the average temperature in the simulated tank increased by 8 °C, and the steam-affected area increased by 14%, which further confirms that the Marangoni convection generated by ethanol can enhance heat and mass transfer.

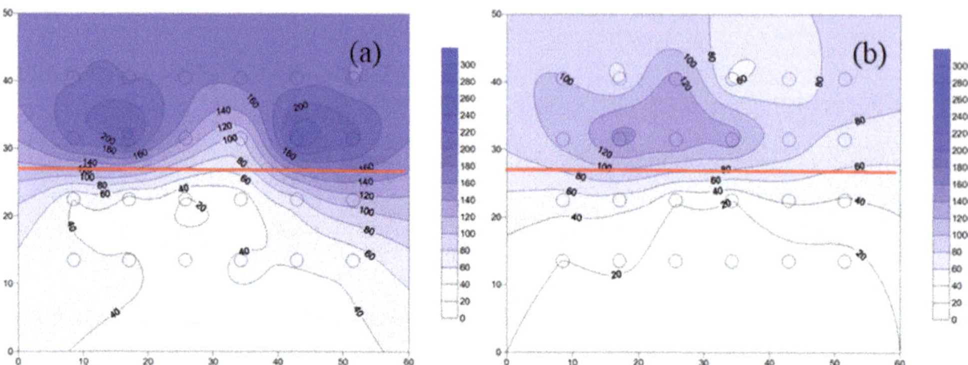

Figure 9. Isograms of NB concentration without ethanol (**a**) and with ethanol (**b**) at 9 h.

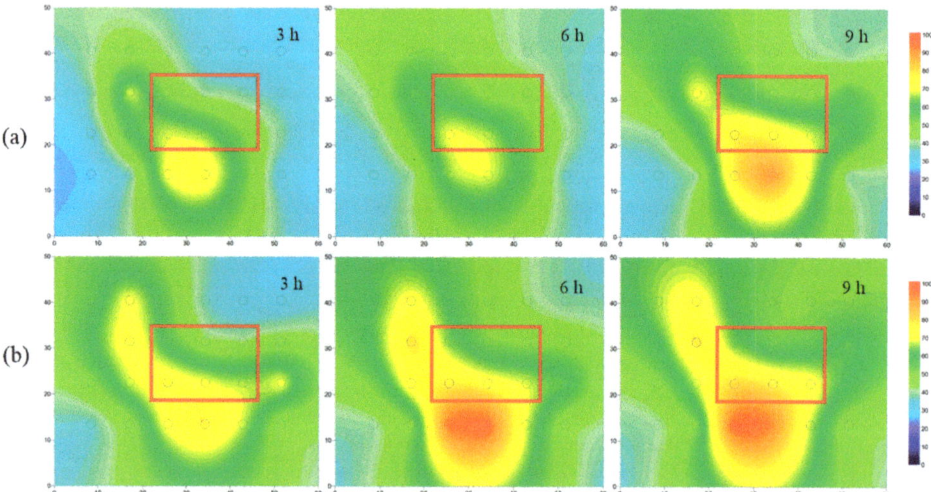

Figure 10. Temperature distribution without ethanol (**a**) and with ethanol (**b**) in the lens heterogeneous aquifer.

Figure 11 shows the isolines of NB concentration in the heterogeneous aquifer with the lens. In the lens heterogeneous aquifer, the remediation effect of steam at the edge of the simulation tank and in the lens was weakened due to the preferential flow. After ethanol was added, the mass transfer and heat transfer were enhanced, and the boiling area increased, resulting in an increase in the removal efficiency of NB. The removal efficiency of the tank was increased by 10%, and the NB concentration in the lens was reduced by approximately 58%. Under the same steam flow rate, ethanol addition enhanced the remediation effect of the lens heterogeneous aquifer and reduced the remediation time and cost.

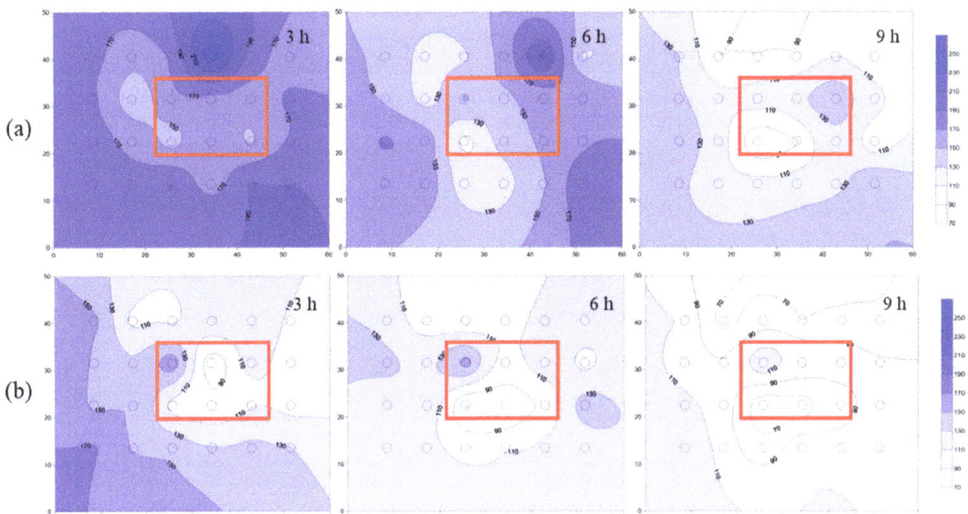

Figure 11. Isograms of NB concentration without ethanol (**a**) and with ethanol (**b**) in the heterogeneous aquifer.

5. Conclusions

(1) The four alcohols reduced the surface tension of the solution from 72 mN/m to less than 30 mN/m; however, among the four alcohols, only ethanol reduced the azeotropic temperature of NB by 15 °C. Therefore, ethanol can be used as a remediation agent to enhance the steam remediation of heterogeneous aquifers.

(2) When steam entered the layered heterogeneous aquifer, the interface between the low-and high-permeability strata blocked the steam transmission, resulting in poor remediation effect of the low-permeability stratum. After adding ethanol, the average temperature increased, and the area affected by steam increased by 13%, which proved that ethanol addition increased the steam transmission. Moreover, the removal effect of NB was significantly improved, and the average concentration of NB in the simulation tank decreased by 51%.

(3) In the lens heterogeneous aquifer, steam flowed around the lens, forming a preferential flow that resulted in the reduced removal of NB from the lens. After adding ethanol, the steam-affected area increased by 14%, the removal rate of NB increased by 10%, and the NB concentration in the lens decreased by 58%.

(4) Ethanol-enhanced steam injection remediation of heterogeneous aquifers not only enhances the remediation effect and increases the remediation range, but also reduces the remediation costs.

Author Contributions: R.L.: Conceptualization, Investigation, Formal analysis, Writing-original draft. X.Y.: Investigation, Validation. J.X.: Investigation, Validation. X.L.: Investigation, Validation. Y.Z.: Conceptualization, Resources, Writing-review & editing. All authors have read and agreed to the published version of the manuscript.

Funding: This work was supported by the National Nature Science Foundation of China (Grant No. 41530636), the National and Local Joint Engineering Laboratory for Petrochemical Contaminated Site Control and Remediation Technology of Jilin University, the Jilin Provincial Key Laboratory of Water Resources and Environment of Jilin University, and the Key Laboratory of Groundwater Resources and Environment of Ministry of Education (Jilin University).

Institutional Review Board Statement: Not applicable.

Informed Consent Statement: Not applicable.

Data Availability Statement: Some or all data, models, or codes that support the findings of this study are available from the corresponding authors on request.

Acknowledgments: This study was financially supported by the Key Project of the National Natural Science Foundation of China (Grant No. 41530636). The authors thank the Key Laboratory of Groundwater Resources and Environment, Ministry of Education for providing financial assistance.

Conflicts of Interest: The authors declare no conflict of interest.

References

1. Al Yousef, Z.; AlDaif, H.; Al Otaibi, M. An overview of steam injection project in Fractured Carbonate Reservoirs in the Middle East. *J. Pet. Sci. Res.* **2014**, *3*, 101–110. [CrossRef]
2. Davis, E.L. Steam injection for soil and aquifer remediation. *Environ. Prot. Agency* **1998**, *505*, 1–16.
3. Udell, K.S. Application of in situ thermal remediation technologies for DNAPL removal. *Groundw. Qual. Remediat. Prot.* **1998**, *250*, 367–374.
4. Gudbjerg, J.; Sonnenborg, T.O.; Jensen, K.H. Remediation of NAPL below the water table by steam-induced heat conduction. *J. Contam. Hydrol.* **2004**, *72*, 207–225. [CrossRef]
5. Basel, M.D.; Udell, K.S. *Effect of Heterogeneities on the Shape of Condensation Fronts in Porous Media*; American Society of Mechanical Engineers: New York, NY, USA, 1991; pp. 63–70.
6. Schmidt, R.; Gudbjerg, J.; Sonnenborg, T.O.; Jensen, K.H. Removal of NAPLs from the unsaturated zone using steam Prevention. *J. Contam. Hydrol.* **2002**, *55*, 233–260. [CrossRef]
7. Heron, G.; Christensen, T.H.; Enfield, C.G. Henry's law constant for trichloroethylene between 10 and 95 °C. *Environ. Sci. Technol.* **1998**, *32*, 1433–1437. [CrossRef]
8. Schwardt, A.; Dahmke, A.; Köber, R. Henry's law constants of volatile organic compounds between 0 and 95 °C—Data compilation and complementation in context of urban temperature increases of the subsurface. *Chemosphere* **2021**, *272*, 129858. [CrossRef]
9. Sleep, B.E.; McClure, P.D. The effect of temperature on adsorption of organic compounds to soils. *Can. Geotech. J.* **2001**, *38*, 46–52. [CrossRef]
10. Philippe, N.; Davarzani, H.; Colombano, S.; Dierick, M.; Klein, P.-Y.; Marcoux, M. Experimental study of the temperature effect on two-phase flow properties in highly permeable porous media: Application to the remediation of dense non-aqueous phase liquids (DNAPLs) in polluted soil. *Adv. Water Resour.* **2020**, *146*, 103783. [CrossRef]
11. Tzovolou, D.N.; Aggelopoulos, C.A.; Theodoropoulou, M.A.; Tsakiroglou, C.D. Remediation of the unsaturated zone of NAPL-polluted low permeability soils with steam injection: An experimental study. *J. Soils Sediments* **2010**, *11*, 72–81. [CrossRef]
12. Heron, G.; Vanzutphen, M.; Christensen, H.T.; Enfield, G.C. Soil heating for enhanced remediation of chlorinated solvents_a laboratory study on resistive heating and vapor extraction in a silty, low-permeable soil contaminated with trichloroethylene. *Environ. Sci. Technol.* **1998**, *32*, 1474–1481. [CrossRef]
13. Ochs, S.O.; Class, H.; Färber, A.; Helmig, R. Methods for predicting the spreading of steam below the water table during subsurface remediation. *Water Resour. Res.* **2010**, *46*, 1–16. [CrossRef]
14. Class, H.; Helmig, R. Numerical simulation of non-isothermal multiphase multicomponent processes in porous media. 2. Applications for the injection of steam and air. *Adv. Water Resour.* **2002**, *25*, 551–564. [CrossRef]
15. Nilsson, B.; Tzovolou, D.; Jeczalik, M.; Kasela, T.; Slack, W.; Klint, K.E.; Haeseler, F.; Tsakiroglou, C.D. Combining steam injection with hydraulic fracturing for the in situ remediation of the unsaturated zone of a fractured soil polluted by jet fuel. *J. Environ. Manage.* **2011**, *92*, 695–707. [CrossRef] [PubMed]
16. Davis, E.; Akladiss, N.; Hoey, R.; Brandon, B.; Nalipinski, M.; Carroll, S.; Herom, G.; Novakowski, K.; Udell, K. Steam Enhanced Remediation Research for DNAPL in Fractured Rock. *EPA* **2005**, *10*, 1–190.
17. Baker, R.S.; Heron, G. In-Situ Delivery of Heat by Thermal Conduction and Steam Injection for Improved Dnapl Remediation. In Proceedings of the 4th International Conference on Remediation of Chlorinated and Recalcitrant Compounds, Monterey, CA, USA, 24–27 May 2004; Battelle Press: Monterey, CA, USA, 2004.
18. Heron, G.; Carroll, S.; Nielsen, S.G. Full-Scale Removal of DNAPL Constituents Using Steam-Enhanced Extraction and Electrical Resistance Heating. *Ground Water Monit. Remediat.* **2005**, *25*, 92–107. [CrossRef]
19. Chao, H.P.; Hsieh, L.C.; Tran, H.N. Increase in volatilization of organic compounds using air sparging through addition in alcohol in a soil-water system. *J. Hazard. Mater.* **2018**, *344*, 942–949. [CrossRef]
20. Chang, Y.; Yao, M.; Bai, J.; Zhao, Y. Study on the effects of alcohol-enhanced air sparging remediation in a benzene-contaminated aquifer: A new insight. *Environ. Sci. Pollut. Res. Int.* **2019**, *26*, 35140–35150. [CrossRef]
21. Abd Rahman, N.; Azizan, N.A.; Kamaruddin, S.A.; Chelliapan, S.; Mohd Jaini, Z.; Yunus, R.; Rahmat, S.N. Steam-Enhanced Extraction Experiments, Simulations and Field Studies for Dense Non-Aqueous Phase Liquid Removal: A Review. *MATEC Web Conf.* **2016**, *47*, 05012. [CrossRef]
22. Yuan, Z.; Herold, K.E. Surface tension of pure water and aqueous lithium bromide with 2-ethyl-hexanol. *Appl. Therm. Eng.* **2001**, *21*, 881–897. [CrossRef]
23. Biscay, F.; Ghoufi, A.; Malfreyt, P. Surface tension of water-alcohol mixtures from Monte Carlo simulations. *J. Chem. Phys.* **2011**, *134*, 044709. [CrossRef]

24. Moraveji, M.K.; Sajjadi, B.; Davarnejad, R. Gas-Liquid Hydrodynamics and Mass Transfer in Aqueous Alcohol Solutions in a Split-Cylinder Airlift Reactor. *Chem. Eng. Technol.* **2011**, *34*, 465–474. [CrossRef]
25. Chen, F.; Freedman, D.L.; Falta, R.W.; Murdoch, L.C. Henry's law constants of chlorinated solvents at elevated temperatures. *Chemosphere* **2012**, *86*, 156–165. [CrossRef] [PubMed]
26. Koproch, N.; Dahmke, A.; Kober, R. The aqueous solubility of common organic groundwater contaminants as a function of temperature between 5 and 70 °C. *Chemosphere* **2019**, *217*, 166–175. [CrossRef]
27. Komninos, N.P.; Rogdakis, E.D. Geometrical investigation and classification of three-suffix margules binary mixtures including single and double azeotropy. *Fluid Phase Equilib.* **2019**, *494*, 212–227. [CrossRef]
28. Maalem, Y.; Zarfa, A.; Tamene, Y.; Fedali, S.; Madani, H. Prediction of thermodynamic properties of the ternary azeotropic mixtures. *Fluid Phase Equilib.* **2020**, *517*, 112613. [CrossRef]
29. Imaishi, N.; Suzuki, Y.; Hozawa, M.; Fujinawa, K. Interfacial turbulence in gas-liquid mass transfer. *Int. Chem. Eng.* **1982**, *22*, 659–665. [CrossRef]
30. Marangoni, C. Ueber die Ausbreitung der Tropfen einer Flüssigkeit auf der Oberfläche einer anderen. *Ann. Der Phys. Und Chem.* **1871**, *219*, 337–354. [CrossRef]
31. Bénard, H. *Les Tourbillons Cellulaires—Dans une Nappe Liquide Propageant de la Chaleur par Convection, en Régime Permanent*; Thèse Présentée à la Faculté des Sciences de Paris; Par, M., Bénard, H., Eds.; Gauthier-Villars: Paris, France, 1901.
32. Rayleigh, L. LIX. On convection currents in a horizontal layer of fluid, when the higher temperature is on the under side. *Lond. Edinb. Dublin Philos. Mag. J. Sci.* **1916**, *32*, 529–546. [CrossRef]
33. Mao, Z.-S.; Chen, J. Numerical simulation of the Marangoni effect on mass transfer to single slowly moving drops in the liquid–liquid system. *Chem. Eng. Sci.* **2004**, *59*, 1815–1828. [CrossRef]
34. Ma, Y.; Yu, G.; Li, H.Z. Note on the mechanism of interfacial mass transfer of absorption processes. *Int. J. Heat Mass Transf.* **2005**, *48*, 3454–3460. [CrossRef]
35. Bodenschatz, E.; de Bruyn, J.R.; Ahlers, G.; Cannell, D.S. Transitions between patterns in thermal convection. *Phys. Rev. Lett.* **1991**, *67*, 3078–3081. [CrossRef]
36. Vidal, A.; Acrivos, A. Effect of Nonlinear Temperature Profiles on Onset of Convection Driven by Surface Tension Gradients. *Ind. Eng. Chem. Fundam.* **1968**, *7*, 53–58. [CrossRef]
37. McTaggart, C.L. Convection driven by concentration- and temperature-dependent surface tension. *J. Fluid Mech.* **1983**, *134*, 301–310. [CrossRef]
38. Buffone, C.; Sefiane, K.; Easson, W. Marangoni-driven instabilities of an evaporating liquid-vapor interface. *Phys. Rev. E* **2005**, *71*, 056302. [CrossRef]
39. Sha, Y.; Chen, H.; Yin, Y.; Tu, S.; Ye, L.; Zheng, Y. Characteristics of the Marangoni Convection Induced in Initial Quiescent Water. *Ind. Eng. Chem. Res.* **2010**, *49*, 8770–8777. [CrossRef]
40. Liu, D.; Tran, T. Vapor-Induced Attraction of Floating Droplets. *J. Phys. Chem. Lett.* **2018**, *9*, 4771–4775. [CrossRef] [PubMed]
41. Alexeev, A.; Gambaryan-Roisman, T.; Stephan, P. Marangoni convection and heat transfer in thin liquid films on heated walls with topography: Experiments and numerical study. *Phys. Fluids* **2005**, *17*, 062106. [CrossRef]
42. Zhang, J.; Oron, A.; Behringer, R.P. Novel pattern forming states for Marangoni convection in volatile binary liquids. *Phys. Fluids* **2011**, *23*, 072102. [CrossRef]

Article

Use of Foamed Cement Banking for Reducing Expressways Embankment Settlement

Yuedong Wu [1,2], Hui Liu [1,2] and Jian Liu [1,2,*]

[1] College of Civil and Transportation Engineering, Hohai University, Nanjing 210098, China; hhuwyd@163.com (Y.W.); lxiao_hui@hhu.edu.cn (H.L.)

[2] Key Laboratory of Ministry of Education for Geomechanics and Embankment Engineering, Hohai University, Nanjing 210098, China

* Correspondence: 20170053@hhu.edu.cn; Tel.: +86-183-6296-0323

Abstract: Expressways are often built on soft ground, the foundation of which is not processed adequately during the construction period. Consequently, the traffic safety and embankment stability will be seriously affected due to uneven settlement. The technology of holing the embankment and replacing foamed cement banking (FCB) could control the settlement of an embankment without road closure, thus reduce the impact of construction on normal operation of highways. In this paper, the principle of FCB is described. Additionally, a sedimentation ratio calculation method, through the analysis of the settlement load ratio, is proposed for calculating the roadbed replacement thickness. This paper takes the example of the test section EK0 + 323 on Shen-Jia-Hu expressway in Zhejiang Province and combines with site settlement monitoring data to confirm the effectiveness of the calculation method proposed.

Keywords: construction disturbance; replacement of embankment; foamed cement banking technology; post-construction settlement; method of replacement thickness calculation

Citation: Wu, Y.; Liu, H.; Liu, J. Use of Foamed Cement Banking for Reducing Expressways Embankment Settlement. *Appl. Sci.* **2021**, *11*, 11959. https://doi.org/10.3390/app112411959

Academic Editors: Daniel Dias and Bing Bai

Received: 14 October 2021
Accepted: 13 December 2021
Published: 15 December 2021

1. Introduction

With the rapid development of expressway construction, in China, construction on soft ground is sharply increasing, such as Shen-Jia-Hu Expressway that passes through soft soil [1,2]. Owing to shortened construction periods and to the rheological properties of soft soil, among other reasons, the embankment after settlement of many highways opened abroad is too high, which does not meet relevant regulatory requirements (i.e., post-construction should be less than 30.0 cm) [3]. The post-construction settlement of expressways has become a major type of expressway damage, which will not only seriously affect the vehicle's road speed, traffic safety, and driving comfort, but also seriously affect the stability and safety of the highway itself [4–6].

In order to reduce post-construction settlement, the embankment is replaced with light soil to reduce the overburden load and control the settlement [7,8]. At present, the widely used light soils mainly include EPS beads-mixed lightweight soil, EPS block, and foamed cement banking, which include good integrity, light mass, and convenient construction. The lightweight soil's cohesion, global strength, bulk density, and other parameters can be adjusted by adjusting the output of cement and foam [9]. The advantages and disadvantages of various methods are shown in Table 1. Foamed cement banking research is already well underway in numerous countries [10–12], and some results have been achieved in practice. Compared with other materials, the FCB has a shorter construction period, lower curing time, and higher construction quality, thus it has broad application prospects [13,14]. As FCB's bulk density is only conventional soil bulk density of 1/4 to 1/8, its use as a replacement of embankment material [15–17], can not only reduce the base load pressure, but can also reduce the total settlement and differential settlement [18]. Replacing the undisturbed soil with lightweight soil to process the post-construction settlement of expressway under traffic control seriously affected highway traffic operations. Especially, there are no closed roads (ramps) during construction, the lightweight soil is being put down to achieve subgrade compactness, that causes huge economic losses due to the road closure.

Table 1. Comparison of different lightweight fills.

Material Type	Advantage	Disadvantage
EPS beads-mixed lightweight soil	Low density and compressibility; high strength; environmental protection.	Difficult construction; high material cost; poor corrosion resistance.
EPS block	Low density; stable nature; convenient construction.	High material cost; poor impact resistance; easy to local cavity.
Foamed cement banking	Low density; good integration; high strength and corrosion resistance; short construction period.	High cement hydration heat; high material cost.

In this paper, foamed cement replacement is carried out on both sides of the embankment, which does not affect the normal traffic operation during replacement construction, its use is targeted and flexible, and only small settlement occurs after improvement. Combined with Shen-Jia-Hu EK0 + 323 highway section, this paper verifies the advantages of FCB in dealing with embankment settlement without road closure. Through the analysis of dynamic monitoring data of Shen-Jia-Hu expressway's subgrade settlement, and according to the proportional relationship between settlement and load, the calculation method of embankment replacement thickness is proposed, which can be applied to complex construction processes, this work can thus provide a reference for similar future projects.

2. Design Principles and Calculation Method

2.1. Principle of FCB

FCB is a new technology for reducing embankment settlement while ensuring the normal operation of road transport, its schematic diagram is shown in Figure 1. First, the amount of FCB replacement needed is based on the amount of static load of embankment to be uninstalled, which is, in turn, used to design the transverse hole construction plan. Then, drilling horizontal holes on the embankment continuously, with the hole machinery along the embankment on one side or both sides, next, FCB is injected directly into the holes using perfusion enrichment after the hole formed. The unit weight of regular FCB filling is 5.5–6.0 kN/m^3, which is 1/3 to 1/4 of the general soil bulk density (i.e., 20 kN/m^3), to effectively reduce the embankment permanent load. The undisturbed foundation soil becomes super-solid, and the effect of the improvement is clear. The normal operation of the highway will not be affected during construction.

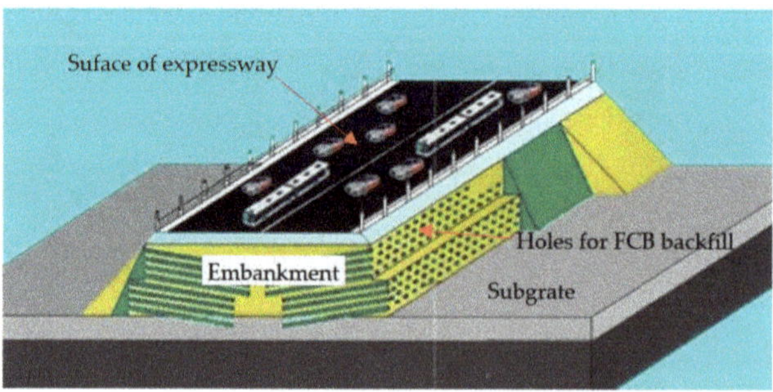

Figure 1. Embankment FCB replacement construction by citing transverse holes.

2.2. Calculation Method

The most important link of FCB displacement control of a subgrade is to identify the replacement thickness of the roadbed. This currently relies heavily on the layerwise summation method for estimation of total deposition, as well as comparative analysis with settlement monitoring data available before improvement, leading to the determination of the replacement thickness [18] in the transverse hole for the filling of the road embankment stage, which mainly depends on the layerwise summation method. When this method is used to calculate the settlement under complex loading and unloading conditions, it produces large errors.

For easier calculation of displacement thickness. Chen [19] had proposed a calculation formula of unloading quantity, but it is necessary to be combined with engineering practice to determine the parameters, which has a high cost. Thus, based on the formula proposed by Chen and the ratio variation between load and settlement, a method for calculating unloading according to the settlement is proposed as follows:

$$P_c = h \cdot (\gamma_1 - \gamma_0) \tag{1}$$

where γ_1 is the embankment unit weight that generally ranges between 19.0 and 21.0 kN/m^3, and γ_0 is the foamed lightweight concrete embankment wet unit weight ranging between 5.5 and 6.0 kN/m^3; P_c is the embankment unloading amount, and h is the replacement thickness.

At present, the existing settlement data is mainly used to predict the settlement curve, and the replacement thickness is determined according to the result. The weight of embankment was reduced by replacing the undisturbed soil, which can reduce settlement. Therefore, when the monitored settlement data does not meet the requirements, the embankment should be improved. The unloading value is mainly determined by the difference between the bulk density of the replacement material and the undisturbed soil mass. Thus, the method proposed in this paper has an advantage of calculating the unloading value of the overload embankment to a certain extent, even though there is a difference in establishing the initial expression. To devise an expression that can calculate the replacement thickness under different conditions, including overload, equivalent load, and under load conditions (i.e., after embankment soil replacement with FCB), further research on the relationship between foundation consolidation settlement and pressure must be conducted.

Based on the assumption that unloading construction is not performed on the embankment, the embankment ultimate settlement under pressure P_0 is $S_\infty^{p+\Delta p}$, and the observed foundation settlement before unloading construction is S_m. If embankment unloading construction is implemented and the unloading amount is P_c, the embankment residual weight P_u^1 induces the ultimate settlement S_∞^h, and a formula for settlement calculation can be given as follows:

$$\frac{S_\infty^{p+\Delta p}}{S_\infty^h} = \frac{P_0}{P_u^1} \tag{2}$$

Owing to the fact that the settlement calculation considers rebound under unloading conditions, the following concepts should be clarified. When unloading P_c in FCB embankment replacement is completed, the load $P_c \cdot S_m / S_\infty^{P+\Delta P}$ had converted into effective load, unloading this part of the load will cause rebound. As a result, this part of the capacity should take the swelling index caused by unloading into account, the swelling index is usually taken as δ times the compression settlement (engineering empirical suggested range of values: 0.05–0.03). Meanwhile, the other part of the load that is not converted is $P_c \cdot (1 - S_m / S_\infty^{P+\Delta P})$, and its settlement computation is suggested to confirm normal compression settlement. Therefore, the settlement caused by unloading is expressed as:

$$S_c = \frac{\left[1 - (1-\delta)\frac{S_m}{S_\infty^{P+\Delta P}}\right]}{P_0} \cdot P_c \cdot S_\infty^{P+\Delta P} \tag{3}$$

After unloading construction, the actual remaining consolidation settlement S_m^1 is expressed as:

$$S_m^1 = S_m - S_c = S_m - \frac{[1-(1-\delta)\frac{S_m}{S_\infty^{P+\Delta P}}]}{P_0} \cdot P_c \cdot S_\infty^{P+\Delta P} \tag{4}$$

Considering the allowable post-construction settlement ΔS in design work, by summing the actual remaining settlement S_m^1 and allowable post-construction settlement ΔS, the total ultimate settlement S_∞^h induced by the residual embankment load after moving the load P_c is obtained:

$$S_m^1 + \Delta S = S_\infty^h \tag{5}$$

$$P_c = \frac{P_0 \cdot (S_\infty^{P+\Delta P} - S_m - \Delta S)}{(1-\delta) \cdot S_m} = \vartheta \cdot P_0 \cdot (\frac{S_r - \Delta S}{S_m}) \tag{6}$$

where S_r is the residual settlement before improvement with FCB, $S_r = S_\infty^{P+\Delta P} - S_m$; S_m is the monitoring data during construction; $S_\infty^{P+\Delta P}$ is the computed ultimate settlement under the embankment load P_0 function; P_0 is the embankment gravity load, and ΔS is the design-allowable post-construction settlement for which a local standard is suggested for determining its value. ϑ is the rebound correction factor expressed as $\vartheta = 1/(1-\delta)$, where δ is the swelling index within the range 0.05–0.3.

Under the condition that the embankment design applies overload pre-loading, if the calculated subgrade unloading confirms that $P_c \le P_0 - P_u$, P_u is the amount of overload, the post-construction settlement requirement can be satisfied after removing the overload pressure. When $P_c > P_0 - P_u$, except for removing overload, it is necessary to apply FCB to process the excess load, $P_c - (P_0 - P_u)$. If equivalent load pre-compression is applied in embankment reinforcement, the computation of FCB replacement thickness should take the unloading value P_c into account. If the embankment design condition is under load pre-compression $(P_0 \le P_u)$, computation of the FCB replacement thickness should confirm the expression: $P_c + P_u - P_0$.

In order to use equation (6) to calculate unloading value, parameters P_0 and S_m can be acquired by measuring before unloading, and δ is determined either by laboratory sample testing or by practical engineering empirical estimation. ΔS is allowable post-construction settlement decided by local standard regulations [20]. The ultimate settlement $S_\infty^{P+\Delta P}$ contributing to the pre-loading effect before unloading, it can be obtained by conventional layerwise summation method or by prediction of measured settlement during preloading construction.

3. Field Study on Shen-Jia-Hu Highway

3.1. Engineering Situation

The practical site for experiments was part of the Zhejiang Shenjiahua Hangzhou Expressway (Lianhang Section) project (hereafter referred to as the Shen-Jia-Hu expressway), which is located on Hang-jia-hu plain, the geographical location of Shen-Jia-Hu is shown in Figure 2. This region has a developed economy, high population density, and high transportation requirements. However, the numerous lakes and rivers cause a large amount of soft soil in this region. Most of its road sections pass through soft soil foundation, which contain soils characterized by high moisture content, high compression, low strength, and low permeability, which will cause greater road settlement.

In view of the engineering geological conditions, in addition to conventional improvement methods such as grout spray pile (pile diameter 500.0 mm, 3.05 million linear meters) and prestressed concrete pipe pile (pile diameter 400.0 mm, wall thickness 60.0 mm, 590 thousand linear meters), Shen-Jia-Hu expressway also adopted equal preloading (2.39 million m^3) along the whole line. The expressway subgrade was filled with earthworks mixture. The foundation improvement mainly adopted stacking load pre-compaction and combined the plastic wick drain. The road section chosen for experiments is the EK0 + 323 ramp bridge road, which has a 10 m wide pavement, a 3.25 m high embankment, and a 1:1.5 slope grade. The soil cohesion and internal friction angle were obtained through direct shear experiment, and the soil water content was obtained through the oven-drying method. The other parameters and specific steps were determined according to the Standard for geotechnical testing method [3], and the parameters are given in Table 2. It should be noted that the soil parameters were obtained after the surcharge preloading method. The cohesion values were relatively higher than other soil.

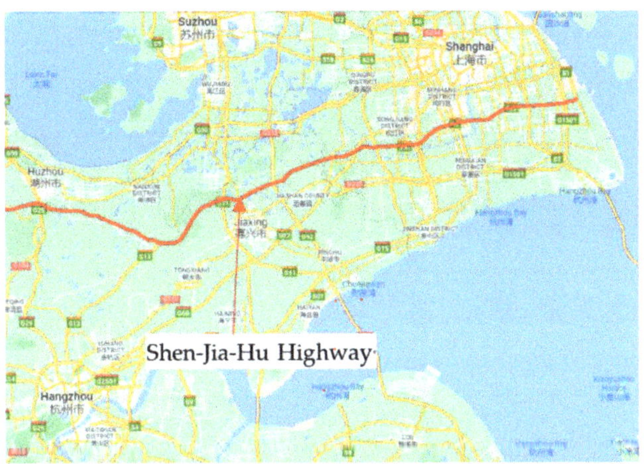

Figure 2. Geographic position of Shen-Jia-Hu expressway.

Table 2. Soil parameters.

Type of Soil Layer	Thickness (m)	Water Content (%)	Natural Density (g/cm^3)	Void Ratio	Compression Modulus (MPa)	Compression Factor (Mpa^{-1})	Cohesion (kPa)	Internal Friction Angle
Planting soil	2.1	-	-	-	-	-	-	-
Silty clay	1.2	22.7	1.95	0.64	7.24	0.22	45.00	15.20
Silt clay	10.4	51.5	1.70	1.40	2.91	0.83	32.50	20.30
Clay	11.2	30.7	1.90	0.86	5.89	0.32	30.30	23.90
Fine sand interlayer	0.4	24.5	1.98	0.70	16.81	0.10	-	-
Silty clay	3.9	23.4	1.99	0.68	10.93	0.15	40.00	26.50

3.2. Construction Process

According to the FCB construction process, the embankment slope was excavated to prepare a construction plane, and then drilling equipment was used to drill lateral holes to a designed depth in the embankment, which would not affect highway traffic operations. PVC plastic tubing or plug gauge was inserted in the hole to test the hole quality, and then the FCB material was cast in the holes, the holes were drilled in the embankment slop by the drilling machine as shown in Figure 3. A construction point worthy of attention is that the FCB pouring process should be implemented immediately after drilling the holes avoid a hole deforming under the overlying embankment load. The slope cover was cast using FCB material after all the holes were filled with FCB. The key point in this construction step is to cast the slope according to design's required proportional grading. Finally, the embankment slope was covered with soil and landscaped with plants.

3.3. Improvement before Analysis of Monitoring Data

Settlement observation was conducted during the period January 2010 to August 2010. In the EK0 + 323 section, the largest settlement, reaching over 129.0 mm, occurred on a road bridge, affecting the speed of traffic and highway safety. In September 2010, asphalt concrete pavement filling was applied, and it was later observed that the roadbed continued sinking at a larger sedimentation rate (i.e., 9.1 mm/month). On 25 March 2011, the settlement reached its maximum, 219.0 mm. Measurements of the settlement before the application of FCB are shown in Figure 4.

Figure 3. Drilling holes on north sides of the embankment. Reprinted with permission from Ref. [18]. Copyright 2021 IEEE Proceedings.

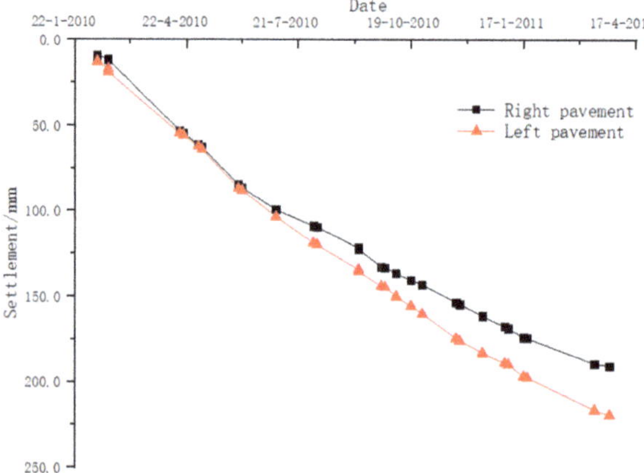

Figure 4. Total embankment settlement before improvement with FCB.

As shown in Figure 4, until 25 March 2011, no obvious convergence trend in the subsidence curves was noted, based on the post-construction monitoring settlement data. The reason may be due to tight deadlines resulting in inadequate preloading foundation, making the work required after settlement too large in scope. A settlement prediction formula [21] for EK0 + 323 settlement was used to demonstrate this method. Based on the bridge sections of EK0 + 323 and two settlement observation sections, the settlement prediction results are shown in Table 3. To simplify the forecasting process and meet the requirements of settlement prediction analysis with prediction process, secondary consolidation settlement was not considered.

The settlement prediction results showed that section EK0 + 323′s settlement will be up to 402.2 mm. Not taking into account the largest remaining secondary consolidation, settlement will again lead to vehicle bumping; thus, holes filled with bubbled and mixed FCB were introduced in section EK0 + 323 for improvement.

Following the above-described unloading calculation method, the FCB replacement quantity was 102.4 m^3, and the total FCB volume for embankment improvement was separated into 774 lateral holes with diameters of Φ 150 mm (including 719 holes that were 6.0 m deep and 55 holes that were 4.0 m deep).

Table 3. Settlement forecast.

Section		Settlement Rate March (mm/month)	Present Settlement (mm)	The Settlement Forecast (mm)	Residual Settlement (mm)
EK0 + 323	left	9.2	189.6	292.9	103.3
	right	11.2	219.3	402.2	182.9

3.4. Effectiveness Analysis of Improvement

For the FCB improvement of section EK0 + 323, the embankment settlement data based on dynamic monitoring data analysis is shown in Figure 5.

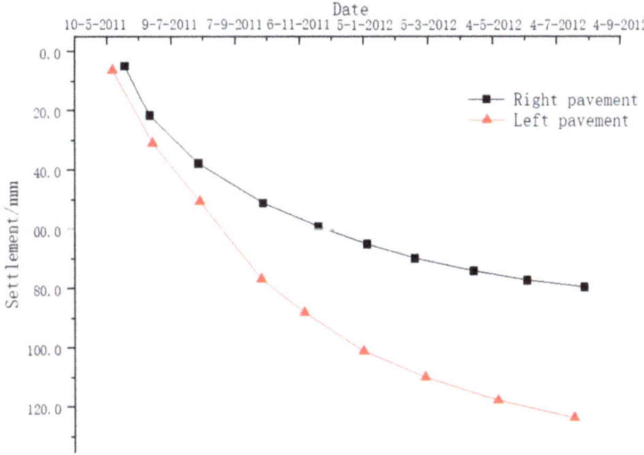

Figure 5. Total embankment settlement after improvement with FCB.

Section EK0 + 323 is located on a bridgehead segment, and according to the pre-construction settlement prediction, the remaining post-construction settlement of the left and right sides were 182.9 and 103.3 mm, respectively, until May 2011. As shown in Figure 5, after improvement with FCB on both the left and right sides, the settlements were 57.1 and 44.2 mm after embankment improvement finished (i.e., October 2011), respectively. Comparing the two sites, after initial construction of the section, the settlement exhibited a relatively sharp decreasing trend. The reason was that during the improvement process, the construction disturbance caused an impact on the embankment. The settlement after improvement was approximately 70% of the predicted settlement. Additionally, the average remaining settlement was 43.2 mm, meeting the residual settlement thresholds (i.e., less than the 30.0 cm) of the expressway embankment after improvement.

The results of EK0 + 323 after replacement improvement using the proposed technology revealed a monthly settlement rate of 50%, which was a reduction from the remaining settlement after construction of 70%. After improvement with the lateral hole replacement technology, the remaining settlement met the threshold for the settlement control after construction standards, and effectively suppressed the recurrence of bumping.

4. Discussion

Using FCB to replace the undisturbed soil of embankment on both sides cannot affect the normal operation of the highway. It can avoid economic losses caused by closure. FCB has very good fluidity, so it can be pumped to the designed depth and has a fast construction speed, so the construction period can be greatly shortened.

By calculating the replacement thickness of the embankment by using the formula proposed in this paper, after improving the embankment with this thickness, the settlement of embankment begins to stabilize. The residual settlement of embankment meets the thresholds (i.e., less than the 5.0 cm). Thus, this calculation formula, proposed in this paper, can be used in embankment improvement projects. During the improvement of the embankment, the settlement has a sharply increasing trend. It may be caused by embankment disturbance during the construction period. Thus, embankments should be replaced by the undisturbed soil in the same section, which is used to analyze the impact of construction disturbance. Additionally, monitoring the road surface settlement without embankment improvement, which is used to prove that the decrease in foundation settlement rate is caused by FCB improvement.

5. Conclusions

In this paper, Shen-Jia-Hu Expressway section EK0 + 323 served as the basis for a project to introduce new FCB design principles and calculation methods. From the results obtained through experiments on the test section as well as via subgrade settlement dynamic monitoring data analysis, the following conclusions can be drawn.

(1) The FCB method is an effective means to reduce highway embankment differential settlement without disturbing normal traffic, which has the following advantages: (1) convenience for continued highway operation, (2) high flexibility, and (3) an ideal improvement effect.

(2) Based on the proportional relationship between unloading amount and settlement value, a calculation method for unloading amount determination was developed herein. This method was used to determine the amount of embankment replacement and apply it to actual projects, which was then verified to be practical and accurate.

(3) According to the theoretical settlement value predicted in the previous article, and from the analysis of field monitoring data from practical highway projects, the pavement monthly settlement ratio was theoretically reduced by 50% and the residual post-construction settlement by 60% when transverse drilling and FCB improvement were completed. Therefore, in practice, FCB improvement is an effective means to control and reduce embankment differential settlement.

Author Contributions: The formal analysis, formula reasoning, and the performance of the experiment, as well as the original draft preparation, were completed by H.L.; Review and editing were completed by Y.W. and J.L. All authors have read and agreed to the published version of the manuscript.

Funding: The research was funded by supported by the Fundamental Research Funds for the Central Universities, number B200204036.

Institutional Review Board Statement: Not applicable.

Informed Consent Statement: Not applicable.

Data Availability Statement: Data sharing not applicable. No new data were created or analyzed in this study.

Conflicts of Interest: The authors declare no conflict of interest.

References

1. Wang, Y.S. Settlement of soft soil foundation of Expressway. *China J. Highw. Transp.* **1993**, *6*, 61–66.
2. Zhou, S.; Wang, B.L.; Shan, Y. Review of research on high-speed railway subgrade settlement in soft soil area. *Railw. Eng. Sci.* **2020**, *28*, 129–145. [CrossRef]
3. CCCC Second Highway Survey and Design Institute. *Specifications for Design of Highway Subgrades*; China Communications Press: Beijing, China, 2015; pp. 69–70, 84–85.
4. Li, J.; He, D.P.; Peng, Y.H. Alleviating bump at bridge-head using geocell flexible approach slab. In Proceedings of the Third International Conference on Transportation Engineering, Chengdu, China, 23–25 July 2011; ASCE: Reston, VA, USA, 2011; pp. 1609–1617.

5. Pan, X.D.; Du, Z.G.; Yang, X.G. Evaluation indexes of the impact of vehicle bumping at bridge-head on driving safety. *J. Tongji Univ. Nat. Sci.* **2006**, *34*, 634–637.
6. Zhao, X.X.; Yang, L.J. Analysis and management methods for the road and bridge transition's sedimentation on the soft clay ground in Shanghai. *J. Lanzhou Jiaotong Univ.* **2008**, *27*, 37–40.
7. Lu, T.; Yang, P.; Li, M.X.; Yi, L.H. Analysis of deformation behavior of foamed lightweight soil subgrade on soft soil foundation. *Subgrade Eng.* **2018**, *46*, 38–41.
8. Liu, H.L.; Yi, F.; Chen, Y.H. Study on method of laterally replacing subgrade of existing expressway by light soil. *J. Highw. Transp. Res. Dev.* **2019**, *36*, 19–27.
9. Huang, T.Y.; Wang, X.Q.; Fang, G.W.; Zhang, H.P.; Diao, H.G. Design and calculation methods of light soil replacement after preloading in the soft soil foundation of the highway. *Urban Rapid Rail Transit* **2019**, *37*, 59–63.
10. Jones, M.R.; McCarthy, A. Utilizing unprocessed low-lime coal fly ash in foamed concrete. *Fuel* **2005**, *84*, 1398–1409. [CrossRef]
11. Wang, Z.J.; Sun, Z.C.; Lu, L.; Li, J.B. Study on the uniaxial compressive mechanical properties of foamed mixture lightweight soil with polypropylene fibers. *Chin. J. Undergr. Space Eng.* **2018**, *16*, 1021–1029.
12. Li, Y.Z. Foamed cement banking for reinforcing soft ground. *Miner. Explor.* **2008**, *11*, 66–68.
13. Gao, Y.L.; Xiao, M.Q.; Guan, H.X. Foamed cement banking and its utilization in highway engineering. *Bull. Chin. Silic. Soc.* **2016**, *35*, 2432–2438.
14. Deng, H.W. Bubble-mixed light soil applied in a metro high-fill roadbed. *Urban Rapid Rail Transit* **2018**, *31*, 88–91.
15. Ding, X.H. Study on design and calculation of foamed cement banking. *Subgrade Eng.* **2017**, *35*, 146–151.
16. Jamnongpipatkul, P.; Dechasakulsom, M.; Sukolrat, J. Application of air foam stabilized soil for bridge-embankment transition zone in Thailand. In Proceedings of the American Society of Civil Engineers GeoHunan International Conference, Changsha, China, 3–6 August 2009; ASCE: Reston, VA, USA, 2009; pp. 181–193.
17. Xiao, L.J. Foamed Cement Banking and Applications in Highway Construction. Master's Thesis, Hunan University, Changsha, China, 2003.
18. Wu, Y.D.; Liu, J. Research on foamed cement banking for non-uniform differential settlement in used super highway. In Proceedings of the International Conference on Electric Technology and Civil Engineering, Lushan, China, 22–24 April 2011; IEEE: Washington, DC, USA, 2011; pp. 5334–5337.
19. Chen, Y.H.; Shi, G.C.; Cao, D.H.; Ying, H.J.; Wang, X.Q. Control of post-construction settlement by replacing subgrade with foamed cement banking. *Chin. J. Geotech. Eng.* **2011**, *33*, 1854–1862.
20. Cao, D.H.; Chen, Y.H.; Wang, X.Q.; Sun, H.; Wei, J. A study on dynamic controlling to embankment settlement of Shen-jia-hu-hang expressway. *Highway* **2010**, *55*, 45–48.
21. Nanjing Hydraulic Research Institute. *Standard for Geotechnical Testing Method*; China Planning Press: Beijing, China, 2019; pp. 16–125.

Article

Analysis of Hydrogeochemical Characteristics and Origins of Chromium Contamination in Groundwater at a Site in Xinxiang City, Henan Province

Wenfang Chen [1], Yaobin Zhang [2,*], Weiwei Shi [1], Yali Cui [2], Qiulan Zhang [2], Yakun Shi [1] and Zexin Liang [2]

1 The First Institute of Geo-Environment Survey of Henan, Zhengzhou 450045, China; chenwf.hndk@hotmail.com (W.C.); shiweiwei.hjyy@hotmail.com (W.S.); shiyakun@yeah.net (Y.S.)
2 School of Water Resources and Environment, China University of Geosciences (Beijing), Beijing 100083, China; cuiyl@cugb.edu.cn (Y.C.); qlzhang919@cugb.edu.cn (Q.Z.); 2105190003@cugb.edu.cn (Z.L.)
* Correspondence: 3005190010@cugb.edu.cn

Abstract: Hexavalent chromium contamination in groundwater has become a very serious and challenging problem. Identification of the groundwater chemical characteristics of the sites and their control mechanisms for remediation of pollutants is a significant challenge. In this study, a contaminated site in Xinxiang City, Henan Province, was investigated and 92 groundwater samples were collected from the site. Furthermore, the hydrogeochemical characteristics and the distribution patterns of components in the groundwater were analyzed by a combination of multivariate statistical analysis, Piper diagram, Gibbs diagram, ions ratio and hydrogeochemical simulation. The results showed that the HCO_3-Cl-Mg-Ca type, SO_4-HCO_3-Na type, and HCO_3-Mg-Ca-Na type characterize the hydrogeochemical composition of shallow groundwater and HCO_3-Cl-Mg-Ca type, HCO_3-Na-Mg type, and HCO_3-SO_4-Mg-Na-Ca type characterize the hydrogeochemical composition of deep groundwater. Ion ratios and saturation index indicated that the groundwater hydrogeochemical characteristics of the study area are mainly controlled by water–rock action and evaporative crystallization. The dissolution of halite, gypsum and anhydrite, the precipitation of aragonite, calcite and dolomite, and the precipitation of trivalent chromium minerals other than $CrCl_3$ and the dissolution of hexavalent chromium minerals occurred in groundwater at the site. The minimum value of pH in groundwater at the site is 7.55 and the maximum value is 9.26. The influence of pH on the fugacity state of minerals was further investigated. It was concluded that the saturation index of dolomite, calcite, aragonite and $MgCr_2O_4$ increases with the increase of pH, indicating that these minerals are more prone to precipitation, and the saturation index of $Na_2Cr_2O_7$, $K_2Cr_2O_7$ and $CrCl_3$ decreases with the increase of pH, implying that $Na_2Cr_2O_7$, $K_2Cr_2O_7$ and $CrCl_3$ are more prone to dissolution. The saturation index of the remaining minerals is less affected by pH changes. The study can provide a scientific basis for groundwater remediation.

Keywords: groundwater; hydrochemistry; hexavalent chromium contamination; water chemistry simulation

Citation: Chen, W.; Zhang, Y.; Shi, W.; Cui, Y.; Zhang, Q.; Shi, Y.; Liang, Z. Analysis of Hydrogeochemical Characteristics and Origins of Chromium Contamination in Groundwater at a Site in Xinxiang City, Henan Province. *Appl. Sci.* **2021**, *11*, 11683. https://doi.org/10.3390/app112411683

Academic Editor: Bing Bai

Received: 29 October 2021
Accepted: 22 November 2021
Published: 9 December 2021

1. Introduction

In many parts of the world, water scarcity has posed a great threat to socio-economic development and the ecological environment [1,2]. Due to the limited surface water resources and precipitation, groundwater resources have become indispensable freshwater resources for domestic drinking, irrigation water and industrial activities [3–6]. With large-scale industrial activities, groundwater pollution has become a serious problem affecting human health and life in many countries and regions, and as of 2018, the number of declared contaminated sites in 31 provincial capitals in China reached 174. In the process of industrial development, chromium has been used in a growing number of industries, and chromium salt production has subsequently developed to a certain extent [7–11]. At the same time, a large amount of chromium slag was produced, and after its long-term open piling, Cr(VI) will seep into the soil and groundwater with surface runoff, thus causing pollution to the surrounding soil and groundwater, and since Cr(VI) is a highly migratory and toxic pollutant, chromium pollution in groundwater has therefore become a serious worldwide problem [12–15]. Cr(VI) has been identified as highly toxic and one of the carcinogens, posing a threat to human health when chromium exceeds drinking water standards.

Groundwater contamination investigation and remediation is urgent, and the analysis of hydrogeochemical parameters are important to grasp the groundwater quality status [16–21]. The study of hydrogeochemical composition and its origins is the basis of groundwater protection and restoration. There are many methods, with the most common ones being multivariate statistical analysis [22,23], the isotope labeling method and the hydrogeochemical simulation method [24–26]. Xiao Yong et al. studied the characteristics and controlling factors of groundwater chemistry in the long-term use area of reclaimed water in Beijing by Piper's trilinear diagram, ionic proportional relationship and saturation index method, and concluded that the changes in the types of groundwater chemistry and water–rock action during the rainy and dry season, and believed that the hydrogeological conditions of the area should be fully considered when reclaimed water was used for agricultural irrigation and ecological landscape [27]. Zhang Jingtao et al. explored the hydrogeochemical characteristics and evolution of groundwater in the Dachaidan area from the pre-mountain alluvial fan to the salt lake by ion ratio analysis, Gibbs plot and hydrogeochemical simulation, and concluded that the water–rock interaction was dominated by the dissolution of halite and gypsum and positive cation exchange [28]. Ashwani used a multivariate statistical and hydrogeochemical approach to analyze Cr(VI) concentration levels, pollutant sources and groundwater geochemistry in an industrial town in Italy to provide an effective aid for water resource management in the region [29]. By testing chemical parameters related to fluoride in groundwater during the periods of abundance, flatness and depletion, D. Laxmankumar concluded that weathering, ion exchange and anthropogenic activities played an important role on the chemical composition of groundwater using Gibbs plot and principal component analysis [30].

Chromium exists in groundwater mainly as trivalent chromium and hexavalent chromium [31]. The migration and transformation of chromium in groundwater is influenced by the redox potential, acidity and concentration [32,33]. Under acidic conditions with low redox potential, chromium mainly exists as Cr(III). As pH increases, the solubility of Cr(III) decreases and $Cr(OH)_3$ precipitation is formed. Under alkaline conditions with high redox potential, chromium is mainly present as Cr(VI) [34,35]. Most of the research focused on the chromium migration and transformation by performing batch and column experiments indoors. The chromium presence pattern was analyzed by varying different pH and concentration values by the controlled variable method. Column experiments usually take a lot of time to complete. In this study, the presence pattern of Cr in representative samples was characterized by numerical simulation using saturation index.

Xinxiang City is a significant industrial base in the North China Plain. To date, there are many left industrial contaminated sites. Thus, it is necessary to get insight into the influence of contaminants on groundwater. Although it has been determined that the groundwater quality at the contaminated site exceeds the Class III water standard, the groundwater hydrogeochemical characteristics and their governing mechanisms are still poorly known. The purpose of this work was to investigate the level of contamination in groundwater, hydrogeochemical characteristics and their origins at a contaminated site in Xinxiang by using mathematical statistics, correlation analysis, ion ratio analysis and saturation index, and to study the hydrogeochemical processes controlling the evolution of groundwater chemistry and the influencing factors of chromium-containing mineral by means of hydrogeochemical simulation. This study will provide effective assistance for groundwater remediation in the region.

2. Description of the Study Area

The study area is located in the northwestern part of Xinxiang City, Henan Province (Figure 1), with a temperate continental monsoon climate and a multi-year average temperature of 14 °C. The multi-year average precipitation is 586.32 mm, with more than 70% occurring in the rainy season spanning June to September. The multi-year average evaporation is 1772.62 mm, which is three to four times the precipitation with the strongest evaporation in June.

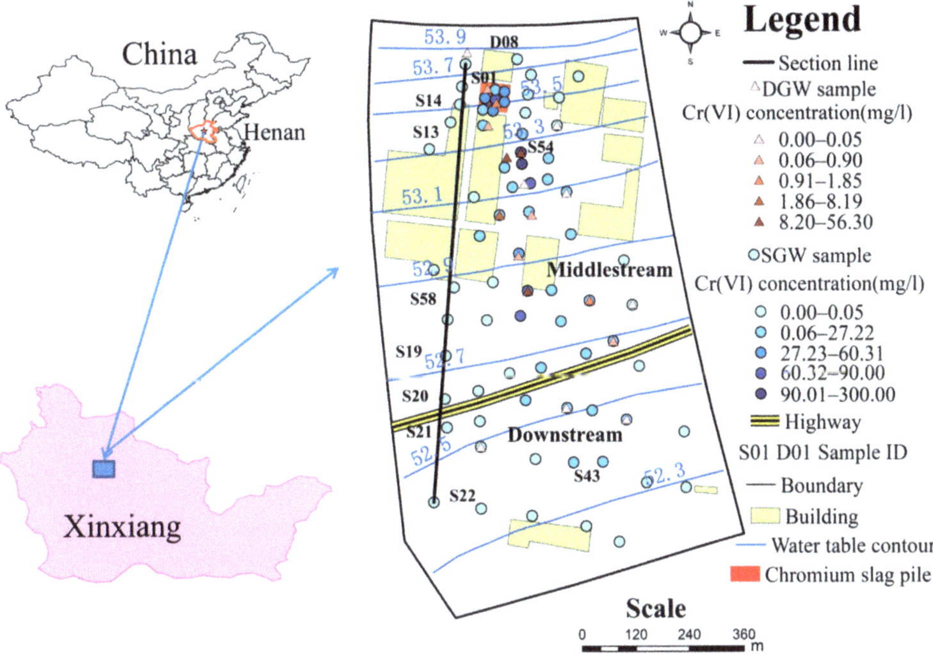

Figure 1. Location of the study area and the sampling sites. DGW represents deep groundwater. SGW represents shallow groundwater. Dots represent shallow sampling points. Triangles represent deep sampling points. The boundary represents the boundary of the numerical model. Henan is located in the central part of China. Xinxiang City is located in the north of Henan Province.

Xinxiang City is located at the southeastern foot of Taihang Mountains and the northern edge of the alluvial plain in the middle and lower reaches of the Yellow River. The terrain generally inclines from northwest to southeast. The regional landform types are divided into three types of landforms: alluvial valley, alluvial sloping plains and alluvial plains. The geomorphological type of the study area is alluvial sloping plain. The surface of the study area reveals the Quaternary river-phase sedimentary layer with a thickness of 50–60 m, and the lithology is mainly clay and sand. The groundwater type is pore water of loose rock of the Quaternary, and the media of the aquifer are mainly medium sand and fine sand, with coarse sand locally. According to the profile, it can be seen that the lithology from 56 m to 49 m is silty clay and silt. The lithology from 59 m to 49 m is silty clay and silt. This is followed by a shallow aquifer with a thickness of 3 m. However, the aquifer thickness increased significantly at borehole S22. The lithology from 46 m to 36 m is mainly silty clay. The bottom lithology is mainly fine sand. The aquifer is mainly recharged by atmospheric precipitation, lateral runoff and irrigation recharge, and groundwater discharge is mainly exploitation, evaporation and lateral runoff.

Chromium salt chemical plants existed in the study area during the historical period, and chromium slag from the production process brought serious contamination to the site soil and groundwater (Figure 2). Test results showed that the average concentration of total chromium in the soil was 12,100 mg/kg, and work has been carried out to remediate the soil and groundwater.

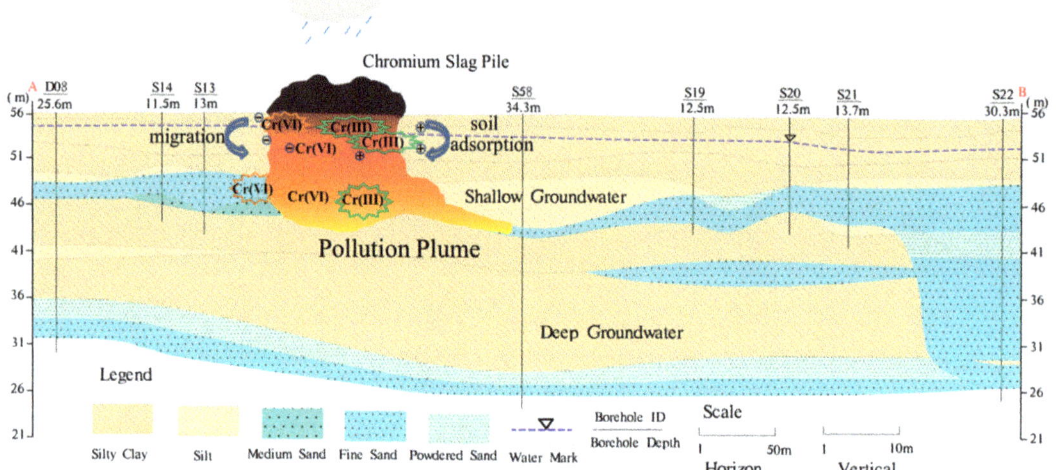

Figure 2. The hydrogeological cross section along the A–B and pollution conceptual model. In the pollution plume, the red color represents the high concentration value. Yellow color represents low concentration value.

3. Materials and Methods

3.1. Sample Collection and Analysis

In groundwater contamination investigation sampling, determining the location of sampling sites was a key factor. This sampling was mainly based on the following principles. Firstly, samples should uniformly along the groundwater flow direction. Secondly, the sampling points are encrypted in the seriously polluted areas. Thirdly, the general shape of the contamination plume can be determined according to the sampling points.

In December 2020, field investigations and sampling were conducted in the study area, and 92 groundwater samples were collected, including 73 shallow groundwater (SGW) samples and 19 deep groundwater (DGW) samples, with the locations of the sampling sites as shown in Figure 1. The total dissolved solids (TDS) and pH were measured in the field using a multiparameter tester. Other chemical indicators (Na^+, Ca^{2+}, Mg^{2+}, K^+, Cr(VI), HCO_3^-, SO_4^{2-}, Cl^-, NO_3^-) were measured in the laboratory of Henan Province Rock and Mineral Testing Center. The charge balance errors of all samples were within 5%, and the accuracy of each indicator met the quality requirements.

3.2. Multivariate Statistical Analyses

Statistical methods are an effective tool in groundwater quality assessment. Groundwater contamination can be determined by graphical and multivariate statistical methods, such as, Piper diagram, Gibbs diagram and Ion ratio. Multivariate statistical methods play an important role in the identification of pollution sources, such as correlation analysis. It can determine the degree of correlation between different parameters. Pearson's correlation coefficient was used to analyze the degree of dependency of one parameter to the other [26,36].

4. Results and Discussion

4.1. Hydrochemistry

The analysis of groundwater chemistry data can help to determine the geochemical parameters of groundwater and the distribution characteristics of contaminants. The statistics and analysis of the test results of groundwater samples were shown below.

The minimum value of pH of shallow groundwater in the study area was 7.55 and the maximum value was 9.26, which was alkaline overall and favorable to the precipitation and oxidation of trivalent chromium. The average distribution of the main cation concentrations: $Na^+ > Ca^{2+} > Mg^{2+} > Cr > K^+ > Fe > Mn > NH_4^+ > As > Pb$, and the average distribution of the main anion concentrations: $HCO_3^- > SO_4^{2-} > Cl^- > NO_3^- > CO_3^{2-} > NO_2^-$. The order of triple nitrogen content in groundwater in the study area was $NO_3^- > NO_2^- > NH_4^+$, which mainly polluted groundwater in the form of nitrate nitrogen [37].

Among the ionic components of shallow groundwater, the coefficients of variation of SO_4^{2-}, Cl^-, NO_3^-, Na^+, Ca^{2+}, Mg^{2+}, HCO_3^- and Fe were relatively small, and the contents of these ions in shallow groundwater were relatively stable. In contrast, the coefficients of variation (C.V.) of NO_2^-, Mn, NH_4^+, As, Pb, Cr(VI), CO_3^{2-} and K^+ all exceeded 100%, which belonged to a strong degree of variation, indicating that the contents of these ions in shallow groundwater were highly variable, and at the same time these ions were sensitive factors that vary with environmental and anthropogenic factors.

The pH of deep groundwater in the study area was 7.8 at minimum and 8.34 at maximum, which was alkaline overall. The distribution of the main cation concentration: $Na^+ > Ca^{2+} > Mg^{2+} > Cr(VI) > K^+ > Fe > Mn > NH_4^+$, the distribution of the main anion concentration: $HCO_3^- > SO_4^{2-} > Cl^- > NO_3^- > NO_2^-$, and As, Pb, CO_3^{2-} were not detected in the deep aquifer. The distribution of the remaining ions was similar to that of the shallow layer.

As can be seen from Table 1, the coefficients of variation of SO_4^{2-}, Cl^-, NO_3^-, Na^+, Ca^{2+}, Mg^{2+}, HCO_3^-, Fe and K^+ in the deep groundwater were relatively small, and the contents of these ions in deep groundwater were relatively stable. The coefficients of variation of NO_2^-, Mn, NH_4^+, Cr(VI) and CO_3^{2-} all exceeded 100%, which belonged to strong degree of variation.

Table 1. Statistical descriptions of chemical parameters.

Index	Shallow Groundwater					Deep Groundwater				
	Min	Max	Mean	SD	C.V./%	Min	Max	Mean	SD	C.V./%
pH	7.55	9.26	8.03	0.32	3.99	7.80	8.34	8.19	0.19	2.27
TDS	738.36	2532.60	1206.26	389.76	32.31	504.00	1198.26	798.04	185.94	23.30
SO_4^{2-}	100.94	1963.05	413.64	403.01	97.43	8.61	475.94	160.72	125.68	78.20
Cl^-	83.05	461.59	204.31	72.86	35.66	7.81	206.98	109.56	45.42	41.46
NO_3^-	0.56	322.92	117.82	81.63	69.28	3.95	200.17	70.75	57.29	80.97
NO_2^-	0.00	2.43	0.39	0.61	156.41	0.00	1.29	0.25	0.31	123.05
Na^+	77.34	1372.00	292.15	266.48	91.21	63.94	322.5	129.35	62.76	48.52
Fe	0.04	4.88	1.09	1.05	96.33	0.01	1.64	0.49	0.45	91.63
Mn	0.01	7.17	0.53	1.05	198.11	0.00	1.36	0.26	0.35	132.48
NH_4^+	0.00	1.80	0.12	0.30	250.00	0.00	0.50	0.07	0.15	230.90
As	0.00	0.10	0.005	0.02	400.00	0.00	0.00	0.00	0.00	0.00
Pb	0.00	0.02	0.002	0.004	200.00	0.00	0.00	0.00	0.00	0.00
Cr(VI)	0.00	299.99	18.34	50.86	277.32	0.00	56.30	8.31	17.85	214.79
Fe^{3+}	0.04	2.55	0.70	0.65	92.86	0.01	1.01	0.31	0.30	96.93
CO_3^{2-}	0.00	40.76	1.17	5.50	470.09	0.00	0.00	0.00	0.00	0.00
HCO_3^-	298.39	1052.65	589.11	140.58	23.86	389.56	687.95	518.25	83.32	16.08
K^+	0.38	14.48	1.45	2.04	140.69	0.45	3.23	0.80	0.61	75.94
Ca^{2+}	15.57	231.90	143.59	34.95	24.34	38.54	148.10	101.48	30.01	29.58
Mg^{2+}	26.65	246.90	117.38	32.88	28.01	52.27	143.50	95.11	24.25	25.49
Cr	0.00	332.90	21.81	57.36	263.00	0.00	63.28	10.41	21.69	208.33

Note: The units of minimum, maximum and average values in the table are mg/L except for pH. SD: Standard Deviation. C.V.: Coefficient of Variation.

4.2. Spatial Distribution Characteristics of Groundwater Chemistry

4.2.1. Hydrogeochemical Facies

A Piper diagram can help to evaluate the geochemical relationships between different dissolved ions and dominant types of water chemistry in groundwater [38]. The TDS of shallow groundwater was 738.4–2532.6 mg/L [39]. The shallow groundwater was highly variable with HCO_3^-, SO_4^{2-}, Cl^-, NO_3^- as the main anions and Na^+, Ca^{2+}, Mg^{2+} as the main cations. The NO_3^- concentration in the water samples was high and exceeded the quality standard for domestic drinking water, probably due to the inappropriate use of fertilizers making NO_3^- enter the groundwater with rainfall and irrigation [40]. The HCO_3-Cl-Mg-Ca type, SO_4-HCO_3-Na type and HCO_3-Mg-Ca-Na type characterized the hydrogeochemical composition of shallow groundwater (Figure 3). The groundwater below the chromium slag pile (CSP) was SO_4-HCO_3-Na type. The reason was related to the production process of chromium salt, and the specific chemical reaction was $2Na_2CrO_4 + H_2SO_4 = Na_2Cr_2O_7 + Na_2SO_4 + H_2O$, so SO_4^{2-} and Na^+ will enter the groundwater under the leaching effect of chromium slag. The downstream was basically uncontaminated, and the groundwater was HCO_3-Mg-Ca-Na type. The groundwater of middle reaches and east of the chromium slag piles was HCO_3-Cl-Mg-Ca type.

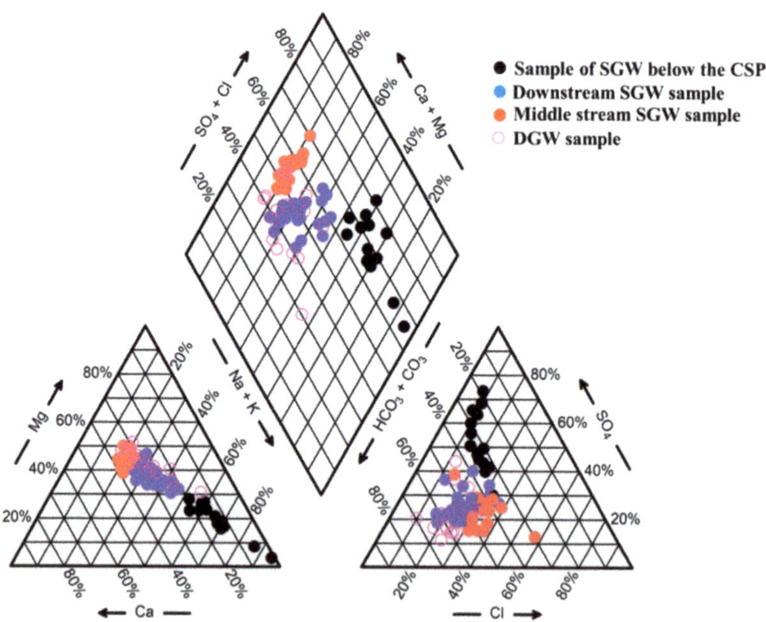

Figure 3. Piper diagram for shallow and deep groundwater. CSP represents chromium slag pile.

The TDS of deep groundwater in the site was 504–1198.26 mg/L. The anions were mainly HCO_3^-, SO_4^{2-}, Cl^-, NO_3^-, and the cations were mainly Na^+, Ca^{2+}, Mg^{2+}. The NO_3^- concentration in the deep groundwater was lower than that in the shallow groundwater, but still exceeded the quality standard of domestic drinking water. It may be that the shallow groundwater overflowed into the deep groundwater, resulting in abnormal NO_3^- concentration in the deep groundwater. The deep hydrogeochemical composition was simpler than the shallow ones, and the groundwater was mainly HCO_3-Cl-Mg-Ca type, HCO_3-Na-Mg type and HCO_3-SO_4-Mg-Na-Ca type.

4.2.2. Spatial Distribution Pattern of Ions

Spatial Distribution Pattern of Ions in Shallow Groundwater

Based on the analytical results of shallow groundwater samples, the spatial distribution pattern of ions was analyzed to study the spatial contamination characteristics of each ion component in groundwater (Figure 4a). The pollutant correlation coefficient matrix was plotted using R (Figure 4b). The Pearson's correlation coefficient was used to characterize the correlation between different pollutants.

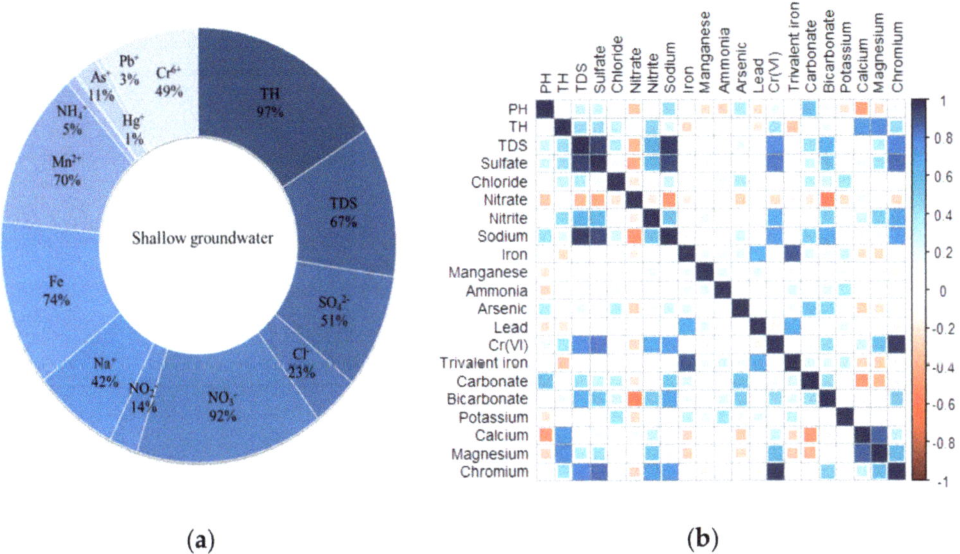

Figure 4. (**a**) Exceedance rate of each component of groundwater, (**b**) pollutant correlation coefficient matrix.

The TDS of 67% of the shallow groundwater samples exceeded three types of groundwater standards, especially the TDS concentration below the chromium slag pile and near the slag pile was the highest, and the water quality was poor, which was seriously affected by the chromium slag pile. Among the tested samples, the total hardness of 97% of the groundwater samples exceeded three types of groundwater standards, and the distribution of total hardness was similar to that of Ca^{2+} and Mg^{2+} concentration distribution, and the correlation coefficient reached 75%, indicating that the total hardness was mainly affected by the Ca^{2+} and Mg^{2+}. The $SO_4{}^{2-}$ concentration of 51% of the groundwater samples exceeded three types of groundwater standards, and the distribution of sulfate concentration was extremely similar to that of TDS, and the correlation coefficient reached 95%. Generally, $SO_4{}^{2-}$ ions came from the dissolution of gypsum or sulfate sedimentary rocks. However, it can be seen that a large amount of sulfuric acid was used in the production of chromium salt, so it can be concluded that the $SO_4{}^{2-}$ exceeded the standard in groundwater and was closely related to the production of chromium salt, and the correlation coefficient with Cr(VI) reached 83%. The $NO_3{}^-$ concentration of 92% of the groundwater samples exceeded three types of groundwater standard. Na^+ concentration distribution was similar to that of TDS and $SO_4{}^{2-}$ with a correlation coefficient of 96%, and the correlation coefficient between Na^+ and Cr(VI) was 0.72, mainly due to the fact that a large amount of sodium chromate and sodium sulfate will be produced in the chromium salt production process, resulting in the exceedance of Na^+ concentration in groundwater. The exceedance of Fe and Mn reached 74% and 70%, respectively. The exceedance of Cr(VI) was 49%, which had a certain contribution rate to TDS, and the correlation coefficient reached 78%. Pb and Hg concentrations in groundwater samples had low exceedance rates, which may be related to human activities.

Spatial Distribution Pattern of Ions in Deep Groundwater

According to the analysis results of deep groundwater samples, it can be concluded that the total hardness of 68% of the tested samples exceeded the three groundwater standards, and the distribution of total hardness was similar to the distribution of Ca^{2+} and Mg^{2+} concentrations, with a correlation coefficient of 75% (Figure 5). The over-standard rate of TDS was 16%. For cations, Mg^{2+} had the highest correlation coefficient with TDS, reaching 0.81, followed by Cr(VI), reaching 0.79, indicating that Mg^{2+} contributed the most to the increase of TDS, and for anions, $SO_4{}^{2-}$ had the highest correlation coefficient with TDS, reaching 0.89, indicating that $SO_4{}^{2-}$ contributed the most to the increase of TDS. Na^+ exceeded the standard by 5%, and the correlation coefficient reached 79% with hexavalent chromium. The over-standard rate of $NO_3{}^-$ was 79%. The over-standard rates of Fe and Mn reached 68% and 74%, respectively, which were comparable to those of Fe and Mn in shallow groundwater. The 58% of Cr(VI) concentration exceeded the standard in deep groundwater, and the highest correlation coefficient between Cr(VI) and $SO_4{}^{2-}$ reached 0.92. Deep groundwater was less polluted than shallow groundwater, and the ion concentration was less than that of shallow groundwater, and the overall exceedance rate of each component was lower than that of shallow groundwater. The reason for exceeding the standard was mainly due to the shallow groundwater overflowing into deep groundwater.

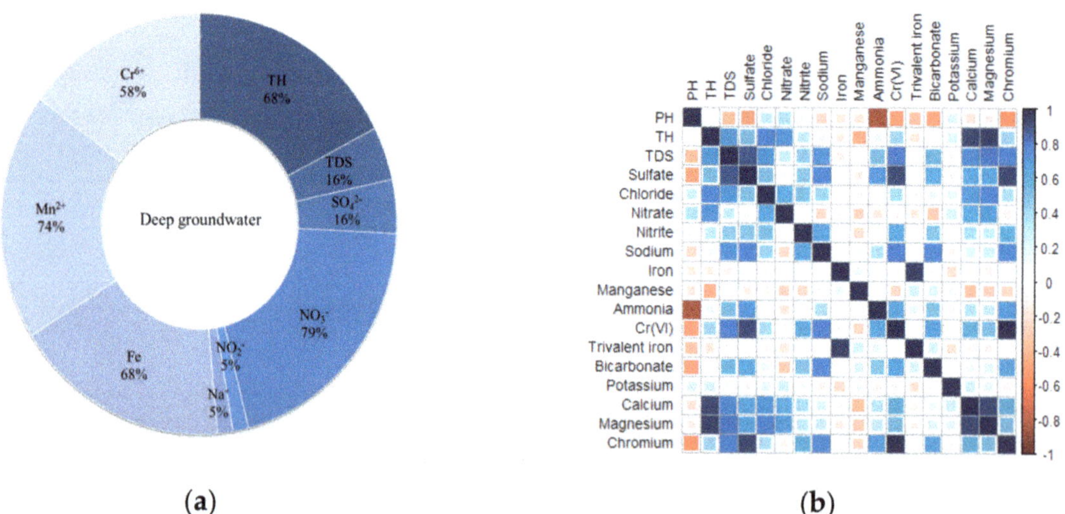

Figure 5. (**a**) Exceedance rate of each component of groundwater, (**b**) pollutant correlation coefficient matrix.

4.3. Analysis of the Causes of Groundwater Chemistry Types

4.3.1. Groundwater Chemistry Control Mechanisms

A Gibbs plot can be used to illustrate the causal mechanism of water chemical composition [41]. In the Gibbs plot, if the TDS value is high and the $Na^+/(Na^++Ca^{2+})$ or $Cl^-/(Cl^-+HCO_3{}^-)$ ratio is close to 1, the ion control mechanism is evaporative crystallization. If the TDS value is medium and the $Na^+/(Na^++Ca^{2+})$ or $Cl^-/(Cl^-+HCO_3{}^-)$ ratio is less than 0.5, the ion control mechanism is water rock effect. If TDS value is low and the $Na^+/(Na^++Ca^{2+})$ or $Cl^-/(Cl^-+HCO_3{}^-)$ ratio is close to 1, the water chemistry component control mechanism is the rainfall effect [42,43]. From the Figure 6, it can be seen that the chemical composition of groundwater in the study area is mainly dominated by water–rock action.

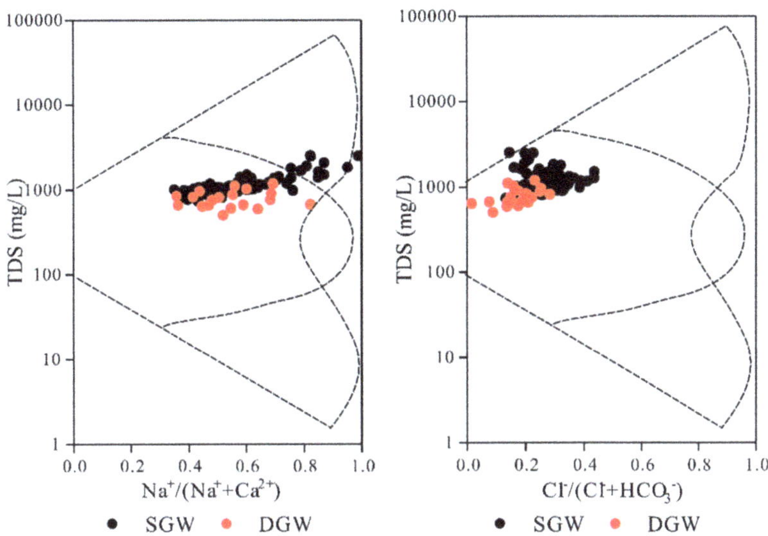

Figure 6. Gibbs diagrams showing the groundwater chemistry controlling mechanisms.

4.3.2. Major Ion Ratio Relationship

In order to further explore the process of site water chemistry, the interrelationship between each major ion was analyzed [42]. As can be seen from Figure 7a, most of the shallow groundwater samples were distributed along the Na/Cl ratio 1:1 line (halite dissolution line), and the Na/Cl ratio of a few samples was greater than 1, indicating that there were other sources of Na^+ in the groundwater in addition to halite dissolution filtration, mainly related to the production process of sodium dichromate, which led to the increase of Na^+ concentration in the groundwater around the chromium slag dumps. Most of the deep groundwater samples were distributed along the Na/Cl ratio 1:1 line, and the Na/Cl ratio values of a few samples were greater than 1, mainly because the shallow contaminated groundwater entered the deep aquifer, leading to the Na^+ concentration to increase.

The sources of Ca^{2+} and $SO_4{}^{2-}$ are commonly the dissolution of gypsum and anhydrite. As can be seen from Figure 7b, most of the shallow groundwater samples were distributed along the Ca/SO_4 ratio 1:1 line (gypsum and anhydrite dissolution line), and a few samples had Ca/SO_4 ratio less than 1, mainly because the plant will use a large amount of sulfuric acid in the production process leading to the increase of $SO_4{}^{2-}$ concentration in groundwater around the chromium slag pile, while the groundwater sample points near the Ca/SO_4 ratio 1:1 line showed most samples with Ca/SO_4 ratios greater than 1, and it may be alternate cation adsorption with constant Ca^{2+} release. The deep groundwater samples were all distributed along the Ca/SO_4 ratio 1:1 line, but most of the samples presented Ca/SO_4 ratios greater than 1, indicating that the dissolution of aragonite, calcite and dolomite was a potential source of Ca^{2+} in the site groundwater.

From the relationship pattern of $(Ca^{2+}+Mg^{2+})/(SO_4{}^{2-}+HCO_3{}^{-})$ [44], it was clear that the majority of water samples were distributed along the 1:1 line, except for groundwater samples with heavily contaminated samples, further indicating that calcite, dolomite, gypsum and anhydrite dissolution were potential sources of major ions in the groundwater mineralization process (Figure 7c).

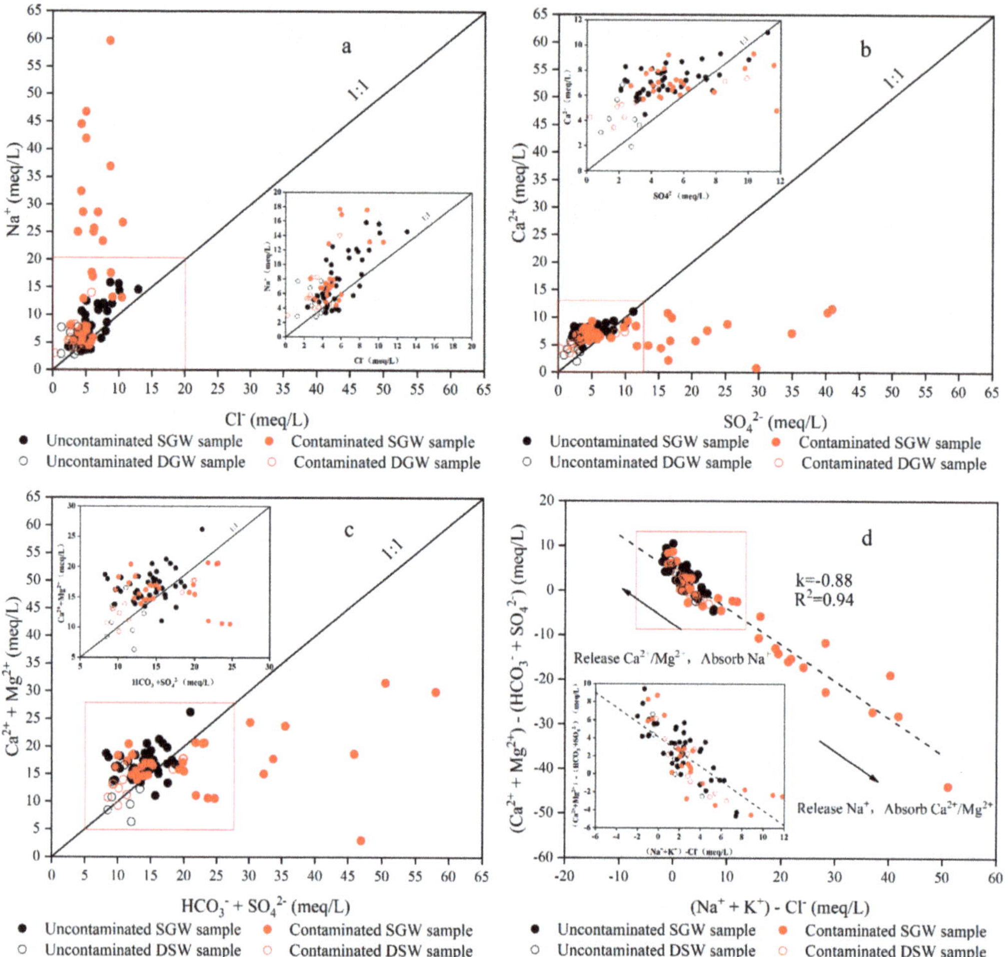

Figure 7. Groundwater main ion relationship diagram. (**a**) represents the ratio of Na^+ and Cl^-, which generally indicates the dissolution of halite. (**b**) rep-resents the ratios of Ca^{2+} and SO_4^{2-}, which generally indicates the dissolution of gypsum and an-hydrite. (**c**) indicates the dissolution of calcite and dolomite. (**d**) can indicate the presence or ab-sence of ion exchange interaction.

In addition, the relationship between (Ca^{2+}+Mg^{2+})-(SO_4^{2-}+HCO_3^-) and (Na^++K^+)-Cl^- can usually be used to determine whether there is an ion alternate adsorption reaction in the aquifer. If the two are linear and the slope is close to −1, it indicates the existence of ion exchange, where the sample points distributed along the fitted line on the right side of the origin are the cation alternate. The sample points distributed along the fitted line to the left of the origin were the reverse reaction of alternate cation adsorption, and R^2 = 0.94 as shown in Figure 7d. This indicated that ion exchange existed in the groundwater of the site, which further illustrated that the part of groundwater sample points continuously released Ca^{2+} under the reverse reaction of alternate cation adsorption. It verified the result that the Ca/SO_4 ratio was greater than 1.

4.4. Hydrogeochemical Simulation and Analysis

The potential hydrogeochemical processes in the groundwater were investigated through ion ratio relationships. In this section, the minerals saturation index (SI) was calculated by Phreeqc 2.8 [45] to further determine the forms of the various minerals present and the geochemical interactions.

4.4.1. Mineral Saturation Index Analysis

The saturation index of minerals can visualize the dissolved equilibrium state of components in groundwater. Three representative water samples were selected for this study: the severely contaminated water sample S54 with hexavalent chromium concentration of 257 mg/L, the moderately contaminated water sample S01 with hexavalent chromium concentration of 60.3 mg/L, and the lightly contaminated water sample S43 with hexavalent chromium concentration of 0.729 mg/L. For the water sample S54, the saturation index of hexavalent chromium minerals was $CaCrO_4 > Na_2CrO_4 > MgCrO_4 > K_2CrO_4 > Na_2Cr_2O_7 > K_2Cr_2O_7 > CrO_3$, and the saturation index was less than 0, which indicated that hexavalent chromium minerals were easily dissolved in the groundwater, and the proportion of chromium oxide was the largest among chromium minerals. The saturation index of trivalent chromium minerals was ordered as $Cr_2O_3 > MgCr_2O_4 > Cr(OH)_3 > CrCl_3$, and the saturation index of Cr_2O_3, $MgCr_2O_4$ and $Cr(OH)_3$ were greater than 0, while the saturation indices of $CrCl_3$ were less than 0. It can be concluded from the saturation index of trivalent chromium minerals that most of the trivalent chromium minerals were insoluble in groundwater, and $CrCl_3$ made the greatest contribution to the concentration of trivalent chromium ions. For groundwater sample S01, the saturation index of hexavalent chromium minerals and trivalent chromium minerals showed the same variation pattern with S54. For groundwater sample S43, the saturation index of hexavalent chromium minerals was ranked as $CaCrO_4 > Na_2CrO_4 > MgCrO_4 > K_2CrO_4 > CrO_3 > K_2Cr_2O_7 > Na_2Cr_2O_7$, and the saturation index of hexavalent chromium minerals was less than 0, indicating that hexavalent chromium minerals were easy to dissolve into groundwater, but the order of mineral saturation index had changed, and the $Na_2Cr_2O_7$ component had the largest proportion in chromium-containing minerals. The saturation index of trivalent chromium minerals showed the same change pattern as S54 and S01(Figure 8a).

By comparing the saturation index of chromium-containing minerals at S54, S01 and S43, and it can be concluded that the higher the chromium ion concentration in the aquifer, the greater the saturation index of chromium-containing minerals, the smaller the tendency for hexavalent chromium minerals to dissolve and the greater the tendency for trivalent chromium minerals to precipitate.

Figure 8b showed the simulation results of the saturation index of the main minerals. The results showed that anhydrite, gypsum and halite were under unsaturated state in groundwater, and halite was the most prone to dissolution, and calcite, aragonite and dolomite were under saturated state, and dolomite was the most prone to precipitation.

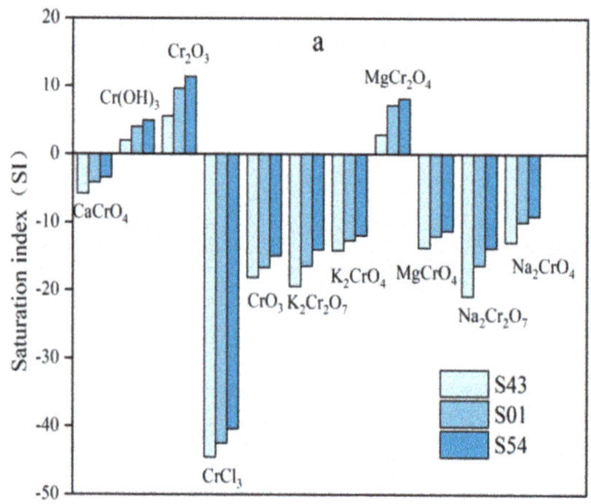

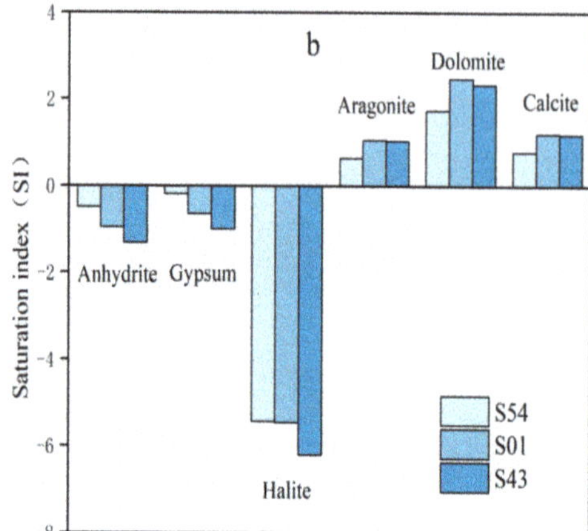

Figure 8. Comparison of saturation index (SI) at three points, (**a**) refers to chromium-containing minerals, including trivalent chromium and hexavalent chromium. (**b**) contains anhydrite, gypsum, halite, aragonite, dolomite and calcite.

4.4.2. Analysis of the Effect of pH on Minerals

After statistical analysis, the pH variation range of the site groundwater was 7.55–9.26, so the trend of minerals in the aquifer was further determined by analyzing the effect of pH variation on the saturation index of minerals. Taking S01 as an example, the pH was taken as 7.55, 8, 8.26, 9 and 9.26, respectively, which can truly reflect the trend of dissolution and precipitation of minerals in the aquifer.

As can be seen from Figure 9, the saturation index of dolomite, calcite, aragonite and $MgCr_2O_4$ is increasing with the increase of pH, indicating that $MgCr_2O_4$ was more prone to precipitation in the alkaline environment, mainly because the ionic activity product increases with the increase of pH under the condition of constant temperature, which led to the easy precipitation. With the increase of pH, the saturation index of $Na_2Cr_2O_7$, $K_2Cr_2O_7$ and $CrCl_3$ were decreasing, indicating that the dissolution of $Na_2Cr_2O_7$, $K_2Cr_2O_7$ and $CrCl_3$ was more likely to occur in the alkaline environment, mainly because the ionic activity product decreased with the increase of pH under the condition of constant temperature, which led to the dissolution to occur easily, and the saturation index of the remaining minerals was less affected by the change of pH.

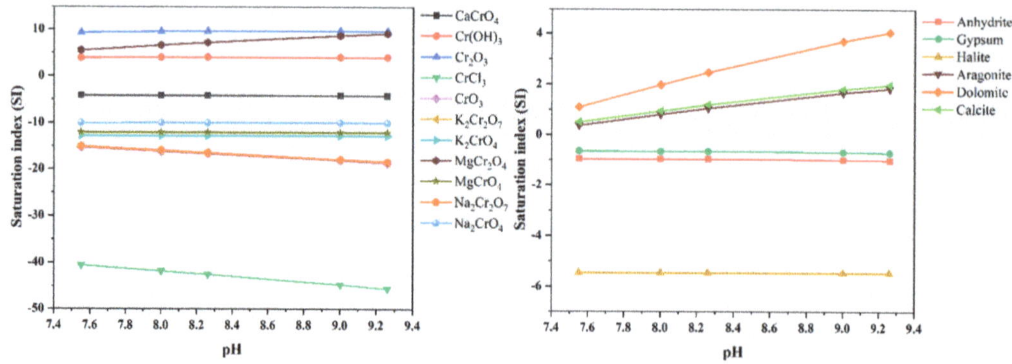

Figure 9. The influence of pH on the SI of minerals.

5. Conclusions

In this study, the distribution pattern of hydrogeochemical characteristics at a contaminated site in Xinxiang was analyzed by combination of multivariate statistical analysis and hydrochemistry, and the following conclusions were drawn.

1. The hydrochemical type of groundwater in the study area have obvious differences in the horizontal and the vertical directions, reflecting the influence of the chromium slag pile on groundwater. In the shallow groundwater, the groundwater below the chromium slag heap is the SO_4-HCO_3-Na type, the hydrogeochemical composition east of the chromium slag heap and in the midstream is the HCO_3-Cl-Mg-Ca type, and the downstream is basically uncontaminated with the HCO_3-Mg-Ca-Na type. The deep groundwater shows the HCO_3 type.
2. Gibbs plots, ion ratios and saturation index indicated that the hydrogeochemical characteristics of the study area are mainly controlled by water-rock action and evaporative crystallization, with dissolution of halite, gypsum and anhydrite, precipitation of aragonite, calcite and dolomite, and precipitation of trivalent chromium minerals other than $CrCl_3$ and dissolution of hexavalent chromium minerals occurring in the groundwater of the site.
3. The groundwater in the study area is generally alkaline with a minimum pH of 7.55, which is conducive to the precipitation of trivalent chromium and the dissolution of hexavalent chromium. With the increase of pH, the saturation index of dolomite, calcite, aragonite and $MgCr_2O_4$ keeps increasing, and these minerals are more likely to precipitate. The saturation index of $Na_2Cr_2O_7$, $K_2Cr_2O_7$ and $CrCl_3$ is decreasing, and the more easily $Na_2Cr_2O_7$, $K_2Cr_2O_7$ and $CrCl_3$ dissolve, and the saturation index of the remaining minerals is less affected by the change of pH.
4. Currently, the groundwater at the site is heavily contaminated and the groundwater chemistry type is altered. First, we should analyze the effects of other contaminants on chromium remediation in conjunction with indoor experiments. Second, we should maintain the concept of joint management of soil and groundwater. Third, the site groundwater runoff conditions are relatively poor and can be remediated by permeable reactive barriers technology and in situ injection of pharmaceuticals.

Author Contributions: Formal analysis, Y.C. and Q.Z.; investigation, Y.Z., Y.S., W.S. and Z.L.; writing-original draft preparation, W.C.; writing-review and editing, W.C. and Y.Z. (Co-first authors); supervision, Y.C. and Q.Z. All authors have read and agreed to the published version of the manuscript.

Funding: This research is funded by Science and Technology Research Project of Henan Provincial Department of Natural Resources (Yu Natural Resources Letter [2019] No. 373-8); National Key R&DProgram of China (No. 2019YFC1804801).

Institutional Review Board Statement: Not applicable.

Informed Consent Statement: Not applicable.

Data Availability Statement: Data is contained within the article.

Acknowledgments: We sincerely thank the reviewers for useful comments and suggestions.

Conflicts of Interest: The authors declare that we do not have any commercial or associative interest that represents a conflict of interest in connection with the work submitted.

References

1. Gu, X.; Xiao, Y.; Yin, S.; Shao, J.; Pan, X.; Niu, Y.; Huang, J. Groundwater level response to hydrogeological factors in a semi-arid basin of Beijing, China. *J. Water Supply Res. Technol.* **2017**, *66*, 266–278. [CrossRef]
2. Elgallal, M.; Fletcher, L.; Evans, B. Assessment of potential risks associated with chemicals in wastewater used for irrigation in arid and semiarid zones: A review. *Agric. Water Manag.* **2016**, *177*, 419–431. [CrossRef]

3. Esmaeili-Vardanjani, M.; Rasa, I.; Amiri, V.; Yazdi, M.; Pazand, K. Evaluation of groundwater quality and assessment of scaling potential and corrosiveness of water samples in Kadkan aquifer, Khorasan-e-Razavi Province, Iran. *Environ. Monit. Assess.* **2015**, *187*, 53. [CrossRef] [PubMed]
4. Xiao, Y.; Gu, X.; Yin, S.; Shao, J.; Cui, Y.; Zhang, Q.; Niu, Y. Geostatistical interpolation model selection based on ArcGIS and spatio-temporal variability analysis of groundwater level in piedmont plains, northwest China. *SpringerPlus* **2016**, *5*, 425. [CrossRef]
5. Lu, S.; Zhang, X.; Liang, P. Influence of drip irrigation by reclaimed water on the dynamic change of the nitrogen element in soil and tomato yield and quality. *J. Clean. Prod.* **2016**, *139*, 561–566. [CrossRef]
6. Bai, B.; Jiang, S.; Liu, L.; Li, X.; Wu, H. The transport of silica powders and lead ions under unsteady flow and variable injection concentrations. *Powder Technol.* **2021**, *387*, 22–30. [CrossRef]
7. Megharaj, M.; Avudainayagam, S.; Naidu, R. Toxicity of hexavalent chromium and its reduction by bacteria isolated from soil contaminated with tannery waste. *Curr. Microbiol.* **2003**, *47*, 51–54. [CrossRef]
8. Yang, G.; Xia, J. Chromium Contamination Accident in China: Viewing Environment Policy of China. *Environ. Sci. Technol.* **2011**, *45*, 8605–8606.
9. Priti, S.; Vipin, B.; Agarwal, S.K.; Vipin, V.; Kesavachandran, C.N.; Pangtey, B.S.; Neeraj, M.; Pal, S.K.; Mithlesh, S.; Goel, S.K. Groundwater Contaminated with Hexavalent Chromium [Cr (VI)]: A Health Survey and Clinical Examination of Community Inhabitants (Kanpur, India). *PLoS ONE* **2012**, *7*, e47877.
10. Hori, M.; Shozugawa, K.; Matsuo, M. Hexavalent chromium pollution caused by dumped chromium slag at the urban park in Tokyo. *J. Mater. Cycles Waste Manag.* **2015**, *17*, 201–205. [CrossRef]
11. Kumari, B.; Tiwary, R.K.; Srivastava, K.K. Physico-Chemical Analysis and Correlation Study of Water Resources of the Sukinda Chromite Mining Area, Odisha, India. *Mine Water Environ.* **2017**, *36*, 356–362. [CrossRef]
12. Bai, B.; Nie, Q.; Zhang, Y.; Wang, X.; Hu, W. Cotransport of heavy metals and SiO2 particles at different temperatures by seepage. *J. Hydrol.* **2021**, *597*, 125771. [CrossRef]
13. Guo, H.; Chen, Y.; Hu, H.; Zhao, K.; Li, H.; Yan, S.; Xiu, W.; Coyte, R.M.; Vengosh, A. High Hexavalent Chromium Concentration in Groundwater from a Deep Aquifer in the Baiyangdian Basin of the North China Plain. *Environ. Sci. Technol.* **2020**, *54*, 10068–10077. [CrossRef]
14. Bai, B.; Rao, D.; Chang, T.; Guo, Z. A nonlinear attachment-detachment model with adsorption hysteresis for suspension-colloidal transport in porous media. *J. Hydrol.* **2019**, *578*, 124080. [CrossRef]
15. Bai, B.; Long, F.; Rao, D.; Xu, T. The effect of temperature on the seepage transport of suspended particles in a porous medium. *Hydrol. Process.* **2017**, *31*, 382–393. [CrossRef]
16. Lockhart, K.M.; King, A.M.; Harter, T. Identifying sources of groundwater nitrate contamination in a large alluvial groundwater basin with highly diversified intensive agricultural production. *J. Contam. Hydrol.* **2013**, *151*, 140–154. [CrossRef]
17. Alam, F. Evaluation of hydrogeochemical parameters of groundwater for suitability of domestic and irrigational purposes: A case study from central Ganga Plain, India. *Arab. J. Geosci.* **2014**, *7*, 4121–4131. [CrossRef]
18. Yang, J.; Graf, T.; Ptak, T. Impact of climate change on freshwater resources in a heterogeneous coastal aquifer of Bremerhaven, Germany: A three-dimensional modeling study. *J. Contam. Hydrol.* **2015**, *177–178*, 107–121. [CrossRef]
19. Li, P.; Wu, J.; Qian, H. Hydrochemical appraisal of groundwater quality for drinking and irrigation purposes and the major influencing factors: A case study in and around Hua County, China. *Arab. J. Geosci.* **2016**, *9*, 15. [CrossRef]
20. Nematollahi, M.J.; Ebrahimi, P.; Razmara, M.; Ghasemi, A. Hydrogeochemical investigations and groundwater quality assessment of Torbat-Zaveh plain, Khorasan Razavi, Iran. *Environ. Monit. Assess.* **2016**, *188*, 2. [CrossRef]
21. Tiwari, A.K.; Pisciotta, A.; Maio, M.D. Evaluation of groundwater salinization and pollution level on Favignana Island, Italy. *Environ. Pollut.* **2019**, *249*, 969–981. [CrossRef]
22. Rezaei, A.; Hassani, H.; Hayati, M.; Jabbari, N.; Barzegar, R. Risk assessment and ranking of heavy metals concentration in Iran's Rayen groundwater basin using linear assignment method. *Stoch. Environ. Res. Risk Assess.* **2018**, *32*, 1317–1336. [CrossRef]
23. Rezaei, A.; Hassani, H.; Jabbari, N. Evaluation of groundwater quality and assessment of pollution indices for heavy metals in North of Isfahan Province, Iran. *Sustain. Water Resour. Manag.* **2019**, *5*, 491–512. [CrossRef]
24. Rezaei, A.; Hassani, H. Hydrogeochemistry study and groundwater quality assessment in the north of Isfahan, Iran. *Environ. Geochem. Health* **2018**, *40*, 583–608. [CrossRef]
25. Rezaei, A.; Hassani, H.; Hassani, S.; Jabbari, N.; Fard Mousavi, S.B.; Rezaei, S. Evaluation of groundwater quality and heavy metal pollution indices in Bazman basin, southeastern Iran. *Groundw. Sustain. Dev.* **2019**, *9*, 100245. [CrossRef]
26. Rezaei, A.; Hassani, H.; Tziritis, E.; Fard Mousavi, S.B.; Jabbari, N. Hydrochemical characterization and evaluation of groundwater quality in Dalgan basin, SE Iran. *Groundw. Sustain. Dev.* **2020**, *10*, 100353. [CrossRef]
27. Xiao, Y.; Gu, X.; Yin, S.; Pan, X.; Shao, J.; Cui, Y. Investigation of Geochemical Characteristics and Controlling Processes of Groundwater in a Typical Long-Term Reclaimed Water Use Area. *Water* **2017**, *9*, 800. [CrossRef]
28. Zhang, J.; Shi, Z.; Wang, G. Hydrochemical characteristics and evolutionary laws of groundwater in Dachaidan area, Qaidam Basin. *Earth Sci. Front.* **2020**, *28*, 194.
29. Tiwari, A.K.; Orioli, S.; De Maio, M. Assessment of groundwater geochemistry and diffusion of hexavalent chromium contamination in an industrial town of Italy. *J. Contam. Hydrol.* **2019**, *225*, 103503. [CrossRef]

30. Laxmankumar, D.; Satyanarayana, E.; Dhakate, R.; Saxena, P.R. Hydrogeochemical characteristics with respect to fluoride contamination in groundwater of Maheshwarm mandal, RR district, Telangana state, India–ScienceDirect. *Groundw. Sustain. Dev.* **2019**, *8*, 474–483. [CrossRef]
31. Zhu, F.; Liu, T.; Zhang, Z.; Liang, W. Remediation of hexavalent chromium in column by green synthesized nanoscale zero-valent iron/nickel: Factors, migration model and numerical simulation. *Ecotoxicol. Environ. Saf.* **2021**, *207*, 111572. [CrossRef] [PubMed]
32. Hausladen, D.M.; Fendorf, S. Hexavalent Chromium Generation within Naturally Structured Soils and Sediments. *Environ. Sci. Technol.* **2017**, *51*, 2058–2067. [CrossRef] [PubMed]
33. Zhou, L.; Zhang, G.; Wang, M.; Wang, D.; Cai, D.; Wu, Z. Efficient removal of hexavalent chromium from water and soil using magnetic ceramsite coated by functionalized nano carbon spheres. *Chem. Eng. J.* **2018**, *334*, 400–409. [CrossRef]
34. Wang, X.; Li, L.; Yan, X.; Meng, X.; Chen, Y. Processes of chromium (VI) migration and transformation in chromate production site: A case study from the middle of China. *Chemosphere* **2020**, *257*, 127282. [CrossRef]
35. Hausladen, D.M.; Alexander-Ozinskas, A.; Mcclain, C.; Fendorf, S. Hexavalent Chromium Sources and Distribution in California Groundwater. *Environ. Sci. Technol.* **2018**, *52*, 782–792. [CrossRef]
36. Rezaei, A.; Hassani, H.; Mousavi, S.B.F.; Jabbari, N. Evaluation Of Heavy Metals Concentration In Jajarm Bauxite Deposit In Northeast Of Iran Using Environmental Pollution Indices. *Malays. J. Geosci.* **2019**, *3*, 12–20. [CrossRef]
37. Esmaeili, A.; Moore, F.; Keshavarzi, B. Nitrate contamination in irrigation groundwater, Isfahan, Iran. *Environ. Earth Sci.* **2014**, *72*, 2511–2522. [CrossRef]
38. Bai, B.; Xu, T.; Nie, Q.; Li, P. Temperature-driven migration of heavy metal Pb^{2+} along with moisture movement in unsaturated soils. *Int. J. Heat Mass Transf.* **2020**, *153*, 119573. [CrossRef]
39. Wagh, V.M.; Panaskar, D.B.; Varade, A.M.; Mukate, S.V.; Gaikwad, S.K.; Pawar, R.S.; Muley, A.; Aamalawar, M.L. Major ion chemistry and quality assessment of the groundwater resources of Nanded tehsil, a part of southeast Deccan Volcanic Province, Maharashtra, India. *Environ. Earth Sci.* **2016**, *75*, 1418. [CrossRef]
40. Karagüzel, R.; Irlayici, A. Groundwater pollution in the Isparta Plain, Turkey. *Environ. Geol.* **1998**, *34*, 303–308. [CrossRef]
41. Gibbs, R.J. Mechanisms Controlling World Water Chemistry. *Science* **1970**, *170*, 1088–1090. [CrossRef]
42. Kanagaraj, G.; Elango, L. Hydrogeochemical processes and impact of tanning industries on groundwater quality in Ambur, Vellore district, Tamil Nadu, India. *Environ. Sci. Pollut. Res. Int.* **2016**, *23*, 24364–24383. [CrossRef]
43. Xiao, Y.; Shao, J.; Cui, Y.; Zhang, G.; Zhang, Q. Groundwater circulation and hydrogeochemical evolution in Nomhon of Qaidam Basin, northwest China. *J. Earth Syst. Sci.* **2017**, *126*, 26. [CrossRef]
44. Wang, L.; Dong, Y.; Xie, Y.; Song, F.; Wei, Y.; Zhang, J. Distinct groundwater recharge sources and geochemical evolution of two adjacent sub-basins in the lower Shule River Basin, northwest China. *Hydrogeol. J.* **2016**, *24*, 1967–1979. [CrossRef]
45. Zhu, G.; Wu, X.; Ge, J.; Liu, F.; Wu, C. Influence of mining activities on groundwater hydrochemistry and heavy metal migration using a self-organizing map (SOM). *J. Clean. Prod.* **2020**, *257*, 120664. [CrossRef]

MDPI
St. Alban-Anlage 66
4052 Basel
Switzerland
Tel. +41 61 683 77 34
Fax +41 61 302 89 18
www.mdpi.com

Applied Sciences Editorial Office
E-mail: applsci@mdpi.com
www.mdpi.com/journal/applsci

www.ingramcontent.com/pod-product-compliance
Lightning Source LLC
LaVergne TN
LVHW071029170726
843515LV00006B/1235

* 9 7 8 3 0 3 6 5 6 1 7 7 6 *